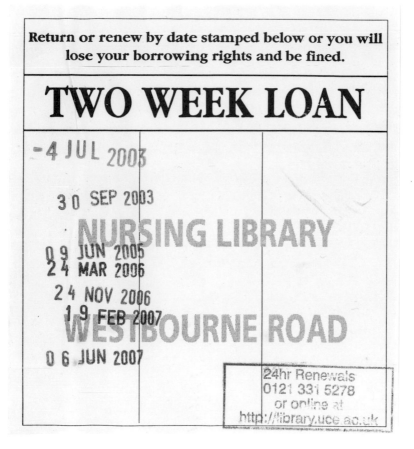

SELF-AWARENESS

Self-Awareness

ITS NATURE AND DEVELOPMENT

Edited by

MICHEL FERRARI
ROBERT J. STERNBERG

THE GUILFORD PRESS
New York   London

© 1998 The Guilford Press
A Division of Guilford Publications, Inc.
72 Spring Street, New York, NY 10012
http://www.guilford.com

Printed in the United States of America

This book is printed on acid-free paper.

Last digit is print number: 9 8 7 6 5 4 3 2 1

Library of Congress Cataloging-in-Publication Data
Self-awareness : its nature and development / edited by Michel
  Ferrari, Robert J. Sternberg.
       p.   cm.
  Includes bibliographical references and index.
  ISBN 1-57230-317-4
  1. Self-perception.   I. Ferrari, M. D.   II. Sternberg, Robert J.
BF697.5.S43.S434   1998
126—dc21   97-52300
                                                         CIP

For Adrien Pinard

Contributors

✦

Thérèse Bouffard, PhD, Département de psychologie, Université du Québec à Montréal, Montréal, Québec, Canada

Raymond Bradley, PhD, Center for Brain Research and Informational Sciences, Radford University, Radford, VA

Jerome Bruner, PhD, Department of Psychology and School of Law, New York University, New York, New York

Jeremy I. M. Carpendale, PhD, Department of Psychology, University of British Columbia, Vancouver, British Columbia, Canada

Michael J. Chandler, PhD, Department of Psychology, University of British Columbia, Vancouver, British Columbia, Canada

Michel Ferrari, PhD, Learning Research and Development Center, University of Pittsburgh, Pittsburgh, PA

Kurt W. Fischer, PhD, Department of Human Development, Harvard University, Cambridge, MA

Martin E. Ford, PhD, Graduate School of Education, George Mason University, Fairfax, VA

Jerome Kagan, PhD, Department of Psychology, Harvard University, Cambridge, MA

David A. Kalmar, PhD, Department of Psychology, New York University, New York, NY

Michelle A. Maher, MA, Department of Psychology, George Mason University, Fairfax, VA

Michael F. Mascolo, PhD, Department of Psychology, Merrimack College, North Andover, MA

Thomas Natsoulas, PhD, Department of Psychology, University of California, Davis, Davis, CA

Sue Taylor Parker, PhD, Department of Anthropology, Sonoma State University, Rohnert Park, CA

Daniel J. Povinelli, PhD, New Iberia Research Center, University of Southwestern Louisiana, New Iberia, LA

Karl H. Pribram, PhD, Center for Brain Research and Informational Sciences, Radford University, Radford, VA

Christopher G. Prince, PhD, Center for Advanced Computer Studies, University of Southwestern Louisiana, New Iberia, LA

John R. Searle, PhD, Department of Philosophy, University of California, Berkeley, Berkeley, CA

Louise Spear-Swerling, PhD, Department of Special Education, Southern Connecticut State University, New Haven, CT

Robert J. Sternberg, PhD, Department of Psychology, Yale University, New Haven, CT

Carole Vezeau, PhD, Département de psychologie, College Joliette—De Lanaudiere, Joliette, Québec, Canada

Preface

✧

The chapters in this volume were specially commissioned to discuss the complex and fascinating topic of self-awareness. Contributors were invited to discuss self-awareness from the perspective of their own disciplines. In this respect, the volume before you offers the chance to see how writers in very different fields converge and diverge when addressing major themes that are central to any study of self and self-awareness, like the relation between biology and culture or between self and others.

The chapter by John R. Searle explores how his views on consciousness and the mind influence individual education. Searle gives a masterly rendition of two key concepts in philosophy of mind—consciousness and intentionality—both of which are foundational to any study of self-awareness. Specifically, he presents five insightful theses to explain how these concepts relate to each other, and their consequences for education.

Thomas Natsoulas follows with a very subtle discussion of how common definitions of the word "consciousness," found in the *Oxford English Dictionary,* can be used to clarify the relationship between the concepts of consciousness and self-awareness. Natsoulas integrates his discussion into an intellectual tradition that includes such illustrious figures as John Dewey, George Herbert Mead, C. S. Lewis, and William James. He pays particular attention to "being conscious to one's self," and argues that such consciousness is intimately related to James's notion of the spiritual self and to self-awareness.

Daniel J. Povinelli and Christopher G. Prince present an extremely thought-provoking chapter in which they cite numerous studies in their lab that compared chimpanzee to children's reactions on theory-of-mind tasks. They show that while human beings may share certain very primitive forms of self-awareness with other higher primates, we may be alone in appreciating that others have a theory of mind.

Sue Taylor Parker follows with a chapter that covers some of the same ground as the previous chapter, but this time from the point of view of the development of emotions. Building on the work of other theorists such as James Mark Baldwin, Michael Lewis, and Michael Mascolo and Kurt Fischer, she traces the evolution and functional significance of self-conscious emotions. She argues that, unlike basic emotions like fear and anger, only self-conscious emotions are uniquely human and may have been selected because of the evolutionary advantage they conferred to early hominids.

In the next section, Jerome Kagan asks whether one is theoretically and empirically justified in postulating that infants have a concept of self. He argues that there is virtually no reason to believe that infants have any abstract representation of themselves as selves. Only by age 1½ or 2 do children begin to give any indication that they are reflexively aware of themselves as agents, as shown by several new ways children have of referring to themselves and of relating to adults.

Michael J. Chandler and Jeremy I. M. Carpendale provide an ambitious attempt to review recent literature on theory of mind and to show how this literature relates (and in many ways does not relate) to our understanding of children's self-awareness. They present a sharp critique of the current theory-of-mind literature, and argue for a more developmental view of the emergence of a mature theory of mind that allows children to increasingly appreciate their own and others' subjectivity. They also describe some innovative studies conducted in their lab that support their theoretical claims.

Martin E. Ford and Michelle A. Maher present an interesting discussion of the role of self-awareness in their Living Systems Framework. They pay particular tribute to the role of attention in directing awareness and controlling behavior in various contexts, especially how such attentional navigation reflects one's social intelligence, including one's ability to coordinate and regulate successful attainment of different types of social goals.

Robert J. Sternberg and Louise Spear-Swerling contribute an interesting chapter built around the notion of personal navigation. Personal navigation refers to how individuals can use self-awareness to exert control over their life course. As the authors point out, the construct of personal navigation cuts across many different areas of psychology, including the study of intelligence, emotion, personality, and others—and does so in a way that looks at the whole person in context.

The notion of control is key to the chapter by Thérèse Bouffard and Carole Vezeau. These authors focus on the impact of children's self-beliefs on their academic achievement. In particular, they describe a longitudinal study in which fourth, fifth, and sixth graders' beliefs about their own abilities were related to their practice and performance of a reading task. They suggest that self-beliefs are an important mediating variable in the success of any educational training or evaluation program.

Karl H. Pribram and Raymond Bradley make an ambitious attempt to examine the neurological underpinnings of affective, conative, and cognitive aspects of self as experienced at individual, dyadic, and communal levels. One of their central themes is that structural principles present meaningful parallels between very different levels of individual and social organization. They suggest that operations at the neurobiological, neuropsychological, and social-psychological levels all show correspondences in the organization of behavior that produce a competent, stable self.

Jerome Bruner and David A. Kalmar provide an exceptionally clear and wide-ranging chapter on the self and self-awareness, paying particular attention to how self-indicators, as expressed through narrative, both individuate us and binds us to our community. They suggest that self-narratives, such as autobiographies, are comparable to other cultural narratives, and are sometimes shaped by them through personal reflection.

Michael F. Mascolo and Kurt W. Fischer bring a very different body of developmental literature to bear on the issue of the development of self and self-awareness. They contrast two views of the self, the Cartesian (rationalist) and the Bakhtinian (dialogical) view. Building on Fischer's theory of skill development, they argue that the self develops along multiple (initially semi-independent) dimensions, in contexts that require agentative control. As children develop, their control over and awareness of themselves become increasingly coordinated and integrated—both in their self-knowledge and in their emotional reactions to events.

The volume concludes with a critical integrative chapter by Michel Ferrari, who looks at the common strands and the essential differences between the various chapters. The aim of this chapter is not to make a definitive statement about self-awareness, but simply to coordinate what has been said by these authors into a more unified story. As I speak with my own voice in this chapter, I allow some of my own biases and preferences among the different positions taken by the authors to show through.

All in all, the volume provides an impressive breadth of opinion on the topic of self-awareness from leading scientists working in a wide range of fields. If the reader learns even half as much, and has even half as much enjoyment, from reading these chapters as we have had in our role as editors, then they will find that this volume is worth every minute they put into it.

MICHEL FERRARI
ROBERT J. STERNBERG

Contents

✧

PART 4
INTEGRATION AND CONCLUSION

SELF-AWARENESS

PART ONE

DEFINITIONS AND UNDERPINNINGS OF SELF-AWARENESS

✧

CHAPTER ONE

The Mind and Education

✧

JOHN R. SEARLE

I have been asked to discuss my work in the philosophy of mind and cognitive science with a view to its relevance to education. I must say at the outset that I am not sure how much relevance my views have for the actual practice of educating young minds. However, if I am right in my criticisms of certain prevailing views about the mind, we may at least avoid making certain mistakes in that practice, mistakes that derive from holding false philosophical and psychological theories about the mind. In this talk, I am going to summarize some of my more controversial views and then make some speculations about their possible relevance to education. For the sake of brevity and clarity, I will state my views as a set of theses.

FIVE THESES ABOUT THE MIND

Thesis 1. The essence of the mental is consciousness. Descartes almost ruined this sentence forever when he said it (or at least its Latin and French versions) to mean something else. What Descartes meant was that there are two kinds of substances in the world, mental and physical, and each has an Aristotelian essence, an essential trait that makes the substance the kind of substance it is. The essential trait of mental substance is that it is always and forever conscious. It is impossible for a mind to exist without being conscious.

In contrast to Descartes, I construe mental states as either conscious or unconscious, and the unconscious ones are understandable as mental because they are the sorts of things that are at least in principle accessible to consciousness. So when I say that the essence of the mental is consciousness, what I mean by that is that our ordinary notion of an unconscious mental state is understandable as mental only in terms of its accessibility to consciousness. *All mental states are either actually or potentially conscious.* A person might not be able to bring his or her unconscious belief to consciousness because of some impediment. These impediments range all the way from brain lesions, to the tip-of-the-tongue phenomenon, to Freudian style repression, but we have no notion of a belief state except in terms of conscious mental states. Our paradigms of mental states are conscious states such as thinking that Bill Clinton is President of the United States, or trying to solve a mathematical problem, or worrying about the size of the deficit, or wondering whether our income tax will be audited this year.

When I say consciousness is the essential trait of the mind, I am presenting that thesis in contrast to a number of other views that say things such as that the essential thing about the mind is that it is behavior, or functional adaption, or perhaps worst of all, that the mind is just a computer program. I will come back to this view later.

It is usually said that "consciousness" is especially difficult to define. I don't really think that "consciousness" is hard to define, if what we are seeking is a commonsense definition rather than the sort of analytic definitions that typically come at the end of a scientific analysis. If we just want a commonsense definition that will identify the phenomenon in question, then we can say that consciousness consists in subjective states of awareness or sentience that begin when one wakes from a dreamless sleep and continue until one goes to sleep again, or falls into a coma, or dies, or otherwise becomes, as one would say, "unconscious."

So construed, consciousness is a biological phenomenon like any other. On my view, we need to stop thinking of the mind as something mysterious and ethereal, and recognize it as an ordinary biological phenomenon, on all fours with digestion, mytosis, meiosis, or the secretion of bile.

Thesis 2. Consciousness is caused by processes in the brain and is a higher-level feature of the brain system. This thesis is an answer to the traditional "mind–body problem." The problem of the relation of consciousness to physical processes, such as brain processes, has obsessed philosophers for the past 350 years. I think, in fact, that if this problem did not have such a particularly sordid history, we would not find it very difficult. I believe that the mind–body problem has a rather simple solution given what we know about how the world actually works.

Mental states, and by that I mean all of our mental states—thoughts, pains, tickles, itches, feelings, worries, anxieties, angst, love, hate, and so on, all of our mental states—are caused by lower-level neuronal processes in the brain and are themselves higher-level features of the brain. Formally speaking, these types of relationships are very common in nature. For example, the liquidity of the water in this glass or the solidity of the wood in this table are both caused by the behavior of lower-level elements of the system, in these cases, by the behavior of the molecules. But at the same time, neither liquidity nor solidity are separate entities. They are, rather, the states that the systems composed of the molecules happen to be in. It is similarly the case with the mind and the brain. Consciousness is caused by the behavior of the lower-level elements, neurons and synapses, but it is not thereby a separate substance or entity. Consciousness is just the state the brain happens to be in at certain times, those times when the neurons and synapses are behaving appropriately. The solidity of the table and the liquidity of the water are not separate things distinct from the molecules that make up the water or the table. Analogously, we should not think of consciousness as a separate thing, distinct from the neuronal systems that make up the brain. Rather, consciousness and other mental phenomena are just states that the brain happens to be in at any particular point, in the same sense that the water in this glass is in a liquid state at present, or the wood in this wooden table is in a solid state.

There are immense empirical mysteries about how the brain works in fact, but I believe there is no deep mystery about the overall relationship between the mind and the brain.

Thesis 3. If we construe the mind as a biological phenomenon, the most important functional feature of the mind is intentionality. Intentionality is simply that feature of the mind by which it is capable of representing objects and states of affairs in the world other than itself. The word "intentionality" is an unfortunate word for English speakers because it seems to imply some special connection with intending, in the sense in which I intend to go to the movies, or intend to get a haircut sometime in the next week. But "intentionality" in the technical sense does not imply intending. Intending is just one kind of intentionality along with perceiving, believing, desiring, hoping, fearing, wondering whether, wishing that, falling in love, or being disappointed. In short, any intentional state that is *about* something other than itself is said to be intentional. In this sense, pains, tickles, itches are not intentional, nor are undirected feelings of elation and anxiety. In this respect, they are unlike intentional states such as believing, desiring, hoping, fearing, and so on.

Intentional states and processes function to relate us to our environment by way of representations that act on the environment and are in turn acted on by the environment. Typical intentional phenomena that direct us to act on the environment are desires, intentions, and intentional actions. Typical intentional states that are caused by the environment and represent the environment are beliefs, memories, and perceptual experiences.

It should be obvious from all this that there is a close connection between intentionality and consciousness, but they are not coextensive. Most conscious states are intentional, but not all of them. If I have an undirected feeling of anxiety, or simply feel an unpleasant sensation in my elbow, these conscious states are not intentional. Furthermore, I have many intentional states that at any given point in my life are not conscious. The man who is sound asleep, for example, can literally be said to believe that Bill Clinton is President of the United States. Thus, not all intentional states are conscious; and not all conscious states are intentional. However, this point needs to be reemphasized: Any unconscious intentional state is understood as an intentional mental state only in terms of its accessibility to consciousness—only in terms of its potentially being conscious.

I guess it is obvious to everyone here that education is for the most part about intentionality. Our aim is to produce knowledge, and knowledge consists in having beliefs that meet certain conditions; that is, they must be true and they must be justified in certain ways.

Thesis 4. The Background: All intentional states, conscious or unconscious, only function given a Background of abilities, capacities, tendencies, dispositions, and general know-how that do not consist in more intentional states. All of our mental life is conditioned by a set of presuppositions that are not part of our conscious awareness. It is a mistake to think of these as a set of unconscious beliefs that we could bring to awareness at will. Rather, it seems to me best to think of them as a set of abilities that we have for coping with the environment and with other people. For the most part, the Background is not a matter of "knowing that" but of "knowing how."

Any educator has the following sort of experience: Some things you say are systematically misunderstood, because the people to whom you are speaking have a different Background. They just bring a different mode of sensibility to bear on the problem.

The most obvious proof of the thesis of the Background is to consider ordinary sentences. If you consider sentences containing the word, "cut," for example, as in "Bill cut the grass," or "Sally cut the cake," we understand the word "cut" as having the same meaning, yet it has a different application. What counts as cutting the grass is quite different from what counts as cutting the cake, not because the meaning of the word is different, but rather because we have a set of abilities and practices derived from our biology and our culture, and these are what I am calling "Background."

Background abilities are not simply cultural; many of them are biological. If you read in a book the sentence "Bill ate the hamburger," you know that he ate it by putting it in his mouth and not by, let's say, stuffing it in his ear or ingesting it through capillary action in his skin. All of that is pretheoretical, preintentional Background. Much education is about shaping the Background. I will return to this point later.

Thesis 5. It is a mistake to suppose that the mind is a computer program. Perhaps the biggest single mistake we make about the mind today is to suppose that the brain is a digital computer and the mind is simply the program of that computer. This mistake is the latest in a long series of reductionist attempts to get rid of consciousness and intentionality and replace them with something else. If you are as old as I am, you will remember certain earlier versions of this reductionist tendencies, such as behaviorism.

Behaviorism says that the essential thing about the mind is publicly observable behavior. I think that view is obviously false, as we all know from our own experiences. There is no essential connection between our mental states and our behavior, though of course, our mental life evolved to enable us to cope better with our environment, and that coping is done through our behavior. But you can have mental states without exhibiting behavior, and you could, at least in principle, build systems that would exhibit the behavior but not have the mental states. I think behaviorism is pretty much dead, but occasionally there are efforts to revive it.

The more interesting and currently more popular view is that mental states are essentially computational, that a computer, if it had the right program and the right inputs and outputs, would have to have the same mental states we have, and indeed, that our mental states consist entirely in computer programs that are capable of mediating the relations between input and output. I will not spend much time on this view, because I believe it can be refuted decisively in a couple of minutes. Briefly, here is the refutation:

What is a computer program? A computer program is a set of rules for formal symbol manipulation. These are usually defined in terms of zeros and ones. A computer manipulates zeros and ones. In fact, on Turing's original definition of a Turing machine, a Turing machine can perform exactly four operations: It can move its head one square to the left, or move it one square to the right; it can erase a zero and print a one, or erase a one and print a zero. That is all a computer does. It does those things very rapidly—several million operations per second, and it does them in accordance with rules of the form: Under condition C, perform act A; if C, then A. It is one of the great intellectual achievements of the 20th century that we are able to do so much with such a simple apparatus, but equally it is one of the great intellectual mistakes of the latter part of the 20th century to suppose that that is what is going on in our minds. Now, we know that the mind could not be just an implemented computer program. This view could not be right for the following simple reason: The operation of the

computer is defined purely in terms of meaningless symbols such as the zeros and ones. But what goes on in our minds is more than the manipulation of meaningless symbols; rather, our mental states actually have a mental content. They have something more than just zeros and ones. If I wish I
had a cold beer or am planning a trip to Seattle, I actually have thoughts
about Seattle or beer, and not just about zeros or ones.

I gave a simple proof of this over 15 years ago. If it were true that a
computer program were sufficient for mental states, then just imagine that
you are carrying out the computer program for some mental state that you
do not in fact have. I do not understand Chinese, so I imagine myself carrying out a computer program that simulates the understanding of Chinese. I imagine I receive a series of inputs in Chinese. I look up in the rule
book, that is to say, the program, I am supposed to manipulate these symbols, and I give a set of outputs in Chinese—I give back Chinese symbols to
those people who have been giving me Chinese symbols. Unknown to me,
it might be the case that the program is enabling me to give the correct Chinese answers to the Chinese questions that are presented to me, but I know
nothing of that. I am simply doing what a computer does, manipulating
meaningless symbols. Now, the point of this whole parable can be stated
quite simply: If I do not understand Chinese on the basis of implementing a
digital computer program for the understanding of Chinese, then neither
does any other digital computer solely on that basis, because that is all a
computer has. It has a set of rules for manipulating formal symbols. There
is a much simpler summary of this argument—it can be stated in two slo-
gans of four words each. Syntax is not semantics; and simulation is not du-
plication. The syntax of the implemented computer program is not sufficient for the semantics of ordinary human minds; and simulating a process
is not the same as duplicating it.

Intellectual historians will regard it as one of the most amazing features of the 20th century that the computational theory of the mind,
though it is preposterous on its face, survived as long as it did.

I think, incidentally, that these two mistakes are important for education in that behaviorist theories of education and computational theories
of education have both had deleterious consequences.

The upshot of my discussion so far can be stated quite simply: What is
going on inside our skulls consists exactly of two sorts of things. First,
there are brain processes with all their enormous complexity, and these
have many levels of description ranging all the way from quarks and
muons up through synapses, neurons, and cell assemblies to large neuronal
circuits and even gross molar organs such as the hypocampus and the thal-
amus. Second, there is consciousness with all of its color and variety. But
that is it. There is not, in addition, a whole lot of mental models, or the
language of thought, or computer programs; there is just a brain, and

sometimes it is conscious. Consciousness functions against a biological and cultural Background of nonconscious capacities that are realized in the brain.

CONSEQUENCES FOR EDUCATION

What are the practical applications to education, if any, of what I just told you? Well, as I said at the beginning, I am not at all sure that there are any practical applications, and my own instinct about education is to think that if we have been doing it more or less successfully, both systematically and unsystematically, for the past 2,500 years, we must have worked out some effective strategies for doing it, and we ought to take our own know-how as superior to anything some theoretician might tell us. But, anyway, there are perhaps a few lessons, so here goes.

One of the most surprising things about educational theories in the 20th century is how certain common expectations were disappointed. When radio was first invented, it was assumed it would revolutionize education. We had similar expectations for the movies and then for television. Instead of having ordinary fellows trying to educate you, you could be educated by the world's leading superstars, because you would be able to see them on television and nowadays see them on videotape. Education, in short, was going to be revolutionized by the electronic media. Why did it not work? Every study that I have seen shows that the retention rate from material acquired while watching television is vastly lower than material acquired while listening to a live lecture in the classroom. Why? What is so magical about actual classrooms inhabited by living people? Well, no doubt, the answer is complex, but there is a rather simple aspect that I want to call to your attention. If you have to get up, put on your overcoat, go to the classroom, sit in the classroom, look at an actual human being and take notes, you have a much greater intensity of commitment and involvement than you do if you are sitting at home in front of a television set. Live education, in short, involves a degree of *conscious* involvement and commitment that is not present in the passive reception of visual and auditory images. Furthermore, not only does it take a greater commitment on the part of a student, but also there is actually a self-awareness of the student's interaction with the teacher. When I am a student in a classroom, the teacher is not only there to be an object of my scrutiny, but also I am aware that the teacher is looking at me.

It may be that the most educationally effective of the electronic media will be those interactive forms of electronic technologies that involve active participation by the student.

The points I am making here are not especially confined to education.

I think that you are aware that you are much more affected by a drama if you see it live on stage or even if you have to get up and go to a movie house to see it, than if you see the same thing at home on television. At home, when you see it on television, you are busy carrying on a conversation with your spouse, your mind is on other things, and at various points you will get up and go to the refrigerator to get a beer, and so on. So the first lesson of what I have been saying is not very surprising. *The key to successful education is total commitment of the consciousness.* To the extent that you succeed in doing that, you will have an impact on the brain of the recipient; to the extent that you fail, you will not. So let's turn and talk about brains a bit.

The most important advice anyone can give about the relevance of brain research to education is not to take this research too seriously. We just do not know enough about how the brain works to use it as guide to education. However, as far as we know, the brain is losing neurons throughout most of our lives. This is not necessarily a bad thing. As far as we now know anything about how it works, it seems best to think of the brain more as a selectional rather than as an instructional mechanism. What I mean by that is that, in general, learning does not take place through reinforcing existing circuitry in an undifferentiated fashion. Rather, especially among the young, learning is often a matter of selecting certain circuitry and allowing other circuits to become inactive or simply to die. It is very important to reinforce the right circuitry at the right age. This is obvious to us in the case of physical skills. The child who does not learn ice-skating, musicianship, or skiing at an early age will never become a world-class performer in these activities. If you start as an adult, it is too late. Any classroom teacher knows that this is most obvious in the case of the pronunciation of foreign languages. Children exposed to a foreign language prior to adolescence will be able to speak that language without any trace of an accent. The older you get, the harder it is. This is not because of middle-aged laziness or lack of attention; it is simply that the circuitry is no longer able to accommodate the perception and the production of the appropriate phonetic shapes. There are some wonderful experiments that show this in the case of the contrast between Japanese children and Japanese adults, where the perception of the distinction between the English "r" and the "l" is concerned. The adults simply cannot hear a difference. The children have no problem hearing the difference. And just as they have no problem hearing the difference, they have no problem producing the difference, in contrast to their elders.

The moral of this is obvious: Teach the child as much as you can as early as you can. One of the tragedies of American education is the idea that bright children should not proceed faster than the rest of the class. The aim of education is to develop each child to the maximum of his or her potential, and though that is an ideal we will seldom realize in practice, all the same, we should not adhere to policies that go dead against that ideal. In his autobiography, Santayana says that the only really well-educated people he knows were educated privately at home by tutors and not in schools, neither public nor private schools. One of the advantages of private tutoring is that the student can go as fast and as far as he or she is able and as fast and as far as the tutor can teach.

Now I want to say a bit about the Background. Any teacher knows that much of what is important in what you teach is not the explicit content of the information that the child learns, but the attitude, the cast of mind, the ways of responding to knowledge and information, and indeed the whole mode of sensibility that the successful teacher can only convey by exemplifying it and by exposing the student to works that exemplify it. The French have a saying that translates as: You forget the information you learned, but the education remains. I take it that the point of that remark is that education is not just a matter of acquiring information, but the effect of acquiring the information is to change pretty much every aspect of your life, in ways that are typically not represented in the information you acquire.

Please do not misunderstand the point I am making. I am not denigrating the educational value of acquiring large amounts of sheer information. You will not learn the history of the United States without memorizing a great many names and dates, and you will not learn how to speak French without memorizing the meaning of a lot of French words. If anything, there is not enough memorization of large amounts of information going on in the schools at present. The point is that the information gives you a much deeper understanding of American history, and learning French—or any other foreign language—well, will change your whole way of responding to language and experience. But that understanding and those ways of responding are not additional items of information. They are matters of what I call the Background.

The real disaster produced by television is not that children waste so many hours watching it, but that it alters their Background. It alters their whole mode sensibility as they get a conception of how adults are normally supposed to behave in responding to each other. I do not need to tell this audience that the model of human behavior presented in most popular television programs is violent, brutish, and stupid.

CHAPTER TWO

Consciousness and Self-Awareness

✧

THOMAS NATSOULAS

Before embarking on a scientific discussion of the topic of consciousness, a psychologist does well to consider the several ordinary meanings of the word "consciousness" (Dewey, 1906; Natsoulas, 1983; cf. Husserl, 1900/1970, pp. 535–536). Attention to these nontechnical concepts should reduce the risk of difficult-to-avoid referential displacements, namely, from consciousness to phenomena that are more amenable to investigation by means of the methodologies and theoretical approaches currently in favor among psychologists (cf. Koch, 1980, p. 45; Miller, 1990, p. 7; Wittgenstein, 1947/1980, p. 180e).

One might suppose that in the years since the beginning of the cognitive revolution, the disciplinary factors largely responsible for such referential displacements have been mitigated in major part. However, note this recent astute observation:

> But the basic issue, I think, is never really addressed. . . . I have the same feeling about consciousness because whenever consciousness comes up— and it usually comes up in meetings of philosophers and psychologists— it's the same kind of phenomenon, where people are prepared to give a theory of consciousness but what they give you is a theory of something else. And the relationship between consciousness and that other thing is never made explicit. (Pylyshyn, 1990, p. 201)

I too have observed similar intellectual behavior among present-day psychologists: a continued, albeit anachronistic, championing of operational definitions as providing the true path for achieving the conceptual advances that all of us in this area of psychological science seek; and an advocacy of empirical research as itself constituting a way by which to bootstrap ourselves somehow—how exactly is not made clear—to a more faithful and complete conceptual grasp of the phenomena that fall commonsensically under the name "consciousness."

Quite germane to the purposes of the present chapter is the ordinary meaning of "consciousness" that is listed second in the *Oxford English Dictionary* (1989; henceforth, the *OED*). The *OED* describes as follows the kind of psychological state picked out when one properly uses "consciousness" in its second sense. At one time, the phrase "consciousness to oneself" was also used to make the same reference: "Internal knowledge or conviction; knowledge as to which one has the testimony within oneself; esp. of one's own innocence, guilt, deficiencies, etc." As the present chapter proceeds, it will be seen repeatedly that consciousness in the second *OED* sense involves reference to oneself as such in every case; and self-awareness is a necessary part of any instance of this kind of consciousness—as is not the case for every other kind.

The preceding dictionary definition expresses the oldest ordinary sense that continues to find expression with the word "consciousness." Use of the word in a different early sense—which the *OED* defines simply as "joint or mutual knowledge"—is now obsolete (Natsoulas, 1991a).[1] Both of these senses date from the 17th century, which is when "consciousness" first came into the English language. Later, four additional meanings of the word were added. I soon consider two of these four, together with the second sense. The two remaining senses of "consciousness" have no role to play in the present chapter. However, here is how the *OED* expresses them: "The totality of impressions, thoughts and feelings, which make up a person's conscious being." "The state of being conscious, regarded as the normal condition of healthy waking life." For detailed discussion of all six *OED* concepts, one may want to consult two previous publications of mine (Natsoulas, 1983, 1986–1987).

The late entry of "consciousness" into the English language does not signify that, prior to that time, ancestors of ours who knew only English could not refer to what we can refer to with the word. They were in firm possession of, among other linguistic means, the cognate word "conscience." From the 14th century onward, "conscience" had to serve, among its other functions, some of the functions that "consciousness" was later to assume. Indeed, one might wonder: Given the availability of "conscience" to users of English, why should there have emerged a word "consciousness" at all? This interesting question lies beyond the scope of the present chapter. However, one might begin to answer it along the following lines: After its poetical introduction into the English language in imitation of Latin authors (see Lewis, 1967), "consciousness" caught on and its number of uses expanded because "conscience" could thereby become more specialized and less ambiguous (cf. Engelberg, 1972).

What more exactly is consciousness in the *OED*'s second sense? What is the character of the kind of psychological state to which the dictionary is adverting with its second definition? The definition leaves one wanting to know more, especially if one is a psychologist who is seeking to address consciousness itself, rather than something else that may be related to it. The *OED*'s definition fails to provide adequate information to enable the discernment of instances of the indicated kind of consciousness. If one attempts to decipher the second meaning on the basis of the explicit definition alone, questions such as the following come to mind, and they are not answered:

1. Does "having the testimony within oneself" amount simply to being in a position to give testimony about something that one has witnessed? What are the limitations, if any, on the possible objects of consciousness in the *OED*'s second sense? Does one possess consciousness in this sense of everything to which one can attest on a firsthand basis?

2. Can a happening or state of affairs external to one, even independent of one, be an object of one's consciousness in the second sense? Can a crime, for example, be an object of one's consciousness in the second sense whether or not one has participated in it, so long as one actually witnessed it? If it can, is the crime an object of one's consciousness only whenever it comes back to mind? Or does it suffice that, possessing the knowledge, one is in a position to remember the crime?

3. How is consciousness in the second sense "internal"? Is there a counterpart phenomenon that qualifies as external to oneself yet is not "internal" to someone else? Does the *OED* describe consciousness in the second sense as internal because the object of this consciousness, whenever one instantiates the latter, is something that no one else can apprehend and give testimony about on a firsthand basis, something that is private to oneself in principle? Is the knowing, or awareness, that defines the *OED*'s second kind of consciousness in some way directed round upon the mind itself?

4. Or is consciousness in the second sense internal to oneself merely because, in the particular instance, one happens to be or to have been the only witness to the particular matter that one now knows or believes? Is it simply that the knowledge or conviction to which the *OED*'s definition refers is personal, at most secret, rather than held in common?

It is such questions regarding consciousness in the *OED*'s second sense that the present chapter seeks to address. I shall call the psychological referent of the second sense "consciousness$_2$," which has been my term for it in previous publications as well (e.g., Natsoulas, 1983, 1986–1987, 1991b). With the help of numerical subscripts, I have been distinguishing the six kinds of consciousness that the *OED* describes in its series of entries under the word. Comparisons between kinds of consciousness will again be useful as I proceed with making explicit what consciousness$_2$ is. Quite soon, the distinction between consciousness$_2$ and consciousness$_3$ enters the picture that I am drawing. Later, I also show how consciousness$_4$ is involved, always, in consciousness$_2$.

Moreover, as I proceed, I bring out the relation between each of three kinds of consciousness and self-awareness. In summary:

1. A case of consciousness$_3$ may or may not be a case of self-awareness; consciousness$_3$ can take place although one is not self-aware at the time.
2. Consciousness$_4$ is a form of self-awareness; it is a way in which one is occurrently aware of certain occurrent parts of oneself, typically, of them as being such.
3. Every instance of consciousness$_2$ involves, as a feature of itself, self-awareness in several ways.

My comments on the difference between consciousness$_2$ and consciousness$_3$, and on the relation between consciousness$_2$ and consciousness$_4$, serve a further purpose, in addition to helping to spell out what consciousness$_2$ is. They show that I am not engaged in advocacy for a particular sense of "consciousness." I am not suggesting that, for purposes of their research, psychologists should adopt the concept of consciousness$_2$ to the exclusion of all other technical or nontechnical concepts of consciousness. Nor am I engaged in the process—which has become traditional in psychology—of constructing a technical concept to be expressed by means of a familiar, ordinary word.

So too, the reader should not draw from my bringing out, perforce, the ethical dimension of consciousness$_2$, that I consider the *OED*'s other referents of "consciousness" to be less worthy as topics of psychological investigation. Actually, I believe even consciousness$_1$ is an intriguing and worthwhile topic, notwithstanding that the word "consciousness" is no longer used to speak of the interpersonal cognitive relation that consciousness$_1$ is (Natsoulas, 1991a).[2]

Thus, I simply hold that how we commonly use ordinary language reflects differentiations of proven utility over a considerable period of time. These differentiations may well be more refined than the distinctions that psychologists, at this early point in the history of their science, are finding useful or attractive. It will become evident as I proceed that the concept of consciousness$_2$ has reference to phenomena that should be of special interest to those psychologists, and others, who are currently rediscovering consciousness and want to advance their understanding of self-awareness.

THE CHARACTER OF CONSCIOUSNESS₂

More Than Being Conscious₃

Whenever someone is, as the *OED* puts it, "mentally conscious or aware *of* anything," that is, whenever there takes place an occurrent awareness of anything at all, consciousness₃ is therein instantiated by definition (Natsoulas, 1992b). Regarding consciousness₃, John Dewey (1906) accurately wrote: "'Conscious' means aware: 'consciousness' the state of being aware. This is a wide, colourless use; there is no discrimination as to contents, as to what there is awareness of,—whether mental or physical, personal or impersonal, etc." (p. 40). A case of consciousness₃ may fail to qualify as a case of each of the five other kinds of consciousness that are described under "consciousness" in the *OED*.

Note also that a psychological phenomenon does not need to occur in any special way for it to qualify as a case of consciousness₃. Its being a case of consciousness₃ does not depend on how an occurrent "awareness-of" comes to take place. Paranormal and strange futuristic means would be all right, if only they existed. In contrast, consciousness₂ requires witnessing or having witnessed evidence bearing on what, along certain lines, one is now occurrently aware of regarding oneself. Consciousness₂ always includes reference to oneself, and involves self-awareness in several ways, as I explain further on. An instance of consciousness₃, however, may be an instance of self-awareness, but often it is not. It is often an awareness of something else, which can be something completely unrelated to oneself. As the *OED* states, instances of consciousness₃ can be "of anything."

Mead

I must acknowledge that, according to George Herbert Mead's (1934) influential and enlightening conception of mind and self, every instance of consciousness₃ that one produces, however simple that instance may be, includes a reference to oneself (see Natsoulas, 1985). Whenever one is occurrently aware of anything at all, one is therein indicating it (and, often, characteristics of it) to oneself as though one were another person. Although Mead does introduce, in addition, a kind of experience entirely lacking the feature of self-reference, such experience is claimed by Mead not to possess any cognitive aspect. Assuming this noncognitive kind of experience does in fact take place,[3] it would not fall under the category of consciousness₃. Mead's proposed involvement of self-awareness in every case of consciousness₃ implies that all cognitions are founded on other cognitions, which in turn require self-awareness, and so on (see Natsoulas, 1991b). In any case, Mead's understanding of consciousness₃ is not the common one—which amounts simply to being aware of anything at all, and does not necessarily include oneself as the object of awareness.[4]

Being Conscious to Oneself

In its entry for consciousness$_2$, the *OED* calls our attention to a corresponding entry under the adjective "conscious." We are told to see *"conscious to oneself* (*of* anything, *that*, etc.)." The parallel meaning turns out to be very much the same. To be conscious to oneself of something in particular is, with regard to the latter: "Having the witness of one's own judgement or feelings, having the witness within oneself, knowing within oneself, inwardly sensible or aware." Whenever one has consciousness of . . . or that . . . in the second *OED* sense, one is conscious to oneself of . . . or that . . . , or, as I shall also say, conscious$_2$ of . . . or that. . . . Both of the latter, "conscious to oneself" and "conscious$_2$," will serve here as adjectival descriptors of someone who is, was, or will be instantiating consciousness$_2$.

Is there an external kind of knowing which, in the previous definition of "conscious to oneself," the *OED* is implicitly contrasting with the internal kind that is consciousness$_2$? The previous definition of "conscious to oneself" implies that the witnessing could be external to oneself; the testimony could be within someone else, instead of being within oneself, although it is not external to oneself in those cases picked out by the proper use of "conscious to oneself." Indeed, one may partake of the witness of someone else's judgments or feelings—if he or she reports about them to one. However, so to partake is not equivalent to being conscious to oneself of what the other person has witnessed. In the very process of being conscious to oneself (i.e., in being conscious$_2$ of something), the witness cannot be someone else; it is oneself alone—one knows within oneself.

Thus, the previous definition of "conscious to oneself" suggests not only (1) the exclusivity of the personal knowing involved in consciousness$_2$, but also (2) to be conscious to oneself of . . . , or that . . . , is to know something that one may want to keep to oneself, something about which one would keep one's own counsel. This is consistent with the original source of the words "consciousness" and "conscious," as well as the word "conscience." Near the start of its entries for "conscience," the *OED* helpfully identifies the Latin source as follows:

> L. *conscientia* privity of knowledge (with another), knowledge within oneself, consciousness, conscience, f. *conscient-* [present participle] of *conscire*, f. *con-* together + *scire* to know; thus *conscire alii* to know along with another, to be privy with another to a matter, thence, *conscire sibii* to know with oneself only, to know within one's own mind.

"Consciring"

C. S. Lewis (1967) coined the word "consciring" to refer to knowing together, or sharing the knowledge of something with someone, that is, to the state of affairs that the Latin *conscire* was primarily employed to indicate. Lewis explained that his new English word would have a use only in cases of shared knowledge between two or a few people in a secret together with respect to that knowledge. Also, just as the Latin allowed one to be one's own *conscius* or *conscia*, that is, to be the one who knows along with oneself, so Lewis generalized the meaning of his word "consciring" to the reflexive, or intrapersonal, case. Lewis justified his latter move by means of a number of quotations from literature. And, drawing upon common knowledge, he eloquently stated,

> Man might be defined as a reflexive animal. A person cannot help thinking and speaking of himself as, and even feeling himself to be (for certain purposes), two people, one of whom can act upon and observe the other. Thus he pities, loves, admires, hates, despises, rebukes, comforts, examines, masters or is mastered by, "himself." Above all he can be to himself in the relation that I have called consciring. He is privy to his own acts, is his own *conscius* or accomplice. And of course this shadowy inner accomplice has all the same properties as an external one; he too is a witness against you, a potential blackmailer, one who inflicts shame and fear. (p. 187)

What more specifically is involved in the makeup of the psychological state, relation, or activity that is Lewis's consciring with oneself? Clearly from the aforementioned statement, consciring is not to be equated with self-observation, self-awareness, or self-referential behavior, nor with emotion, thought, or action that is directed on oneself. Therefore, are intrapersonal consciring and the *OED*'s consciousness$_2$ the same? In a deeper sense, what is it for a person to be his own *conscius* or her own *conscia*? What is it for someone to have the testimony or witness within himself or herself? Having learned to distinguish particular cases of consciousness$_2$ from similar though different psychological phenomena, we will still want to know: In what does consciousness$_2$ consist? What more exactly is going on when one is conscious$_2$?

Relation to Consciousness qua Guilty Awareness of Wrongdoing

The *OED*'s definitions for "conscious to oneself" (quoted at the beginning of the preceding subsection) and "consciousness" in its second sense (quoted near the beginning of the present chapter) are, I believe, too broadly drawn. Surely, the compilers of the *OED* did not intend to include under the second concept so large a variety of psychological phenomena. The two definitions—on their own (i.e., ignoring etymology and examples of usage)—allow for too many possible kinds of items that one might attest to, or be in a position to attest to, and thereby be conscious$_2$ of . . . or that. . . .

There are at least two grounds for thinking that the definitions are too broad for capturing the true second sense of consciousness. I develop the first ground in this subsection, the second in the subsection right after the next one. The first ground pertains to a further meaning of "conscious," which the *OED*'s compilers considered to be very closely related to the meaning of "conscious to oneself."[5] The second ground is provided by the *OED*'s totality of quotations taken from literature for the purpose of illustrating either the use of "consciousness" in its second sense or the corresponding uses of "conscious to oneself" and "conscious."

The entry under "conscious" immediately after the one for "conscious to oneself" is comprised of two subentries. The first subentry attributes to "conscious" on its own the same meaning as for "conscious to oneself." The second subentry gives the following, distinct definition of "conscious": "Having guilty knowledge (*of* anything); *absol.* inwardly sensible of wrong-doing, guilty." This use of "conscious" is no longer current; yet the connection between being conscious$_2$ and being conscious in this old sense is no less pertinent.

Guilty awareness of wrongdoing is not a feature of every case of consciousness$_2$. However, it is such a feature often enough for "conscious" to have meant, in certain contexts, one's feeling guilty concerning something about which, being conscious$_2$ of it, one is in a position to attest. Compare the *OED*'s several illustrative quotations for the old sense of "conscious"; here are two of them: (1) "She being conscious, did of her own accord . . . make confession of her wickedness;" (2) "*conscious*, inwardly guilty, privy to ones self of any fault or errour." The reference of "conscious" in this sense was not simply to the fact that someone instantiated consciousness$_2$. Feelings of guilt played a necessary role in this further way in which an individual could be conscious.

Both quotations clearly implicate consciousness$_2$. It was not by hearsay that "she" knew of her wickedness, faults, or errors; rather, she had within herself the witness to the personal facts about which she felt guilty. However, to have guilty awareness of wrongdoing is more than to be conscious$_2$ that one has done something wrong. Guilty feelings are not essential to being conscious$_2$. This can be seen from the *OED*'s relevant definitions, and especially from the examples of usage accompanying them.

Note too that any case of guilty awareness of wrongdoing is also a case of being conscious$_3$ of something. Yet the respective concepts are not as closely related as the pair under discussion. The uniquely close relation between being conscious$_2$ and having guilty awareness of wrongdoing strongly suggests that what one can be conscious$_2$ of is not just anything at all; nor can it be just anything at all about oneself. As Lewis (1967, pp. 188–189) points out, one would fail to grasp some of what was going on if one construed the statement "I am conscious to myself of many failings" as though its author was making reference to a psychological state that did not involve intrapersonal consciring, that is, that did not involve his being "privy to himself, in his own secret" as regards many of those failings.

Distinct from Consciousness$_4$

The *OED* provides a total of 17 illustrative quotations that are proposed to be instances of exercising the concept of consciousness$_2$. This total does not include, of course, the quotations illustrating the use of "conscious" to refer to guilty awareness of wrongdoing. Nevertheless, many of the 17 statements imply that the individual referred to is not merely in a position to give testimony; also, he or she feels guilty regarding the witnessed state of affairs—not for having witnessed it, of course, but for what that state of affairs is or indicates about him or her. Moreover, except for one case out of the 17, the author is speaking of someone's being aware of something personal to him or to her that possesses ethical significance as defined broadly (to be explained).

Before I develop the latter point, let me address the single aberrant example. This example turns out to be useful for the present purpose because it has reference to a basic kind of consciousness that is distinct from consciousness$_2$. It is the kind that I have been calling "consciousness$_4$" following the *OED* (e.g., Natsoulas, 1983, 1986–1987, 1994a). Also, I have been calling consciousness$_4$ "inner awareness," because consciousness$_4$ is the direct, noninferential awareness that we all have of some of our own mental-occurrence instances. Although there is disagreement among theorists concerning the process by which inner awareness is accomplished (e.g., Brentano, 1911/1973; Dulany, 1991, p. 105; Humphrey, 1987; Natsoulas, 1993b; Rosenthal, 1993; Woodruff Smith, 1989), it is the rare psychologist who will deny that we possess the ability to be directly aware of some of our own mental-occurrence instances (e.g., Hebb, 1972, 1982; cf. James, 1890/1950, pp. 304–305).[6]

The stray *OED* example of usage is none other than John Locke's (1706/1975) application of "conscious to oneself" in the first sentence of this passage from *An Essay Concerning Human Understanding*:

> If they say, That a man is always conscious to himself of thinking; I ask
> How they know it? Consciousness is the perception of what passes in a

Man's own mind. Can another man perceive, that I am conscious of any thing, when I perceive it not my self? No Man's knowledge here, can go beyond his Experience. (p. 115)

Clearly, Locke is not speaking here of consciousness$_2$. His foregoing use of "conscious to oneself" should have been included in a different *OED* entry. In the previous passage, Locke is using both "consciousness" and "conscious to himself" in a very important psychological sense. He is referring to a kind of consciousness that is more basic than consciousness$_2$. He is clearly referring to inner awareness.[7]

To say, as I just did, that consciousness$_4$ is more basic than consciousness$_2$ is not to side, so to speak, with one kind of consciousness against another, in the fashion of some psychologists. Among much else, the *OED*'s six kinds of consciousness are *all of them* our subject matter, and *in no part* to be denied. If consciousness$_4$ is the more basic of the two, then it is essential to study consciousness$_4$ in order fully to understand consciousness$_2$. However, study of consciousness$_4$ cannot substitute for study of consciousness$_2$, except on pain of omitting psychological phenomena of great human importance.

But how did Locke's statement find its way into the *OED* entry for "conscious to oneself"? The form of words must have been responsible; as was, no doubt, the fact that every case of consciousness$_2$ instantiates the kind of "inwardness" that Locke was adverting to in the previously quoted passage. The latter fact will be seen in the final subsection of the present chapter, which bears the title The Self-Awareness Involved.

Locke's kind of consciousness is described later in the *OED*'s two lists of entries. The second sentence of the quoted passage from Locke serves as one of the illustrative quotations in the first half of the *OED*'s fourth entry under "consciousness." Indeed, the compilers of the *OED* could have adopted "Consciousness is the perception of what passes in a Man's own mind" for their definition of "consciousness" in the fourth sense.[8]

One's being conscious$_4$, or having inner awareness (cf. Brentano, 1911/1973) of one's thinking now going on, or of any other kind of mental activity or mental state or event now taking place within one, is a distinct kind of consciousness. It is distinct with two qualifications:

1. Consciousness$_4$ is part of what essentially occurs whenever one of certain of the other five *OED* kinds of consciousness happens to be instantiated.
2. Every instance of consciousness$_4$ is also an instance of consciousness$_3$—which is, as already identified, occurrent awareness of anything at all by whatever means, including however it may be that, whatever the process by which, we have inner awareness.

The familiar distinction between conscious and nonconscious mental-ocurrence instances amounts to whether or not a particular mental-occurrence instance is an object of inner awareness. Also, we now commonly speak of conscious mental occurrences and nonconscious mental occurrences, meaning, respectively, mental occurrences that can and cannot be the objects of inner awareness; that is, to qualify for the epithet "conscious," a mental occurrence need not be an object of inner awareness on every occasion of its taking place, but it must be capable of being such an object.[9]

Like consciousness$_3$, consciousness$_4$ is instantiated with very great frequency in everyday life (Natsoulas, 1993a; contrast Delius, 1981). And consciousness$_4$ certainly does not require that one be instantiating consciousness$_2$ at the time. Typically, consciousness$_2$ is, as it were, a more reflective process; that is, consciousness$_4$ occurs most often merely in passing, as one orients oneself with reference to what is taking place in one's "stream of consciousness" (James, 1890/1950, 1892/1984).[10] However, as already mentioned and as will be seen in some detail, in every case of being conscious$_2$, inner awareness of mental-occurrence instances does take place as a feature of consciousness$_2$ itself.

At the start of the preceding paragraph, I contradicted a conviction that would seem to be held by many psychologists. Consciousness$_4$ is supposed to take place under special circumstances, namely, when one has a task or a problem requiring that one determine what exactly one is now experiencing. Otherwise, one's awarenesses are directed on the environment and on oneself as inhabiting the environment, but not on the stream of consciousness itself. In this erroneous view, although the stream of consciousness always has the potential to be consulted (i.e., it is open to one's consciousness$_4$), there is normally no need to perform such a consultation.

I disagree with this view frequent among psychologists. They so hold, I believe, because they construe consciousness$_4$ as a process of reflecting on one's mental life, that is, as a matter of turning one's thoughts back upon one's mental life. They fail to realize that, in the absence of inner awareness, we would be ignorant as we go along of our seeing, feeling, thinking, or intending whatever we may be seeing, feeling, thinking, or intending (Natsoulas, 1993a). Absent all inner awareness, we would be, with respect to all of our mental life, like people who possessed only "blindsight" over the entirety of their field of view (Weiskrantz, 1993; cf. Natsoulas, 1996). We would be "mindblind." All of our self-awarenesses would result either from perceiving our bodies and behaviors by means of our senses or from apprehending ourselves as objects of thought.[11] The stream of consciousness could not be, in any part, its own direct object (cf. James, 1890/1950, pp. 304–305).

The Possible Objects of Consciousness$_2$

With the exception of that stray statement quoted from Locke, the *OED*'s 17 illustrative quotations for "consciousness" in the second sense, for "conscious to oneself," and for "conscious" in the same sense characterize one or more persons, respectively, as being conscious$_2$ of his, her, or their own

> wants; ignorance; guilt; wrongdoing; immeasurable superiority to others; well-spent life; great weakness; great defectiveness; useless medicines; having done everything possible to warn the nation; offence; highest worth; errors of omission; defects and vices; unfitness; or having engaged in adultery.

Whoever was said to be conscious$_2$ of any one of most of these items probably felt guilty about instantiating it. However, the exceptions to the latter concomitance are useful in helping us to determine the range of items of which one can be said to be conscious$_2$. The following five are among the exceptions:

1. Happy in the consciousness$_2$ of a well-spent life, one is likely to have applied moral standards to one's past actions and activities in coming to this consciousness. In judging one's life to have been well spent, one may also have applied other kinds of standards, in addition to or instead of the moral. In any case, one's favorable judgment implies an absence of feelings of guilt at least concerning the general way in which one has spent one's life.

2. If one were conscious$_2$ of being immeasurably superior to other people and drew support from being so—just as the respective illustrative quotation states about someone called Bentley—one would not feel guilty about this state of affairs, which one judged to be the case, although one would likely keep one's judgment to oneself.

3. In *A Journal of the Plague Year*, Daniel Defoe (1722/1960) states, "Abundance of quacks too died, who had the folly to trust to their own medicines, which they must needs be conscious to themselves were good for nothing" (p. 43). These quacks could not but be conscious$_2$ of the uselessness of the medicines that they were dispensing. Thus, they could not but be conscious to themselves of their own powerlessness against the plague. Nevertheless, Defoe goes on to say, they failed to feel guilty concerning their fraudulent practice. And, curiously, they did not act in such a way as would have avoided the punishment that they deserved. Defoe would seem to be suggesting that these quacks were deeply engaged in self-deception. They placed faith in medicines of which they were conscious to themselves could not protect them from illness, or restore them to health if they contracted the disease, as was very likely.

4. John Milton (1667/1935, II, 428) describes Satan as conscious of

his "highest worth" when Satan announces to his fellows that he alone will take on the great task that the rest of them feared. Satan's "monarchal pride" is mentioned in this connection, but no feelings of guilt are implicitly or explicitly ascribed to him.

5. In a letter, Edmund Burke (1779/1963) wrote, "I am low and dejected at times in a way not to be described. The publick Calamities affect me; and would much more, if I were not conscious to myself of having done every thing in my power to warn the Nation of the Evils that were bringing upon them" (p. 125). Although the government's actions and inactions were endangering the nation's safety from its enemies, Burke did not have to feel guilty, at least, for having remained silent regarding the situation. He was conscious to himself of having done all that he could in this respect.

Now, one's being immeasurably superior to other people, for example, or one's possessing the highest worth, is not obviously the kind of personal characteristic that one might feel guilty about. The potential for guilty feelings concerning the personal characteristic being judged would not seem to delimit the potential objects of consciousness$_2$. If every object of consciousness$_2$ is not to be distinguished by accompanying feelings of guilt regarding one's instantiating it, what do all the possible objects of consciousness$_2$ have in common?

I mentioned earlier that a large number of them clearly do have ethical significance. However, this statement cannot be generalized without a special interpretation of the ethical that includes intellectual powers and achievements within that category. For example, one can be conscious to oneself of being ignorant regarding how parts of one's body specifically function, or conscious to oneself of one's superior ability in treating the sick.

James's Spiritual Self

William James's discussion of "the spiritual self" in *The Principles of Psychology* (1890/1950) can help to explain what all of the possible objects of consciousness$_2$ have in common. It would seem that everything of which one can be conscious$_2$ has direct relevance to those features of a person that James considered to be parts of his or her spiritual self. The possible objects of consciousness$_2$ are either psychological characteristics of the kind constituting the spiritual self, or they are behaviors or mental occurrences that indicate one's having a particular one or more such characteristics.[12]

James (1890/1950) defined the spiritual self as

a man's inner or subjective being, his psychic faculties or dispositions. . . . These psychic dispositions are the most enduring and intimate part of the self, that which we most verily seem to be. We take a purer self-satisfaction when we think of our ability to argue and discriminate, of our moral sensibility and conscience, of our indomitable will, than when we survey any of our other possessions. Only when these are altered is a man said to be *alienatus a se*. (p. 296)

We gather better what James's spiritual self consists of in its entirety when James explains that "spiritual self-seeking" corresponds to "every impulse towards psychic progress, whether intellectual, moral, or spiritual in the narrow sense of the term [i.e., religious]" (p. 309). Thus, a person's spiritual self, when it is considered abstractly, is equivalent to all of the psychological powers, abilities, traits, dispositions, and tendencies that make up the intellectual, moral, and religious dimensions of his or her personality.

When, instead, the spiritual self is theoretically considered by James as something concrete, then it amounts simply to the stream of thought or mental life. In contrast, the aforementioned features, which are said to make up the spiritual self, do so abstractly; that is, although they can be judged or inferred about by the person whose spiritual self it is (and by others as well), they cannot be directly apprehended themselves, whether by inner awareness or in any observational sort of way.

James states that the spiritual self may be theoretically considered both abstractly and concretely. This does not mean that someone's spiritual self amounts merely to how that individual thinks it is in his or her own case, that is, to a kind of conception that one keeps on developing of an important part of oneself. Certainly, one can think about one's spiritual self and come to conclusions concerning each of the features that comprise it. But these features are one thing and how we construe them is another thing, not to be confused with them. The features of the spiritual self are each something objective about the individual and susceptible to being gotten wrong by the individual and others. That certain of an individual's powers or tendencies largely manifest themselves by affecting his or her stream of consciousness does not make them any less objective.

James's inclusion of the intellectual as part of the spiritual self is, of course, intriguing. For we commonly hold that the intellect and its development may serve a variety of purposes, including morally neutral ones, spiritually irrelevant ones, or worse. I want to be sure to emphasize that James considered the intellectual to be part of the spiritual self only insofar as intellectual ends are being pursued for themselves alone, without concern for any material or social advantage—not even God's approval—that may result from the intellect's advancement. According to James and others of his time, the desire to improve one's intellectual functioning derives from the same group of basic instinctive impulses as do moral and spiritual purposes.

The Self-Awareness Involved

Certain kinds of judgments are a part of being conscious$_2$. However, judgments do not comprise all that any instance of consciousness$_2$ essentially involves. Another essential part of being conscious$_2$ is the *basis* of the specific judgments involved. Elsewhere, I have indicated what this basis is in the following words:

> I newly learn or remind myself, on a firsthand basis (not from hearsay), about the kind of person I am in one or another specific respect. . . . I newly learn or remind myself of this, from having witnessed relevant actions I performed or experiences I had, and by now bringing this evidence to bear on how I conceive of myself, in terms of a trait or ability I therefore consider myself to possess, on perhaps other grounds as well. (Natsoulas, 1991b, p. 344)

In any instance of consciousness$_2$ that is based on evidence from the past, several kinds of self-awareness must take place, as described next; one cannot be conscious$_2$ in their absence.

Witnessing of Oneself

Consciousness$_2$ cannot be, as already noted, about just anything at all. Rather, one is, or has been at some point, aware firsthand, on the spot, at the time of its occurrence, of a certain relevant piece of one's own behavior or a certain relevant part of one's stream of consciousness. This behavior or the particular segment of one's stream must be suitable for serving as evidence about one's intellectual, moral, or religious powers, abilities, traits, dispositions, or tendencies. The evidence cannot pertain instead to merely social or material facts about oneself. Evidence pertaining only to one's health, physical attributes, kinship, wealth, social status, sociability, and the like, would not be relevant, because consciousness$_2$ is always about one or more aspects of one's spiritual self. Of course, in evaluating evidence concerning an aspect of one's spiritual self, one may take into account such factors as the aforementioned. One may acknowledge one or more of these factors as determinants of one's intellectual performance for example. However, it is not these factors that one has consciousness$_2$ of or about; rather, it is the relevant outcomes to which they may have contributed, such as a habitual way of thinking about the world or treating other people.

Appropriation to Oneself

The self-witnessing just mentioned cannot be of an alienated sort. In James's term, the self-witnessing must "appropriate" to oneself that which is witnessed. One must not only be aware firsthand of the particular piece of one's behavior or segment of one's stream of consciousness that would serve as evidence, but also one must be aware of that piece or segment as being one's own. In this additional sense, the self-witnessing must be "personal." It is possible for one to have, instead, inner awareness wherein one's stream of consciousness seems to have been taken over by another agent or, at least, seems not to be one's own (Natsoulas, 1979). Also, one may observe someone behaving (e.g., as a result of an arrangement of mirrors or by means of video equipment) and not realize it is oneself whom one is observing. Reed (1972) brings out that actions, as well as temporal sections of one's stream of consciousness, may suffer "a loss of personal attribution," either transiently, as a result of situational stress in the case of normal people, or more consistently, as part of a schizophrenic syndrome.

Retrowareness of Oneself

Instances of consciousness$_2$ very frequently take place at a temporal distance from the specific self-witnessing that furnishes evidence for them. For example, something of yourself that you witnessed at a certain, perhaps early point in your life may repeatedly serve you as evidence for judgments regarding the kind of person you are. Thus, remembering proper is usually a part of being conscious$_2$, and all instances of such remembering, whether or not they take place as part of consciousness$_2$, necessarily involve occurrent awareness now of a past happening or state of affairs. In being conscious$_2$, one now remembers something that one previously did or underwent; that is, consciousness$_2$ usually involves a kind of self-awareness that is a "retrowareness" (Natsoulas, 1986). For example, I am apprehending in thought right now, not for the first time, how I behaved (i.e., a particular action of mine) at the party my parents gave to celebrate the 12th anniversity of my birth.

Inner Awareness

For an occurrent awareness to qualify as an instance of remembering proper, the awareness must be a retrowareness of a special kind. It does not suffice that one was originally aware firsthand of the object of the particular retrowareness. In an act of remembering something in particular, one must be occurrently aware now of oneself as now apprehending that which one had earlier apprehended. Thus, necessarily involved at the point of remembering is inner awareness (consciousness$_4$) of one's present self-retrowareness.

Consciousness₄ Extended Backward

And there is involved, as well, a kind of retrowareness that resembles inner awareness. There occurs a present retrowareness of one's past experience. By past experience, I mean, for example, one's perceiving, emoting over, thinking about, planning, remembering, or expecting something or other in particular that was taking place, had taken place, or was going to take place. Thus, in those many cases of consciousness₂ wherein one remembers witnessing the relevant evidence, there occurs both (1) inner awareness of present components of one's stream of consciousness that are retroware-nesses of the evidence witnessed and (2) inner awareness of present compo-nents of one's stream of consciousness that are retrowarenesses of some of the mental acts that took place as part of one's witnessing the evidence.

Self-Thoughts and Self-Judgments

There is more to consciousness₂ than remembering the right sort of item. One is conscious₂ not simply because one is remembering one's act of wit-nessing something about oneself, however important the latter feature of oneself may happen to be and whatever it may be that one remembers. Only through putting what one remembers to use does it become evidence. It becomes evidence by being brought to bear on something that one al-ready believes or something that one may come to believe. To be con-scious₂, one must put what one remembers to use as evidence regarding the kind of person that one is with respect to the spiritual sphere as James broadly defined it.[13] In addition to one's undergoing retrowareness of the particular past events as being occurrent parts of oneself, one must have thoughts regarding one or more characteristics that may belong to the in-tellectual, moral, or religious dimensions of one's personality, and one must make judgments regarding how the remembered evidence bears on whether those characteristics do so belong.

The necessary emphasis on remembering as a feature of conscious-ness₂ based on past evidence should not serve as a distraction from the fact that consciousness₂ from present evidence also takes place. It can occur on the spot, that is, while one is still engaged in witnessing the relevant behav-ior or segment of one's stream of consciousness. Spelling out the self-awareness involved in contemporaneous consciousness₂, as I have done for consciousness₂ from past evidence, would be a belaboring of the closely analogous features. Indeed, the main points of the present subsection on the self-awareness involved in consciousness₂ can now be, in effect, very briefly reviewed simply by noting how the present analysis extends quite naturally to contemporaneous consciousness₂ as well. Accordingly, whenever any instance of consciousness$_2$ takes place at the time of the occurrence of the evidence on which it is based, there is essentially involved self-awareness in, at the least, all of the following forms:

- One witnesses potential evidence about oneself.
- One has inner awareness of this witnessing.
- One has occurrent awareness in thought of one or more features of one's character or personality.
- One brings self-witnessed evidence to bear in judging of this feature or these features.

NOTES

1. However, see the Cambridge physiologist Horace B. Barlow's (1987) use of the obsolete interpersonal sense of "conscious" to make his case for a certain scientific understanding of consciousness (cf. Barlow, 1980). For some discussion of Barlow's view in relation to the *OED*'s first sense of "consciousness," see Natsoulas (1991a).

2. Nor do we still use "conscious" in the corresponding sense that the *OED* defines as follows: "Knowing, or sharing the knowledge of anything with another; privy to anything with another."

3. Compare the following statement from James's *The Principles of Psychology* (1890/1950):

> No one ever had a simple sensation by itself. Consciousness from our natal day is always of a teeming multiplicity of objects and relations, and what we call simple sensations are results of discriminative attention, pushed often to a very high degree. It is astonishing what havoc is wrought in psychology by admitting at the outset apparently innocent suppositions, that nevertheless contain a flaw. (p. 224; see also p. 478)

James held that all basic durational components of the stream of consciousness—therefore, all mental-occurrence instances—possess the property of intentionality (cf. Brentano, 1911/1973); that is, all are "intellections," all have a cognitive aspect—which does not mean that they are not, each of them, feelings as well according to James.

4. In the present chapter, I abbreviate and simplify what needs to be said regarding consciousness$_3$ (see Natsoulas, 1992b, 1995). Also, I do not discuss views, such as James's (1890/1950), to the effect that every basic durational component of one's stream of consciousness intrinsically involves awareness of one's body or of certain occurrences in one's body. This would mean that every instance of consciousness$_3$ is a kind of self-awareness by involving awareness of a part of oneself (see Natsoulas, 1996–1997).

5. The closeness of meaning is of unusual degree, as will be seen. There are a number of entries listed under "conscious" in the *OED*. But the remaining mean-

ings, beyond those already mentioned in the text, need not be brought into the present discussion of consciousness$_2$.

6. A particular theoretical disagreement that is relevant to the relation between consciousness and self-awareness is the one between (1) those who hold that inner awareness requires the ascription or appropriation of its objects to oneself (e.g., Kihlstrom, 1987; Rosenthal, 1986), and (2) those who allow for impersonal or anonymous inner awareness (James, 1890/1950; Woodruff Smith, 1989). For objections to Rosenthal's view, see Natsoulas (1992a). For discussion of this issue with special reference to James, see Natsoulas (1996–1997).

7. Margaret Atherton (1983/1992) does not agree. I have argued against her position elsewhere (Natsoulas, 1994b). It serves no present purpose, however, to summarize or to consider her understanding of Locke again here.

8. Except for one consideration: Insofar as we know our own mental life firsthand, do we really know it by the efficacy of a perception-like process, by, as it were, "inner spection," or Locke's power of reflection? This is a controversial thesis (see Brentano, 1911/1973, pp. 429–434; Woodruff Smith, 1989, pp. 83–88).

9. Thus, all the basic durational components of James's (1890/1950) stream of consciousness are conscious mental occurrences, although inner awareness of every one of them does not occur. But any one of them that happens not to be an object of inner awareness could have been such an object. All of the basic durational components of the stream are open to inner awareness, as they come into existence one after another. However, James's notion of the specious present renders his understanding of inner awareness somewhat more complex than I have intimated here; see Natsoulas (1992–1993).

10. Compare the phenomenologist Aaron Gurwitsch (1964): "Our mental activity is always [?] accompanied by awareness of facts and data belonging to the following three orders of existence: 1. *The stream of our conscious life;* 2. *our embodied existence;* 3. *the perceptual world*" (p. 415).

11. But what could such self-awareness amount to? Lacking inner awareness, we would have to infer the occurrence of an instance of self-awareness from something about ourselves that we could observe. But we could not have inner awareness of any observations and would have to infer their occurrence as well—and so on.

12. James also discussed other constituents of the self, none of which are mentioned in the text of the present chapter, namely, the material self, the social self, and the pure ego (whose existence James rejected). Their exclusion from the present chapter, however, does not signify a judgment against their importance for understanding self-awareness. Rather, as should be clear, James's notion of the spiritual self serves here as a likely key to comprehending consciousness$_2$.

13. Putting evidence to use is indeed what one must do in being conscious$_2$ of . . . or that . . . ; but this need involve no more than drawing a conclusion concerning oneself, partially or entirely on the basis of that evidence. Thus, I do not mean that potential evidence must be worked over, although this may be necessary in some cases if it is to contribute to one's conclusion. What may serve as evidence can point in more than a single direction regarding the kind of person one is, or who is responsible for what happened.

REFERENCES

Atherton, M. (1992). Locke's theory of personal identity. In V. Chapell (Ed.), *John Locke* (pp. 85–105). New York: Garland. (Original work published 1983)

Barlow, H. B. (1980). Nature's joke: A conjecture on the biological role of consciousness. In B. D. Josephson & V. S. Ramachandran (Eds.), *Consciousness and the physical world* (pp. 81–90). Oxford, England: Pergamon.

Barlow, H. B. (1987). The biological role of consciousness. In C. Blakemore & C. Greenfield (Eds.), *Mindwaves* (pp. 361–376). Oxford, England: Blackwell.

Brentano, F. (1973). *Psychology from an empirical viewpoint.* New York: Humanities Press. (Second German edition published 1911)

Burke, E. (1963). *The correspondence of Edmund Burke* (Vol. 4). Cambridge, England: Cambridge University Press. (Letter cited written 1779)

Defoe, D. (1960). *A journal of the plague year.* New York: Signet/New American Library. (Original work published 1722)

Delius, H. (1981). *Self-awareness.* Munich: Beck.

Dewey, J. (1906). The terms "conscious" and "consciousness." *Journal of Philosophy, Psychology and Scientific Method, 3,* 39–51.

Dulany, D. E. (1991). Conscious representation and thought systems. In R. S. Wyer, Jr., & T. K. Srull (Eds.), *Advances in social cognition* (Vol. 4, pp. 97–119). Hillsdale, NJ: Erlbaum.

Engelberg, E. (1972). *The unknown distance.* Cambridge, MA: Harvard University Press.

Gurwitsch, A. (1964). *The field of consciousness.* Pittsburgh: Duquesne University Press.

Hebb, D. O. (1972). *A textbook of psychology* (3rd ed.). Philadelphia: Saunders.

Hebb, D. O. (1982). Elaboration of Hebb's cell assembly theory. In J. Orbach (Ed.), *Neuropsychology after Lashley* (pp. 483–496). Hillsdale, NJ: Erlbaum.

Humphrey, N. K. (1987). *The uses of consciousness* (Fifty-seventh James Arthur Lecture on the Evolution of the Human Brain). New York: American Museum of Natural History.

Husserl, E. (1970). *Logical investigations* (2 vols.). New York: Humanities Press. (Original work published 1900)

James, W. (1950). *The principles of psychology* (2 vols.). New York: Dover. (Original work published 1890)

James, W. (1984). *Psychology.* Cambridge, MA: Harvard University Press. (Original work published 1892)

Kihlstrom, J. F. (1987). The cognitive unconscious. *Science, 237,* 1445–1452.

Koch, S. (1980). Psychology and its human clientele: Beneficiaries or victims. In R. A. Kasschau & F. S. Kessel (Eds.), *Psychology and society* (pp. 30–53). New York: Holt, Rinehart & Winston.

Lewis, C. S. (1967). *Studies in words* (2nd ed.). Cambridge, England: Cambridge University Press.

Locke, J. (1975). *An essay concerning human understanding.* Oxford, England: Clarendon. (Fifth edition published 1706)

Mead, G. H. (1934). *Mind, self, and society.* Chicago: University of Chicago Press.

Miller, G. A. (1990). The place of language in a scientific psychology. *Psychological Science, 1,* 7–14.

Milton, J. (1935). *Paradise lost.* New York: Odyssey. (Original work published 1667)

Natsoulas, T. (1979). The unity of consciousness. *Behaviorism, 7,* 45–63.

Natsoulas, T. (1983). Concepts of consciousness. *Journal of Mind and Behavior, 4,* 13–59.

Natsoulas, T. (1985). George Herbert Mead's conception of consciousness. *Journal for the Theory of Social Behaviour, 15,* 60–75.

Natsoulas, T. (1986). Consciousness and memory. *Journal of Mind and Behavior, 7,* 463–501.

Natsoulas, T. (1986-1987). The six basic concepts of consciousness and William James's stream of thought. *Imagination, Cognition and Personality, 6,* 289–319.

Natsoulas, T. (1991a). The concept of consciousness$_1$: The interpersonal meaning. *Journal for the Theory of Social Behaviour, 21,* 63–89.

Natsoulas, T. (1991b). The concept of consciousness$_2$: The personal meaning. *Journal for the Theory of Social Behaviour, 21,* 339–367.

Natsoulas, T. (1992a). Appendage theory—pro and con. *Journal of Mind and Behavior, 13,* 371–396.

Natsoulas, T. (1992b). The concept of consciousness$_3$: The awareness meaning. *Journal for the Theory of Social Behaviour, 22,* 199–225.

Natsoulas, T. (1992-1993). The stream of consciousness: II. William James's specious present. *Imagination, Cognition and Personality, 12,* 367–385.

Natsoulas, T. (1993a). The importance of being conscious. *Journal of Mind and Behavior, 14,* 317–340.

Natsoulas, T. (1993b). What is wrong with appendage theory of consciousness. *Philosophical Psychology, 6,* 137–154.

Natsoulas, T. (1994a). The concept of consciousness$_4$: The reflective meaning. *Journal for the Theory of Social Behaviour, 24,* 373–400.

Natsoulas, T. (1994b). The concept of consciousness$_5$: The unitive meaning. *Journal for the Theory of Social Behaviour, 24,* 401–424.

Natsoulas, T. (1995). Consciousness$_3$ and Gibson's concept of awareness. *Journal of Mind and Behavior, 16,* 305–328.

Natsoulas, T. (1996). Blindsight and consciousness. *American Journal of Psychology, 110,* 1–34.

Natsoulas, T. (1996–1997). The stream of consciousness: XII. Consciousness and self-awareness. *Imagination, Cognition, and Personality, 16,* 161–180.

Pylyshyn, Z. (1990). Roundtable discussion. In P. P. Hanson (Ed.), *Information, language and cognition* (pp. 198–216). Vancouver: University of British Columbia Press.

Reed, G. (1972). *The psychology of anomalous experience.* London: Hutchinson.

Rosenthal, D. M. (1986). Two concepts of consciousness. *Philosophical Studies, 49,* 329–359.

Rosenthal, D. M. (1993). Higher-order thoughts and the appendage theory. *Philosophical Psychology, 6,* 155–166.

The Oxford English dictionary. (1989). (2nd ed., 20 vols.). Oxford, England: Clarendon.

Weiskrantz, L. (1993). Unconscious vision. *The Sciences, 32*(5), 23–28.

Wittgenstein, L. (1980). *Remarks on the philosophy of psychology* (Vol. 1). Oxford, England: Blackwell. (Original German typescript dated 1947)

Woodruff Smith, D. (1989). *The circle of acquaintance.* Dordrecht, The Netherlands: Kluwer.

PART TWO
THE EVOLUTION OF SELF-AWARENESS

CHAPTER THREE

When Self Met Other

✧

DANIEL J. POVINELLI
CHRISTOPHER G. PRINCE

"Starting from what I know of the operations of my own individual mind, and the activities which in my own organism they prompt, I proceed by analogy to infer from the observable activities of other organisms what are the mental operations that underlie them." So wrote John George Romanes (1882, pp. 1–2), outlining his method for gaining scientific leverage on the problem of animal minds. In the wake of Darwin's (1871/1982) publication of *The Descent of Man*, Romanes took up the challenge of investigating the evolution of mind with a fervor. Darwin had marshaled an impressive array of observations to suggest that not only were humans descended from other species in bodily structures, but also in mental structures. For Romanes, the agenda for a new science was clear. Just as anatomists "aim at a scientific comparison of the bodily structures of organisms," he observed, "so [comparative psychology] aims at a similar comparison of their mental states" (Romanes, 1883, p. 5). But Romanes faced a problem. Although his analogy between comparative anatomy and a new science of comparative psychology was powerful, it began to break down when it came to the substances to be compared. Anatomists had access to dead bodies, but the stuff of psychology was not so easily available for examination on the laboratory bench. Recognizing this, Romanes offered an interim solution by turning to the only source of material available—the spontaneous behavior of animals. But even here, he was far behind the anatomist in that there was no existing corpus of data. Romanes knew he might be roundly chastised for doing so—and he was—but ultimately he was forced to rely on anecdotes as his database. "If the present work is read without reference to its ultimate object of supplying facts for the subsequent deduction of principles," Romanes apologized, "it may well seem but a small improvement upon the works of the anecdote-mongers" (1882, p. vii).

Although the foundations of Romanes' approach collided with an age-old philosophical problem far more profound than that of inferring mental states in other species—the uncertainty of making inferences even about the minds of our fellow humans—in a very real sense, the core of his method was unassailable. After all, we are not just bodies cohabiting the same physical locations. We are thinking, feeling beings, negotiating a social world teeming with each others' desires, goals, intentions, and emotions. We are linked not just in space, but also in mind. Each of us possesses an intense desire to have our lives understood by others. Indeed, this desire to understand and be understood, to be part of a group—whether it be the village, school, family, or kibbutz—would appear to be as universal a trait of the human species as any.[1] This inextricable connection between self and other even occurs in our most silent moments. Someone looks around a corner, furrows his or her brow, and we effortlessly attribute a visual experience of having just seen something. In this sense, Romanes' solution to the problem of understanding animal minds was simple: If we can make reasonably accurate inferences about what other humans are thinking and feeling by just observing them, why can we not do the same with members of other species? If the method of introspective analogy can bridge the gap that separates the mind of one human from another, why cannot it not likewise span the distance that separates one species from another?

In this chapter, we return to Romanes' problem by examining the evolution of the psychological connection between self and other that appears so characteristic of our species. We explore three related questions. First, when did the ability to conceive of the self and others as mentalistic agents evolve? That is, when and in what lineages did organisms first evolve the capacities to reason about themselves and others in terms of mental states and events such as thinking, knowing, believing, attending, desiring, and perceiving? The data we review suggest that humans may be largely unique in being able to reason about such internal states. Second, we ask whether the evolution of self–other psychology occurred in synchrony, or whether an understanding of self as a mental agent preceded a comparable understanding of others. A main issue we address is whether the fusion of self and other understanding we observe in humans can be dissociated in other species. For example, the data we review are consistent with the hypothesis that although chimpanzees may possess at least a limited objective self-concept, they may lack the ability to conceive of others (and perhaps even themselves) as mentalistic agents. Finally, we address the thorny problem of how it is that nonhuman primates can share with us so many behavioral patterns that in humans are clearly associated with a mentalistic under-standing of others, yet not possess such understanding themselves. Our conclusion is that humans may have evolved these abilities not because they endowed us with scores of novel behaviors per se, but rather because they allowed for the complex reorganization and redeployment of existing behavioral patterns. Thus, although humans may indeed possess unique behavioral capacities (e.g., active pedagogy), the initial utility of interpret-ing behavior in terms of internal, nonobservable mental states may have been at an organizational level—at a higher level of abstraction than any specific set of behaviors. If this general idea is correct, it may mean that traditional efforts to find a coherence between an understanding of self and others as mental agents, and some set of naturally occurring behaviors, will remain largely futile.

In order to set the stage for examining these ideas, we review several aspects of current theorizing and research concerning self and social under-standing. First, by examining certain aspects of the spontaneous social be-havior of primates in nature, we explore how the traditional approach to understanding the evolution of social intelligence has led to the idea that many, if not most, nonhuman primates possess some understanding of the mental states of themselves and others. Second, we explore the evidence concerning the evolution of self-conception and explain why these data have been interpreted as suggesting a qualitative psychological difference between the great apes and humans on the one hand, and most other forms of life on the other. Third, we examine experimental data concerning the development of one aspect of human infants and children's understanding of their own and others' minds, and parallel research with chimpanzees. Our interpretation of these data is that despite their striking similarity to us at the level of specific behavioral patterns, not even chimpanzees possess a theory-of-mind system comparable to that which develops in humans during late infancy and early childhood.

SELF, SOCIALITY, AND INTELLIGENCE

Evolving an Intelligence of Others

Alison Jolly (1966) authored the first careful statement of the idea that the truly remarkable features of primate intelligence had evolved in the context of coping with each other—a kind of social intelligence of others. Return-ing from an early field survey of the prosimian primates of Madagascar, Jolly was left with the impression that primate intelligence was far from a unitary concept. She reflected that intelligence about objects and physical events appeared to have evolved independently from intelligence deployed in the social realm. She minimally differentiated between intelligences "toward objects, including food; toward other active species, including predators; and toward fellow members of one's own species" (p. 504). Some of her own previous laboratory research had suggested that prosimians lagged far behind other primates both quantitatively and qualitatively on standardized tests of intelligence (Jolly, 1964a, 1964b; see also Andrew, 1962). In contrast, the sophistication of their social behaviors seemed comparable to what had been previously observed in various monkey species, including the characteristically long chains of social interactions.

Given that prosimians appeared able to generate the same level of social complexity as the anthropoid primates, but were far less proficient at traditional object-oriented laboratory tests of intelligence, it occurred to Jolly that prosimians might share a different form of intelligence with other primates—a social intelligence. As she explained:

> The social use of intelligence is of crucial importance to all social primates. As the young develop, they depend on the troop for protection and for instruction in their role in life. Since their dependence on the troop both demands social learning and makes it possible, social integration and intelligence probably evolved together, reinforcing each other in an ever-increasing spiral. And although it is very likely that the learned social relations of monkeys are in fact more complex than those of lemurs, our present techniques of description emphasize the similarity between lemur and monkey social interactions. (1966, p. 504)

Jolly's speculations amounted to suggesting a possible dichotomy or compartmentalization between these two types of intelligence. Indeed, experimental evidence from rhesus monkeys reared in social isolation was already hinting that there might be some validity to this claim by demonstrating that despite their inability to learn a variety of species-typical patterns of social behaviors, they performed like normal monkeys on most traditional, object-oriented intelligence tests (Harlow, 1965; Harlow, Schlitz, & Harlow, 1968).

Nicholas Humphrey (1976) independently reached a similar conclusion in the mid-1970s. Like Jolly, he noted that most primates, on the surface, appeared to possess a veritable "surplus" of intelligence: a seemingly unneeded cornucopia of intellectual abilities unrelated to the demands of their way of life in nature. Given his view of nature as an uncompromising optimizer, Humphrey quickly reached the conclusion that the idea that such a surplus truly existed "was most likely to be wrong" (p. 303). The solution, he reasoned, might lie in the function that intellect played in the social arena:

> [The] social primates are required by the very nature of the system they create and maintain to be calculating beings; they must be able to calcu-

late the consequences of their own behaviour, to calculate the likely behaviour of others, to calculate the balance of advantage and loss—and
all of this in a context where the evidence on which their calculations are
based is ephemeral and liable to change, not least as a consequence of
their own actions. (p. 309)

And, also like Jolly, he took special note of the self-generating nature of intellect in this social context. Social intelligence was driven through an evolutionary ratchet effect, "acting like a self-winding watch to increase the
general intellectual standing of the species" (p. 311).

But what about the interconnection of self and other—did the social
intelligence hypothesis (as the Jolly–Humphrey idea came to be known)
have anything to say about the apparent fusion of these intelligences? In
later essays, Humphrey (1980, 1982) traced out the implications of his
ideas for precisely this problem. Marshaling a long-standing philosophical
position, he argued that once self-consciousness had provided our species
with a way of making introspective sense of our own behavior, then our
own experiences and ways of understanding behavior would "immediately
and naturally [be] project[ed] onto other people" (Humphrey, 1982, p.
477). Thus, like others before him, Humphrey concluded that humans
come to understand the inner thoughts and desires of those around them
through an introspective examination of their own mental states and
processes—a kind of mental simulation of what it must be like to be the
other person (for variations on this solution to the problem of other minds
see Hume, 1739; Smith, 1759/1961; Adams, 1928; Russell, 1948; Stewart,
1956; Schutz, 1962; Gallup, 1982; Gordon, 1986; Harris, 1991).

Humphrey's speculations about the evolution of the connection between self and social understanding were largely confined to humans—but
were other species capable of generating inferences about the mental lives
of those around them? Humphrey (1982) offered cautious skepticism on
this point: "I am not yet convinced that any other species has followed the
same path to consciousness as man . . ." (p. 477). But, to be fair, he noted
that "[i]t may turn out there are, in fact, non-human species . . . [that are]
making use of explanatory systems which bear the hallmarks of a mind capable of looking in on the inner workings of the brain" (p. 477). As we
shall see, at least one other comparative psychologist hypothesized that
chimpanzees and some other great apes might be in this very same epistemological position (see Gallup, 1982, 1983).

Social Manipulation and Deception in Primates

If Jolly and Humphrey were right, and the social dimensions of intellect
had been under intense selection due to the burgeoning sociality of the primates, then surely there must be examples of such complex uses of this in-

telligence in the day-to-day interactions of these animals. And indeed, by the late 1960s, the evidence for such maneuverings was already starting to surface. After a lengthy field study of hamadryas baboons, Kummer (1967) reported instances of females presenting and glancing at a dominant male while threatening a rival female. He speculated that this behavior functioned to manipulate the male into attacking the other female. By the mid-1980s, Smuts (1985) was urging primatologists to recreate the common sense of their discipline around emotional and mental states such as ambivalence, flirtation, trust, jealously, affection, and grief. For example, she offered preliminary evidence for a system of "selective retaliation" in savanna baboons by showing that after baboons had an aggressive encounter with more dominant animals, the animals apparently waited to retaliate—surreptitiously and selectively—against the relatively subordinate "friends" of the animals that had initially aggressed against them.

Meanwhile, other researchers had been reporting even more striking episodes of social manipulation—episodes that appeared to qualify as instances of intentional deception (e.g., Menzel, 1974). In Tanzania, for example, Jane Goodall (1971) reported several cases of chimpanzees acting in ways that seemed to suggest that they were capable of suppressing normal behaviors in order to mislead each other. A decade later, Frans de Waal's (1982) *Chimpanzee Politics*—a careful blend of quantitative data and anecdotal accounts—helped to establish chimpanzees' reputation as Machiavellian beasts par excellence (see also, de Waal, 1986). But chimpanzees were not to claim exclusive rights to such status for long, as evidence of deception in other primate (and nonprimate) species was soon to follow. Byrne and Whiten (1985) described examples of apparent deception among familiar baboons. They distinguished between deception well known to ethologists (such as mimicry and camouflage) and what they dubbed "tactical deception." They defined tactical deception as those instances in which an individual used an act from its normal behavioral repertoire in a unique context, so that it served to manipulate another individual. For instance, in one of the episodes they reported, a juvenile male baboon, who was being approached by threatening males, suddenly stopped, stood on his hind legs, and looked into the distance. This gesture was familiar to the team of researchers—it was a signal typically given in the context of an approaching predator or another group of baboons. The other males, apparently reacting to the warning, abandoned their hostile ambitions toward the juvenile and scanned the horizon as well. But when the human observer scanned the horizon, there was no trace of a predator or rival baboon troop. After several years of scouring the literature and soliciting unpublished accounts from other primatologists, Whiten and Byrne (1988) concluded that many different species of primates engage in tactical deception. In formalizing Humphrey's (1980) claim that primates are natural psychologists, Whiten and Byrne were attempting to revive the anecdotal method as a means of gaining insight into the social psychology of nonhuman primates.

As we have seen, Romanes (1882) had also attempted a careful use of anecdotes to infer the kinds of mental states other organisms must be experiencing. Likewise, after categorizing their examples of deception, Whiten and Byrne (1988) aimed to use these anecdotes to "sketch the features of the state of mind that an individual with deceptive intent must be able to represent" (p. 233). Whiten and Byrne noted that their database of anecdotes could never be more than a starting point for "more systematic" work (p. 243). But by "more systematic" they were not necessarily calling for experimental approaches to the question. Instead, they were directly advocating the collection of more detailed observations in natural settings. But could Whiten and Byrne's rebirth of the anecdotal method succeed where Romanes (1882, 1883) had failed? Even more generally, could the spontaneous behavior of free-ranging primates tell us anything at all about whether these animals interpret each other in terms of unobservable mental states?

The skeptics were unconvinced. One set of problems had to do with the ontogenesis of the behaviors. "The plural of anecdote," Irwin Bernstein (1988) quipped, "is not data" (p. 247). Gordon Burghardt (1988) agreed: "How does [Whiten and Byrne's] call for multiple records really counter the problems inherent in interpreting any '*single* observation'? How much problematic data adds up to one conclusive bit of evidence?" (p. 249). David Premack (1988) was more blunt: "How many 'trials' go into producing the anecdotes that are reported from the field? Since this is rarely known, readers are led to indulge their ignorance and to draw romantic conclusions" (p. 171). But a second set of problems hinged on the more fundamental issue of whether, regardless of their ontogenetic origins, such behavior must be based upon a mentalistic interpretive system. Premack put the problem simply: "In calling [an] observation an anecdote, I do not mean to question the reliability of the report or even, for that matter, the general accuracy of the circumstance in which the act occurs" (p. 162). But if many, highly reliable anecdotes could be coded and catalogued, why would the need for experimentation exist? "The need for experiments," Premack replied, "arises with respect to the interpretation we wish to place on the act" (p. 162). Thus, in the view of Premack and other skeptics, reports of the spontaneous actions of nonhuman primates, as intriguing as they are, bring us no closer to answering the fundamental question posed by Edward Thorndike's discovery of the law of effect over a century ago: Do animals understand social interactions in terms of behavior alone, or both behavior *and* mind?

Self-Recognition, Self-Conception, and Theory of Mind

While field researchers were observing the spontaneous behavior of nonhuman primates and speculating about the evolution of self and social understanding, an experimentally minded comparative psychologist, Gordon G. Gallup, Jr., was also reflecting on the relationship between the evolution of self-consciousness and the ability to reason about the minds of others. In a series of ingenious experiments conducted in the late 1960s, Gallup (1970) had uncovered a peculiar quirk of psychological evolution: Humans, chimpanzees, and presumably other great apes, were capable of recognizing themselves in mirrors, whereas various species of monkeys he tested were not. Gallup reported that after several days of mirror exposure, chimpanzees spontaneously learned to use mirrors to explore parts of themselves that they had never (or only rarely) had the opportunity to see before. Thus, although they initially reacted to the mirror as if they had been suddenly confronted with another chimpanzee, his subjects soon altered their disposition toward their images and began using the mirror to manipulate and explore their ears, eyes, noses, teeth, and anogenital areas—all while carefully monitoring the effects in the mirror (see Figure 3.1).

In order to provide more direct experimental evidence for his impression that the subjects had learned to recognize themselves, Gallup anesthetized the chimpanzees and marked them with a bright red, odorless, tactile-free dye on their upper eyebrow ridge and ear. These areas were selected because in absence of a mirror, the apes would not know that parts of their faces had been dyed red. After recovery from the anesthesia, the subjects were observed for 30 minutes in the absence of the mirror, and virtually never touched the marks. In direct contrast, and in clear support for the idea that they had learned to recognize themselves, when the mirror was reintroduced, the chimpanzees reached up to touch these otherwise invisible marks. Various aspects of Gallup's findings with chimpanzees have been replicated both within this species, and in orangutans, but generally not gorillas[2] (Lethmate & Dücker, 1973; Ledbetter & Basen, 1982; Suarez & Gallup, 1981; Swartz & Evans, 1991; Calhoun & Thompson, 1988; Povinelli, Rulf, Landau, & Bierschwale, 1993; Eddy, Gallup, & Povinelli, 1996). More recent research has pointed to striking developmental and individual variability in the phenomenon (Swartz & Evans, 1991; Povinelli et al., 1993; Eddy et al., 1996).

Heyes (1994, 1995) has recently attempted to discount both the spontaneous reactions of chimpanzees to mirrors, as well as the results of the mark tests as evidence of self-recognition. However, these alternative explanations have not fared well against either theoretical or empirical scrutiny (see Figure 3.2A-3.2B; Gallup et al., 1995; Eddy et al., 1996; Povinelli et al., 1997). Although not all great apes display the ability to recognize

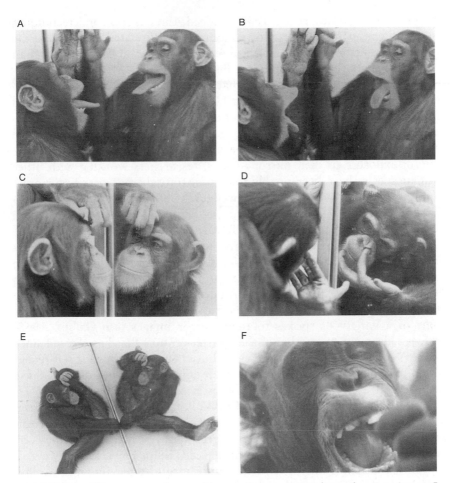

FIGURE 3.1. Chimpanzees show evidence of recognizing themselves in mirrors. In contrast to other species, many chimpanzees engage in self-exploratory behaviors (C–F). They also use mirrors to make and view exaggerated facial displays (A and B). Many chimpanzees that display the types of self-exploratory behaviors that are shown in C–F also will pass a "mark test" (through use of a mirror, these animals often display emphasized inspection of parts of their faces which have been surreptitiously marked red).

themselves in mirrors (especially gorillas), results from our laboratory, using both cross-sectional and longitudinal designs and several new methodological techniques, have converged to support the idea that by about 5–8 years of age, many chimpanzees display unambiguous evidence of using mirrors to gain information about their physical appearance (Povinelli et al., 1993, 1997; Eddy et al., 1996). Indeed, the experimental techniques used with chimpanzees have allowed a more sophisticated diagnosis of the presence of mirror-mediated, mark-directed behavior in chimpanzees than in 18- to 24-month-old children (see Figure 3.2A, B).

Coincident with his discovery that chimpanzees are capable of recognizing themselves in mirrors, Gallup (1970) also uncovered an additional finding that many comparative psychologists have found difficult to accept: the apparent absence of self-recognition in primate species outside the great ape/human group. Despite weeks, months, and in some cases a lifetime of exposure to their mirror images, researchers working with other primate species have been unable to demonstrate the pattern of phenomenon reported for chimpanzees and orangutans (Gallup, 1970, 1977; Lethmate & Dücker, 1973; Benhar, Carlton, & Samuel, 1975; Gallup, Wallnau, & Suarez, 1980; Anderson, 1983; Bayart & Anderson, 1985; Suarez & Gallup, 1986; Itakura, 1987; Anderson & Roeder, 1989; Fornasieri, Roeder, & Anderson, 1991; Gallup & Suarez, 1991). Attempts to demonstrate self-recognition in a wide range of nonprimate species (including elephants and dolphins) have thus far not produced compelling positive evidence (Povinelli, 1989; Marino, Reiss, & Gallup, 1994; for a review of the reactions of a variety of species to their mirror images, see Gallup, 1968, 1975). This is not to say that in the quarter-century that has elapsed since Gallup's original report appeared there have been no claims for self-recognition in one or more monkeys, and even dolphins. To be sure, there have been several (e.g., Boccia, 1994; Thompson & Boatright-Horowitz, 1994; Marten & Psarakos, 1994; Hauser, Kralik, Botto-Mahan, Garrett, & Oser, 1995). However, the best available evidence continues to point to a qualitative functional difference between the responses of many great apes versus other primates to their mirror image (see Gallup, 1994, for a review).

But what about the theoretical significance of these findings? Gallup (1970) saved the final two paragraphs of his original report to account for the phylogenetic difference he had apparently uncovered: "Such a decisive difference between monkeys and chimps," he struggled, "is particularly interesting in view of the fact that most investigators have found only relatively slight quantitative differences on other, more traditional behavioral tasks" (p. 87). His pivotal point was summarized in a single, guarded comment: "Insofar as self-recognition of one's mirror-image implies a concept of self, these data would seem to qualify as the first experimental demonstration of a self-concept in a sub-human form" (p. 87). Gallup concluded that the data suggested that a qualitative psychological difference among primates had been detected, and that it seemed doubtful that the capacity for self-recognition extended "below [sic] man and the great apes" (p. 87).

As efforts to facilitate self-recognition in monkeys continued to yield negative results, and as the number of innovative techniques designed to

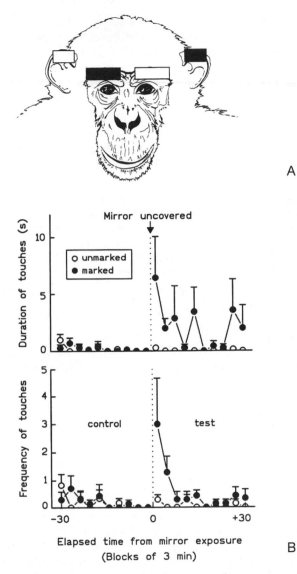

FIGURE 3.2. (A) An advanced experimental technique used to assess mark-directed behavior compares touches to comparable marked (filled areas) and unmarked regions (open areas) on a subject's head. Raters score videotapes for touches to each of these four regions in 30-minute control period (mirror covered) and 30-minute test period (mirror uncovered). (B) Duration (top panel) and frequency (bottom panel) of touches to marked and unmarked facial regions. Subjects selectively touch marked regions in test period (Povinelli et al., 1997). From Povinelli et al. (1997). Copyright 1997 by Academic Press Limited. Reprinted by permission.

rule out simple procedural explanations mounted, Gallup continued to ponder the potential significance of this apparent qualitative difference between monkeys and apes. It struck Gallup that chimpanzees must possess some form of self-awareness. In short, he reasoned that an organism must first know that it exists before it can recognize a reflection of itself (Gallup, 1975, 1977, 1979). If his chimpanzees had no self-concept, he argued, then there would have been no way for them to have known that the marks were on themselves—and hence no reason to search the corresponding areas on their own bodies (see Davis, 1989, for a philosophical defense of Gallup's inference). As Gallup (1977) put it,

> The unique feature of mirror-image stimulation is that the identity of the observer and his reflection are necessarily one and the same. The capacity to correctly infer the identity of the reflection must, therefore, presuppose an already existent identity on the part of the organism making this inference. . . . While many organisms are ostensibly conscious of different features of themselves as the result of visual, tactile, chemical, and proprioceptive feedback, in principle this is quite different from self-consciousness. (p. 334)

Evidently, Gallup reasoned, his chimpanzees had not only grasped the duality that is apparent in any mirrored surface, but they had also possessed a sufficiently well-integrated self-concept to realize that the dualism applied to themselves as well. Indeed, Gallup (1982) later speculated even further. The presence of a self-concept in chimpanzees, he suggested, might reflect a limited capacity to introspect about some subset of their own mental experiences.

An organism's view of itself was one thing, but how would such a creature view other organisms? Gallup (1992) reasoned that self-aware organisms might understand others in much the same fashion as they understand themselves. If they explain their own behavior in terms of emotions, desires, intentions, and beliefs, perhaps they might assume that others have similar mental states as well. Such organisms, like humans, might use their own experiences to make inferences about what other creatures in similar situations might think and feel. Thus, Gallup proposed that evidence of self-recognition in mirrors—because of its possible connection to the presence of self-awareness—might be indirect evidence for an ability to understand the minds of others. His idea, therefore, united self and social intelligence, and in particular, intelligence related to the ability to introspect about one's own mental states (Gallup's self-awareness) and the ability to make inferences about analogous mental states in others.

Gallup's ideas constituted a theory in that they generated what were, at the time, some rather striking predictions about both those species that had displayed evidence of self-recognition in mirrors and those that did not. The theory predicted that whereas the former would succeed on tests designed to tap into their ability to reason about the private, unobservable mental states of others, the latter would not. Indeed, if the common ancestor of the great apes and humans had uniquely evolved a psychological system related to the self-concept—a system that simultaneously allowed them to make inferences about the mental states of others—then there ought to be a class of identifiable social behaviors unique to its living descendants as well (e.g., Gallup, 1982, 1983, 1985). In other words, Gallup (1982) speculated that if the great apes possessed a psychological system that distinguished them from other primates, then there ought to be corresponding behavioral differences as well. To illustrate his argument, Gallup quickly surveyed the primate literature, pulling out examples of empathy, deception, grudging, and the like, to illuminate his ideas. His intent was not to conduct an exhaustive review of the literature on primate social cognition, but rather to highlight a class of behaviors that could make his comparisons explicit. Of course, he realized that some researchers might immediately reject the notion of a dichotomy between great apes and other nonhuman primates because of widespread anecdotes of deceptive, empathic, and other behaviors that would, at first glance, appear to qualify as evidence of reasoning about mental states.

But Gallup drew a more subtle distinction between self-aware instances of these behaviors, on the one hand, and learned or hardwired (innate) ones on the other. As an example, he drew on Gene Sackett's (1966) demonstration that very early in their development, isolation-reared infant rhesus monkeys show different (and appropriate) reactions to photographs of adult male rhesus monkeys who display threatening faces and those that do not—despite the fact that they have never had the social interactions in which they could have learned this distinction. Gallup explained the significance of such findings:

> In the absence of ever having had any experience with adult males there would be no basis upon which to make that distinction. From an adaptive point of view, however, it is not important that [the infants] know, but only that they act as if they knew something about the apparent intentions of threatening adults males. (Gallup, 1982, p. 245)

Thus, nature was far too complex to simply "read into" animals an understanding of intentions, desires, and beliefs, as Romanes had counseled a century earlier. Instead, clear standards were needed for distinguishing between real and learned/hardwired instances of mind. Indeed, as primatologists interested in the mental states of animals began to rely on anecdotes (especially those concerning deception) more and more frequently, Gallup's suggestions for how to distinguish among such cases began to take a different turn. In particular, as his concerns about the difficulties in separating learned/hardwired behaviors from those based on mental state attribution mounted, he began to favor more controlled experimental demonstrations of these abilities (compare Gallup, 1982, 1985, 1991). Indeed, he found it increasingly necessary to explain why the interaction of learning and hardwired developmental mechanisms on the one hand, and anthropomorphic tendencies in humans on the other, meant that naturalistic observations—no matter how reliable—were not well-suited to answer questions of this type. "Mindlessness," Gallup (1985) commented in reference to the illusions of mind that can be produced through these mechanisms, "is not obvious" (p. 634).

As an alternative to using anecdotes, Gallup turned to David Premack and Guy Woodruff's (1978) report concerning the possibility that chimpanzees possessed a theory of mind. By "theory of mind" they were referring to the effortless way in which humans make inferences about the mental states of themselves and others. "A system of inferences of this kind," they observed, "may properly be viewed as a theory because such states are not directly observable, and the system can be used to generate predictions about the behaviors of others" (p. 515). Turning to Premack's star pupil, Sarah, Premack and Woodruff attempted to determine if chimpanzees could reason about the intentions of others. They constructed a series of videotape sequences that depicted familiar humans struggling to solve staged problems, and then gave Sarah ample opportunity to view them. Each videotape was then played for her, placed on pause, and she was provided with pairs of still photographs—one of which represented the solution to the implied problem, the other, which did not. So, for example, on one trial, Sarah witnessed an actor struggling to obtain a banana suspended from the ceiling out of reach. Sarah was presented with two photographs, one in which the actor was stepping onto a box, the other in which the actor was engaging in an irrelevant activity. Sarah's above-chance success in selecting the photographs that depicted the solution to a range of such problems was interpreted as evidence that she "recognized the videotape as a problem, understood the actor's purpose, and chose alternatives compatible with that purpose" (p. 515). Many researchers have challenged Premack and Woodruff's interpretation of their data, but suffice it to say that their conceptualization of the problem of the attribution of mental states, and their experimental approach to the question, impressed Gallup (1982, 1983) as a straightforward means of testing a central prediction of his model: that organisms that displayed clear-cut evidence of recognizing themselves in mirrors ought to be capable of reasoning about the mental lives of others.

On the other hand, Gallup's theory predicted that species incapable of self-recognition did not possess an introspective system, and hence could not use their own experiences to simulate or model the experiences of others. Likewise, in a later developmental application of these ideas, Gallup and Suarez (1986) argued that if the theory was correct, young children should not begin to show signs of self-awareness or of using their own experiences to make sense of the behavior of others until they displayed evidence of recognizing themselves in mirrors. Indeed, subsequent research provided evidence for such correlations with respect to the development of empathy, altruistic behaviors, the self-conscious emotions, and synchronic play (Asendorpf & Baudonniere, 1993; Lewis, Sullivan, Stanger, & Weiss, 1989; Brownell & Carriger, 1990; Johnson, 1982; Bischof-Köhler, 1988, 1991).

Gallup's hypothesis provided a formalism to the idea that primate evolution had been marked by a connection between self and social understanding, and offered a set of specific, testable predictions concerning the evolutionary history of this connection. Like Humphrey (1982), Gallup argued that our indirect, inferential knowledge of others is derived from our direct, primary access to our own mental and emotional states. In this sense, Gallup's proposal was striking not so much for its claims about how knowledge about the minds of others is obtained, but rather for its startling claim that self-recognition in mirrors, because of its purported relationship to introspection, might be a useful phylogenetic (and developmental) marker of the presence of the ability to know such things about others in the first place. What Gallup's model did not do—indeed, did not purport to do—was to provide an explanation for why these forms of self-awareness had evolved in the first place.

Self-Recognition and Self-Conception: Alternative Views

Gallup's model, while making some clear predictions regarding mental-state reasoning in different primate species, left open the question of the exact nature and scope of the self-representation that allows for self-recognition in mirrors in the first place (see Gallup, 1979; Povinelli, 1991; Mitchell, 1993). Gallup's (1970) initial speculation was that "self-directed and mark-directed behaviors would seem to require the ability to project, as it were, proprioceptive information and kinesthetic feedback onto the reflected visual image so as to coordinate the appropriate visually guided movements via the mirror" (p. 87). It was only later, as Gallup began to consider the connection between self and other, that he hypothesized an even more elaborate form of self-awareness in chimpanzees—one that included an ability to introspect about one's own mental states (Gallup, 1982, 1983, 1985). It is important to note that this later formulation was not offered as deductively valid: Nothing forced the conclusion that chimpanzees were self-aware in this latter sense. Rather, Gallup offered a speculative model that contained clear and falsifiable predictions.

Several investigators have proposed alternatives to, or elaborations of, Gallup's ideas about the cause of self-recognition in mirrors (Epstein, Lanza, & Skinner, 1981; Mitchell, 1993; Parker, 1991). Although they differ in a number of ways, most of these models either explicitly or implicitly assume that before organisms can display evidence of self-recognition in mirrors, they must first learn that mirrors provide contingent accurate images of objects in front of them. This line of thinking has led researchers to approach the mark test as a problem in which the organism is searching for the correct location of the red marks they see in the mirror (see Bertenthal & Fischer, 1978; Mitchell, 1993).

We have recently proposed an alternative to this line of thinking about the nature of the self-representation underwriting self-recognition in mirrors (Povinelli, 1995). Although our model shares much in common with Gallup's (1970) original view, and Mitchell's (1993) subsequent elaboration of some of Gallup's ideas (Mitchell's "inductive" model), our explanation begins by denying that self-recognition in mirrors has anything to do with understanding the reflective property of mirrors; that is, we have argued that understanding that mirrors provide an accurate and contingent image of one's own body is not necessary for mirror self-recognition. Indeed, we suspect that despite their capacity to recognize themselves in mirrors, not even 18-month-old human infants know this fact (see Povinelli, 1995).

How, then, is self-recognition in mirrors possible? A majority of human infants display evidence of self-recognition in mirrors (and other forms of live visual feedback) by about 18–24 months (Amsterdam, 1972; Schulman & Kaplowitz, 1977; Bertenthal & Fischer, 1978; Lewis & Brooks-Gunn, 1979; Johnson, 1983; review by Brooks-Gunn & Lewis, 1984). Because far more is known about aspects of cognitive development in human infants, we start by asking about other changes that occur at around 18–24 months of age that might support this ability. Povinelli (1995) has argued that the typical development of a domain-general system of representation in 18- to 24-month-old children provides them a means of forming relations between objects and events in the world and schemes held in mind. Although they differ in important ways, a number of specific proposals are compatible with this view (e.g., Perner, 1991; Olson & Campbell, 1993). Once such a system of representation becomes possible, an organism is able to construct a number of first order relations, but of particular interest to us is their ability to grasp equivalence relations. Elsewhere, we have outlined how the ability to form such equivalence relations allows the organism to understand that the representation of self held in mind (an immediate, on-line representation of the self's actions, movements, and possibly desires) is equivalent to the image they see in the mirror (see Povinelli, 1995). Everything that they can detect about the self's appearance, movements, and desires (as recognized in action) is equivalent to the image in the mirror. This does not require the organism to understand that the mirror is an accurate and contingent image of itself. This fact may or may not be understood, but it is not necessary for self-exploratory and mark-directed activities to occur. Following William James (1890/1950), we have labeled this self-representation the "present self."

One of the central components of our model is that it posits the contingency between the self's actions and the actions in the mirror, is what triggers the formation of an equivalence relation between the organism's internal self-representation and the external stimuli (the mirror image). Thus, if 18-month-old children were able to understand not only the relation between their present self and live visual feedback, but also understood how their present self is related to previous states of the self, our model would be in serious jeopardy. Recently, we have tested whether such a developmental dissociation exists in young children's capacity for self-recognition using contingent visual feedback (e.g., mirrors, live video) and delayed visual feedback (e.g., prerecorded video). In a series of studies, we have shown that although 2- and 3-year-old children will reach up to remove a sticker that was secretly placed on their head if they see themselves in a mirror or in live video, they will not typically do so if they see a delayed videotape that depicts the experimenter placing the sticker on their head 3 minutes earlier (Povinelli, Landau, & Perilloux, 1996). It is not that the children fail to recognize their physical features on the delayed tape; to the contrary, they are quite willing to use their proper name or the first-person pronoun "me" when asked who it is (see also, Brooks-Gunn & Lewis, 1984). Rather, they do not seem to understand that the image has anything to do with their current self—their on-line representation of their actions and desires. Indeed, consistent with the theoretical predictions of our model, it is not until 4–5 years of age that children pass this test of self-recognition using delayed feedback (see also Povinelli & Simon, 1998).

These empirical findings have important implications for assessing the scope of the chimpanzee's self-concept. Although to date we have only conducted pilot tests with chimpanzees using our delayed self-recognition procedures, our theoretical model explicitly argues that organisms such as chimpanzees can recognize themselves in mirrors without necessarily appreciating that they are unique, unduplicated selves with a particular past and possible futures. Passing our tests of delayed self-recognition, in contrast, requires a representational system that allows them to link particular instances of the self into a higher-order self that endures through time. Indeed, there appears to be a relationship between these phenomena and the development of autobiographical memory in children. Our findings of the age at which young children appear to grasp the connection between their current state and briefly delayed video images of themselves, mirrors the age of onset of autobiographical memory (see Nelson, 1992, 1993; Welch-Ross, 1995). Although at present we do not yet know whether chimpanzees develop an ability to pass tests of delayed self-recognition, their ability to recognize themselves in mirrors uniquely counts only as evidence for a limited self-concept restricted to their on-line actions and possibly desires. One curious fact remains, however. As we have seen, the emergence of self-recognition in mirrors in human infants is significantly correlated with the emergence of other abilities that may relate to an understanding of other minds (e.g., empathy, altruism, synchronic play, the self-conscious emotions). If similar correlations held true for apes, Gallup's model would turn out to have heuristic merit. However, as we shall see, such correlations may not be present in chimpanzees.

THEORY OF MIND AND THE SELF–OTHER FUSION

In contrast to the simulation view adopted by Humphrey and Gallup, a very different relation between self and other has recently been advanced by the "theory" theorists. Following Premack and Woodruff's (1978) lead, the theory-theorists have argued that like adults, young children (and possibly even infants) possess a naive, folk "theory of mind" that is used to explain and predict the behavior of others (Wellman, 1990; Perner, 1991; Gopnik & Wellman, 1994; Gopnik & Meltzoff, 1997). These researchers use the term "theory" quite literally, meaning that our beliefs about other minds are acquired, revised, and used in the very same manner as are scientific theories—it is just not as formalized (see Gopnik & Meltzoff, 1996). Thus, in characterizing the development of young children's knowledge about mental states, these researchers argue that the child really and truly is constructing a bona fide (if naive) theory of how the mind works. The development of knowledge about the mind, in their view, can best be understood as a series of revisions of theories about how the mind operates— revisions forced by the accumulation of new evidence (Wellman, 1990; Gopnik & Wellman, 1994; Gopnik & Meltzoff, 1997).

The theory-theory has direct implications for the means by which we come to know the minds of those around us. Alison Gopnik (1993), for example, has maintained that the intuitive, cherished belief that we have direct access to our own mental states, but only indirect access to the mental states of others, is simply wrong (Gopnik, 1993; Gopnik & Wellman, 1994; Gopnik & Meltzoff, 1997). Thus, in direct contrast to the position adopted by Humphrey, Gallup, as well as by Paul Harris (1991) and Robert Gordon (1986), Gopnik has argued that the way in which we know about the mental states of others is the very same way in which we know about our own. She has proposed that from early infancy forward, knowledge of the mental states of self and other are derived from the same psychological system, and in particular, a system that forms theories about how the social world operates. But, just as our knowledge of the mental states of others is indirect and inferential, so too, she claims, is our knowledge of our own mental states. As we develop, we invent and learn about theoretical mental constructs such as goals, desires, intentions, and beliefs that guide our interpretation of both our own behavior and that of others. Thus, Gopnik maintains that mental states are every bit as theoretical when we invoke them to explain our own behavior as when we use them to explain the behavior of others.

The relation between self and other minds proposed by Gopnik (1993) is certainly a strong version of the claim that knowledge of self and other are inextricably connected. Indeed, not all developmental psychologists concur with her characterization of the self–other relation, or with her more general allegiance to the idea that human infants and children are forming "theories" about how the world operates. Although their reasons for doing so vary widely, there are many who find it impossible or difficult to accept the idea that young infants and children actively construct "theories" of mind during their development (e.g., Leslie, 1994; Fodor, 1987, 1992; Hobson, 1993). However, it is important to note that even if we were ultimately to reject the extreme position of the theory-theorists, we would still be left acknowledging that humans possess an uncanny psychological connection to each other. In a particularly rich attempt to understand the developmental origins of our psychological connection to others, Peter Hobson (1993) argues that this psychological identification has "pretheoretical origins," and argues that the earliest manifestations of understanding the psychological attitudes of self and other occur by observing mental states that, in some sense, are directly observable. But placing the particular developmental accounts aside for the moment, an important phylogenetic question remains: Are humans alone in this psychological fusion of self and other? More broadly, are humans alone in possessing a theory of mind?

In the sections that follow, we select a particular case study of theory-of-mind development in humans and examine what is known about comparable abilities in chimpanzees. In particular, we explore their understanding of seeing as a mental event. Ideally, our review would be more broad, including a wide range of abilities associated with the development of theory of mind in human children, and would include data from not just chimpanzees but other nonhuman primates as well. Unfortunately, at present, such a review is impossible given that there have been very few systematic investigations of theory-of-mind abilities in nonhuman primates. Even in the case of chimpanzees, where the most research has been conducted, the kinds of systematic studies that will be necessary to come to closure on these issues are in their infancy (see Tomasello & Call, 1994; Povinelli, 1996a). Nonetheless, by comparing chimpanzees' and children's understanding of seeing as a projection of attention, we hope to shed some light on both the similarities and differences in the cognitive systems of human and nonhuman primates. We hope to show how it is possible that the evolutionary emergence of theory-of-mind abilities forever altered the manner in which humans understand the behaviors they share in common with their primate ancestors.

Some might question the logic of comparing the psychological development of human children and other species, such as chimpanzees, on any number of grounds. Elsewhere, we have explored these objections and have explained the methodological and biological rationale of such comparisons in detail (see Povinelli & deBlois, 1992a; Povinelli & Eddy, 1996a; Premack & Dasser, 1991). In summary, despite the separate evolutionary histories of these species, comparisons of the type we review here can help to establish features of psychological functioning that these species inherited from a common ancestor, as well as unique features that one of more of the species have evolved after their lineage diverged from that ancestor. In short, systematic comparisons of chimpanzee and child, despite their methodological difficulty, can ultimately provide a fair characterization of the unique cognitive abilities of the two species.

DEVELOPING A THEORY OF MIND:
THE CASE OF SEEING-AS-ATTENTION

Before we begin our examination of what is currently known about the development of young children's understanding of the act of seeing as the mental state of attention, it is important to emphasize that this research is really just a small subset of a much broader field of inquiry. Since Premack and Woodruff's (1978) report, investigations of the development of young children's knowledge about mental states have escalated at an unprecedented rate. Currently, researchers are investigating young children's understanding of a wide range of mental states and events such as consciousness, desire, attention, intention, knowledge, belief, false belief, and thinking. A number of developmental progressions have been offered to describe the child's developing theory of mind (Wellman, 1990; Perner, 1991; Gopnik & Meltzoff, 1996). One of the most widely accepted is a scheme offered by Henry Wellman (1990), in which young children are postulated to move from desire–belief reasoners in the second year of life to belief–desire reasoners from about 3 or 4 years on. He argues that by about 2–3 years of age, young children conceive of the mental affairs of others primarily on the basis of desires, although they can and do reason about the beliefs of others to a certain extent. In contrast, he believes that by about 4 or 5 years of age, children have firmly grasped the role that beliefs play in regulating the behavior of others, and use this understanding as the centerpiece of their reasoning about the mental life of others. Other researchers have argued that even prior to this desire–belief reasoning, human infants possess a kind of desire–goal psychology in which they understand the action of those around them in terms of observable goal states that are internally desired (Baron-Cohen, 1991). Indeed, others hold out for a continuous extension of intersubjective experiences into early infancy (Gopnik & Meltzoff, 1997).

Several researchers have attempted to provide a unified account of the development of the child's knowledge about the mind. One approach has been to ignore the messy problem of development altogether by postulating that cognitive development is really just the onset of brain modules, prewired to perform the neural computations necessary to reason about the phenomena in question (Fodor, 1983, 1992; Leslie, 1994; Baron-Cohen, 1994, 1995). A second approach has been to specify domain-general representational systems that the child constructs during development and then to show how changes in these general structures allow for the behaviors identified at various ages (Perner, 1991; Olson, 1993; Frye, Zelazo, & Palfai, 1995; Zelazo & Frye, in press; Gopnik & Meltzoff, 1997).

Although there are clear theoretical points of disagreement between and among these approaches, it is important not to overlook two fundamental points of unity. First, they all start with the notion that humans do, at some point, construct a psychological system that allows them to reason about unobservable mental states of the self and other. Second, they agree that however difficult the task, if the relevant environmental parameters and the transformational processes that act upon them were completely understood, we would be able to characterize the modal developmental progression of the developing child's knowledge about the mind. As we shall see, these seemingly trivial points of agreement have nontrivial implications. We now examine a specific case in point concerning the child's growing understanding of the mind—the development of their understanding of seeing-as-attention.

Understanding Seeing: From Meaningful
Stimuli to Mental State

Many organisms—not just primates—are interested in and sensitive to the presence of eyes and eye-like stimuli (e.g., see Blest, 1957; Burger, Gochfeld, & Murray, 1991; Burghardt & Greene, 1988; Gallup, Nash, &

Ellison, 1971; Ristau, 1991; Perrett et al., 1990; Povinelli & Eddy, 1996a, 1996b, 1996c, in 1997; see review by Argyle & Cook, 1976). A moment's reflection on the impact of predation on both solitary and group-living organisms reveals why the process of natural selection may have favored the preservation of perceptual systems that reacted quickly to the presence and direction of the eyes of other organisms. In the case of group-living organisms, such as many primate species, additional advantages to paying attention to who is looking at whom (or what) can easily be envisioned (Chance, 1967; Fehr & Exline, 1987). In many of these cases, we are not readily tempted to conclude that the organisms in question know anything at all about a psychological state of attention behind the eyes—for example, the eye-like stimuli on the wings of certain moths and butterflies (Blest, 1957). But other situations readily tempt our anthropomorphic inclinations. For example, surely a rhesus monkey who leaps up to threaten you after you look her in the eyes must understand that you looked at her. Surely a chimpanzee who spins around to look where you are looking, must have understood that you were looking at something behind him. Well, maybe—but maybe not.

Let us assume for the moment that some organisms can and do process information about the eyes of other organisms without having any notion at all about seeing. Thus, just as many birds have specific courtship displays in which a behavioral act by one of the participants reliably triggers a response by the other one, so too may the reactions of many species to the eyes and eye direction of others have evolved as useful responses to specific stimuli. So, although many species actively monitor and respond appropriately to the visual attention of others, they may not be aware that others have visual experiences. They simply may not interpret visual perception as a mental event. We have not yet presented any evidence that they do not; rather, we are simply holding open this possibility.

On the other hand, we can easily imagine that some species do understand visual perception as an intentional event linking the perceiver to the external world. By "intentional" we refer to the philosophical notion of "aboutness"—that is, seeing is about (or refers to) organisms, objects, or events in the external world (Brentano, 1874/1960). Indeed, we do not merely have to imagine such species, for we know of at least one in which this kind of understanding is well consolidated: our own. Indeed, the eyes seem to occupy center stage in our human folk psychology. The commonly used metaphor of making "eye contact" with someone highlights this property well; although our eyes do not literally contact the other person, our folk psychology foists this connection upon the interaction. When we see someone turn to look at an object, we automatically register that they are attending to that object.

Developing the Idea That Seeing Is Attention

Are humans born with this mentalistic understanding of the eyes, or does it develop gradually? Even 4- to 5-month-old infants are sensitive to the presence of eyes and will look longer at faces that make direct eye contact with them as opposed to those that do not (Lasky & Klein, 1979; Johnson & Vicera, 1993). Between 6 and 18 months, infants construct an ability to follow the gaze of others by turning to look in the same direction (Scaife & Bruner, 1975; Butterworth & Cochran, 1980; Butterworth & Jarrett, 1991; Corkum & Moore, 1994). George Butterworth and his colleagues have proposed three stages in the development of this ability. Initially, infants only appear able to turn their heads in the same direction as the other person. They limit their visual scans to the space that they directly perceive but cannot yet localize the specific object to which the adult is gazing. Later, by about 12 months, this ability is elaborated, allowing infants to localize the specific target of the adult's gaze, again, provided it is within their own visual field. Finally, by about 18 months or so, infants will turn and search in space outside their own immediate visual field if, for example, the adult glances behind them.

Some researchers interpret gaze following as evidence of an intentional understanding of seeing—a kind of joint visual attention in which the infant and the adult are aware of each other's attentional focus on a particular object or event in the external world. This interpretation implies that an infant's ability to follow another person's line of sight expresses an understanding that the person is looking "at" something (Baron-Cohen, 1994, 1995; Franco, in press). Other theorists, although impressed by the sophistication of the gaze-following system, are far less sanguine about its relation to an intentional understanding of visual perception (Butterworth & Jarrett, 1991; Moore, 1994; Tomasello, 1995; Povinelli & Eddy, 1994, 1996a, 1996b). Simply because infants are interested in the eyes, and can use the gaze direction of others to discover useful information in the world, this does not guarantee they appreciate that seeing subjectively connects the observer to the external world. Thus, although a gaze-following system ensures that mother and infant will look at the same object in unison, it is far less clear that the infant is aware of this fact. In this sense, *joint* attention does not ensure *shared* attention.

If sensitivity to the presence and direction of eyes does not necessarily qualify as evidence that infants understand the underlying attentional significance of visual perception, then what might? First of all, it is important to clarify that infants might develop a general understanding of attention before understanding the exact role that the eyes play in regulating its deployment and direction. Generally, when we attend to something, there are multiple redundant cues indicating our attentional focus, including eye direction, facial orientation, and bodily posture. Indeed, these are merely the cues that concern *visual* attention. Often, we attend to things by listening, touching, or smelling. Thus, an infant might construct the general notion of attention as a mental state long before realizing either that there are specific sensory channels that each offer distinct input, or the factors that govern the deployment of each channel. For example, infants might conclude that people can be attending to an object if their head and body are oriented toward it—even if their eyes are closed! Thus, we must ask two questions: (1) At what point do infants come to appreciate the mental state of attention, and (2) at what point do infants appreciate the eyes in particular as portals through which attention is regulated?

With respect to the first question, there is evidence that by about 18 months (and possibly earlier), infants may be able to represent the mental state of attention. For example, Dare Baldwin and her colleagues have used several procedures in which infants are asked to reason about the attentional focus of an adult. In one of them, an experimenter shook an opaque plastic bucket to demonstrate that something was inside. Next, the experimenter looked into the bucket without letting the infant see, and declared, "It's a *modi*! A *modi*! There's a *modi* in here!" Next, the experimenter picked up an identical bucket, shook it, and then looked inside and pulled out a toy that the child had never seen before and let the child play with it. Finally, the experimenter pulled a comparable, unfamiliar toy from the first bucket and let the child play with it as well. The crucial test came after both toys had been retrieved from the infant and then placed in front of him or her. The experimenter said, "There's a *modi* here. Can you point to the *modi*? Point to the *modi*." Of course, seeing the person peer into the bucket would automatically indicate to us that he or she was talking about whatever was inside. Baldwin's (1993a) results suggest that by about 19 months or so, young children are apparently in this same boat. They correctly select the object that the experimenter was presumably looking at while labeling the object. Numerous control procedures have been conducted to rule out alternative explanations, and, in general, the data have converged to suggest that somewhere in the middle of their second year of life, infants are able to interpret the actions of others in terms of a hypothetical mental state that we call attention (Baldwin, 1991, 1993a, 1993b; Baldwin & Moses, 1994; Mumme, 1993). Whether infants younger than 18 months have a similar, if more circumscribed and fragile, competence in this area remains a matter of controversy (see Baldwin & Moses, 1994).

But what about the eyes? When can we feel confident that infants have uncovered the connection between the eyes in particular, and the mental state of attention? That is, when do they understand seeing-as-attention? In an impressive series of studies spanning over a decade, John

Flavell and his colleagues have asked precisely this question (Masangkay et al., 1974; Lempers, Flavell, & Flavell, 1977; Flavell, Shipstead, & Croft, 1978; Flavell, Flavell, Green, & Wilcox, 1980; Flavell, Everett, Croft, & Flavell, 1981; Flavell, Green, & Flavell, 1989). However, they have distinguished two levels at which seeing can be understood mentalistically (see Figure 3.3A). At Level 1, young children may understand that seeing connects people to the external world; in other words, they can register what people can see and what they cannot see. Their studies have revealed that by about 2½ years of age, young children appear to have localized the eyes as being relevant for seeing. So, for example, if you ask young 3-year-olds to make it so that you cannot see a ball, they know to move a screen between you and it; if you ask them to show the ball to someone else who is facing away, they will hold the object up and place it in the other person's

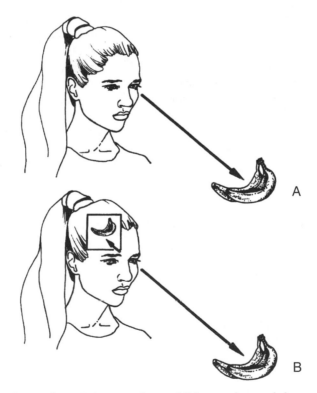

FIGURE 3.3. (A) By about 2½ years of age, children understand that seeing connects people to the external world. For example, they understand that people can see or not see based on the direction of their gaze. (B) In addition, by 4 years of age, children develop an understanding of how visual perception leads to the formation of internal states of knowledge.

line of sight; if you stare at one of several toys and ask them to point to the one you can see, they do so with ease; if you cover your eyes, they know you cannot see them (see Lempers et al., 1977). Povinelli and Eddy (1996a) and Gopnik, Meltzoff, and Esterly (1995) have provided additional evidence for this kind of understanding of seeing-as-attention in 2½- and 3-year-old children. Finally, using cartoon stimuli of faces, Baron-Cohen and his colleagues have shown that 3-year-old children can discern what another person wants based solely on the direction in which the character's eyes are directed (Baron-Cohen, Campbell, Karmiloff-Smith, Grant, & Walker, 1995).[3]

When Seeing Becomes Believing (and Knowing)

Although 2- and 3-year-olds seem to understand visual perception as a mental state, it would be erroneous to conclude that they have achieved an adult-like understanding of seeing. Indeed, it is likely that their understanding of seeing is quite shallow. Flavell and his colleagues have identified a second level of understanding seeing that children of this age do not yet grasp. Although they have no trouble realizing what you can or cannot see, they have striking difficulty understanding how those things appear to you. For example, sit across the table from a young 3-year-old girl and show her a picture of a turtle so that it is right-side-up from her perspective, but upside-down from yours. She will readily assent that you can see the turtle, and indeed, if you cover your eyes, she will readily acknowledge that you can no longer see it. But try as you may, you will have an extraordinarily difficult time getting her to understand that, from your perspective, the turtle appears differently, that is, upside-down (e.g., Flavell et al., 1981). But less than a year later, this same child will demonstrate without hesitation her understanding that although the two of you are visually connected (or attending) to the same thing, your respective mental representations of the object differ considerably (Level 2 understanding, Figure 3.3B). Although one might be tempted to dismiss these findings as being the result of some trivial artifact of the experimental design, Flavell and his colleagues have used procedure after procedure to assist the younger children, generally with little or no success.

Indeed, Flavell's findings seem to tightly parallel other differences discovered between 3- and 4-year-olds related to their understanding of the connection between seeing and knowing. For example, 3-year-olds have great difficulty in understanding the link that perception in general (and seeing in particular) plays in the process of knowledge formation (Wimmer, Hogrefe, & Perner, 1988; Perner & Ogden, 1988; Gopnik & Graf, 1988; Ruffman & Olson, 1989; O'Neill & Gopnik, 1991; Povinelli & deBlois, 1992b; O'Neill, Astington, & Flavell, 1992). When 4- or 5-year-olds see someone lift a lid on a box and look inside, they automatically assume that the other person knows what is there. In contrast, they do not make this assumption if someone just touches the box, or lifts the lid without looking inside. Even though they cannot literally see information entering the person's eyes, traveling to the brain, and resulting in a particular activation of neurons that correspond to the state of knowing the contents of the box, these children appear to assume that internal states of belief have arisen from the act of visual perception (see Figure 3.3B). Yet if the same scenario is rehearsed for 3-year-olds, it rapidly becomes apparent that they are interpreting these events in a very different manner. They appear to have little idea that the one who sees, knows, and that the one who does not see, does not know.

Indeed, 3-year-olds' difficulty is not limited to their understanding of visual information. Their difficulty appears to be a general one in that they have not yet understood that auditory, visual, haptic, and olfactory information lead to, or cause, internal knowledge states. Of course, there is some disagreement on this point, with some researchers offering evidence that 3-year-olds, for example, may possess some understanding of the seeing–knowing relation (e.g., Pillow, 1989; Pratt & Bryant, 1990; Wooley & Wellman, 1993; but see critiques by Povinelli & deBlois, 1992b). However, Lyon (1993) has recently provided evidence that the disagreement could stem from the fact that younger children may conflate the idea of knowledge with interest or desire. His research suggests that young 3-year-olds attribute "knowledge" to those people who display an interest in the contents of a box, more than to those who look inside the box, but express little other evidence of interest. O'Neill (1996) has also recently provided evidence that 2½-year-olds are sensitive to whether their mothers have been visually or attentionally engaged with an event, and if they have not been, will attempt to draw their mothers' attention to the situation. However, she has argued that despite this sensitivity to their mothers' behavioral connection or engagement to a given event, children of this age may still possess little or no appreciation of knowledge states per se.

In parallel to their trouble in understanding the perception–knowledge relation in others, 3-year-olds have great difficulty in explaining how they themselves come to know things. For instance, let a young 3-year-old look into a bag and discover a toy bird inside, then put the bag away. A minute later, if you ask him if he remembers what was inside the bag, he will answer correctly. But then ask him *how* he knows there is a bird in the bag. Here, he will look at you blankly—just before rapidly describing all sorts of irrelevant (if true) details about the dog he has at home that one time tried to bite the mailman. You can rephrase the question, make it as obvious as possible for him, and even give him a choice: "Do you know there's a bird in there because I *told* you or because you *looked* inside?" Try as you may, you will discover that young preschoolers appear to have little idea how, or even when, they came to know such a simple fact (Wimmer et al., 1988; Gopnik & Graf, 1988; O'Neill & Gopnik, 1991; O'Neill et al., 1992; Povinelli & deBlois, 1992b; Taylor, Esbensen, Bonnie, & Bennett, 1994).

Summary: Seeing as a Mental Event

The findings reviewed here suggest that during the earliest months of life, children are sensitive to the head posture, face, and eyes, and by the end of the first year have consolidated a simple gaze-following system that allows them to track where someone else is looking (as long as it is within their own immediate perceptual field). Furthermore, by 18 months or so, children appear to have constructed a mentalistic notion of attention, even if they have not yet sorted out the exact role that the eyes and other sensory modalities play in mediating that attention. By 2½ or 3 years of age, preschoolers seem to be well on the way to understanding the eyes in particular as portals through which attention emanates. Finally, by about 4 or 5 years of age, children come to understand how perception brings information into the mind; that is, they come to understand the role that seeing (along with the other senses) plays in creating the mental states of knowledge and belief.

As we noted earlier, a number of authors have discussed how some of these facts can be explained in more general terms either by implicating changes in a domain-general representational system, the activation of specific brain modules, or the progressive revision of naive theories of how the mind works (Wellman, 1990; Perner, 1991; Leslie, 1994; Olson, 1993; Baron-Cohen, 1994, 1995; Frye et al., 1995; Zelazo & Frye, in press; Gopnik & Meltzoff, 1996). We recognize that not everyone will find our summary in keeping with the majority of the data. However, even if specific proposals about the exact timing of children's development of an understanding of seeing (and their theory of mind in general) are rejected in the face of new data, the general theoretical possibilities we outline should remain tenable.

COMPARATIVE EVIDENCE: WHAT CHIMPANZEES KNOW ABOUT ATTENTION

Chimpanzees: Gaze Following

Until recently, there were only anecdotal reports that chimpanzees and other nonhuman primates follow each other's line of sight (e.g., Byrne &

Whiten, 1985). Indeed, this evidence, combined with a generous "joint-attentional" interpretation of the behavior, led some investigators to conclude that chimpanzees possess an understanding of the intentional aspect of seeing (e.g., Baron-Cohen, 1994). But do chimpanzees (and other nonhuman primates) really follow gaze, and if so, what is the extent of similarity between the gaze-following system that is elaborated between 6 and 18 months in human infants and any similar system in other species?

Over the past several years, our laboratory has conducted a series of studies designed to explore the gaze-following phenomenon in 7 preadolescent chimpanzees. The results of these experiments have uncovered striking similarities between human infants and chimpanzees. First, chimpanzees appear to display gaze following in response to movement of the head and eyes in concert, or just the eyes alone (Povinelli & Eddy, 1996b, Experiment 1; see Figures 3.4 and 3.5). Second, like 18-month-old human infants (but not younger ones), chimpanzees will track the gaze of an experimenter into regions of space not within their immediate perceptual field (Povinelli & Eddy, 1996b). So, if you gaze above and behind a chimpanzee, there is a very good chance he or she will wind up looking there almost immediately. Third, chimpanzees appear to extract specific information about the direction of gaze. Thus, looking behind them does not simply trigger a general visual scanning response on their part; if you look to their right, they will look there first as well (Povinelli & Eddy, 1997; Povinelli, Bierschwale, &

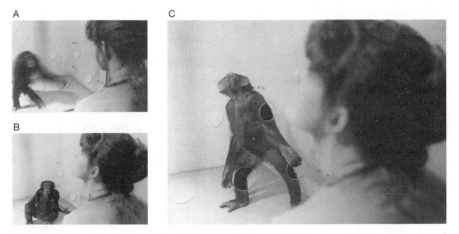

FIGURE 3.4. Like 18-month-old human infants, chimpanzees follow the gaze of an experimenter into regions of space not within their immediate perceptual field. In (A), the chimpanzee enters the test room. In (B), the experimenter turns her head to a region in space not within the immediate visual field of the chimpanzee subject. The subject responds (C) by rapidly orienting head and body toward the same general direction as the gaze of the experimenter.

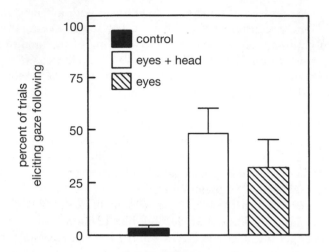

FIGURE 3.5. Chimpanzees turn and look behind themselves in response to an experimenter's eye and head movements in concert (eyes + head) and eye movements only (eyes), but not in response to no movement (control). Each bar represents the group mean for the seven chimpanzee subjects. From Povinelli and Eddy (1996b). Copyright 1996 by Cambridge University Press. Reprinted by permission.

Čech, in press, Experiments 1 and 2). Finally, chimpanzees do not even need to witness the shift in another person's line of sight in order to engage in gaze following (Povinelli & Eddy, 1996a, Experiment 12). Rather, if they merely encounter you with your head and eyes in a fixed orientation, directed above and behind them, this is sufficient to cause them to turn and look where you are looking. People who witness this behavior for the first time are generally deeply impressed. However, not everyone is equally convinced that this behavior qualifies as evidence of an appreciation of the mental state of attention. As we noted in discussing the development of gaze following in human infants, there are good reasons to consider alternative, lower-level explanations of the underlying causes of any such behavior.

In order to gain some empirical leverage on this problem, we attempted to challenge our chimpanzees' gaze-following system in a way that could reveal more about their understanding of gaze. A low-level account posited that they were just automatically registering the sudden shift, or unusual orientation of the eyes/face of others, and then turning along a particular trajectory until their visual system oriented to a novel object or event. A high-level model assumed that they interpreted our line of sight as a projection of attention (see Figure 3.6). We subjected these competing

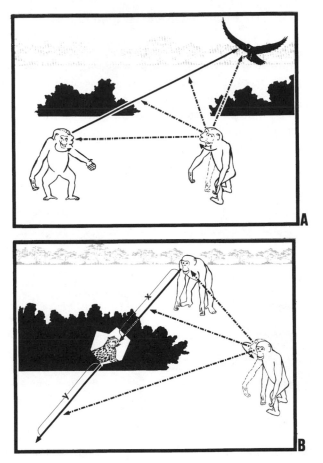

FIGURE 3.6. (A) A prototypical gaze-following event in which the animal on the left first sees an event (a descending raptor), and an observer (on the right) utilizes the first animal's gaze to discover the event. Solid lines represent the line of sight of the gazing animal; dash-and-dot lines represent the line of sight of the animal tracking the gaze. This utilization of gaze may be based on a variety of mechanisms, some being more mechanically based and others being more psychologically based. A low-level model posits that the observer follows the gaze of the first animal without any understanding of the mental state of attention. Rather, it uses a system that projects a trajectory from the face of the first animal out into distant space. Alternatively, a high level model posits that the observer recognizes the change in the first animal's mental state of attention, and this information prompts the gaze-following response. (B) Some naturally occurring situations might assist in distinguishing between the two accounts for a given species. With the animal on the left seeing an event behind a bush, the animal on the right might use the first animal's gaze in one of two ways. The low-level model predicts that without an understanding of attention per se, the "gaze" could be projected directly through the bush, and past it. In contrast, the high-level model expects that the observing animal could ascertain that the first animal must be looking at something on its own side of the bush (vision cannot pass through opaque barriers). From Povinelli and Eddy (1996b). Copyright 1996 by Cambridge University Press. Reprinted by permission.

ideas to a test by having our chimpanzees enter a test unit, approach an experimenter, and use their species-typical begging gesture to request a food item (Figure 3.7). In the crucial test condition, as soon as the ape gestured, the experimenter leaned and looked along a predetermined line of sight that struck an opaque partition that separated the chimpanzee and the experimenter (glance to partition, Figure 3.7). The line of sight was carefully planned so that if projected through the partition, it would strike the back wall of the testing unit. If the apes were just responding to the orientation of our eyes, head, and upper torso, and, as a result, turning to scan along an imaginary trajectory, they should have ignored the interruption in the experimenter's line of sight. They should have projected the experimenter's

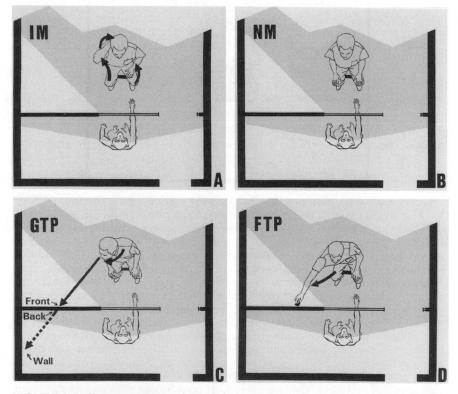

FIGURE 3.7. Four experimental conditions used to differentiate between competing hypotheses regarding chimpanzees' gaze-following behavior. See text for details. From Povinelli and Eddy (1996b). Copyright 1996 by Cambridge University Press. Reprinted by permission.

line of sight through the opaque partition and wound up looking at the back wall of the test unit. In fact, they virtually never did. In support of the predictions generated by the high-level model, in their first two encounters with these conditions, the apes attempted to look around to the experimenter's side of the partition (Povinelli & Eddy, 1996b, Experiment 2; see Figure 3.8). Several control conditions were employed. For example, in one condition, as soon as the ape gestured to the experimenter, the experimenter engaged in an irrelevant movement for 5 seconds; in another condition, the experimenter instead sat still, staring at the subject for 5 seconds before handing over the food (see Figure 3.7). In these conditions, the apes virtually never looked around the partition.

Does this mean that the high-level model is correct—that chimpanzees understand gaze as the projection of the mental state of attention? Again, maybe, but maybe not. The more we considered these results, the more we came to the conclusion that there were really two alternative explanations. One, of course, was that the high-level model was correct. The other was that the low-level model was simply too low-level. In other words, perhaps it failed to consider for how other mechanisms—most notably, general learning—might interact with the gaze-following response. If the low-level account in Figure 3.6 is accurate, then surely with enough experience following the gaze of others, chimpanzees will begin to form measurable anticipatory reactions (e.g., changes in heart rate) when they see someone look off in an unusual direction. Of course, these anticipatory responses will become paired or associated with the changes that occur during the orienting response that results when they see what the other was looking at (see Figure 3.6). In simple terms, then, regardless of what chimpanzees (or young infants) understand about attention as a mental state, when they track someone else's gaze, their neurophysiological systems certainly will anticipate receiving interesting or novel stimulus information. Thus, it does not require understanding gaze as a projection of the mental state of attention to become conditioned to look on the other side of an opaque barrier. After all, in the world of the subjects' previous experiences, this is where the interesting events or objects have always been located. With enough experience in situations of this sort, the apes will be conditioned to look on the other side of opaque obstacles. Thus, the diagnostic response on our test may be accomplished through the interaction of the orienting reflex, general learning mechanisms, and a "simple" gaze-following mechanism. Is the resulting behavior that emerges from these fairly simple mechanisms unimpressive and trivial? Absolutely not. But this should not dupe us into prematurely heralding these results as unambiguous evidence of a theory of attention in chimpanzees. Indeed, as will be evident shortly, we have additional reasons to be extremely skeptical.

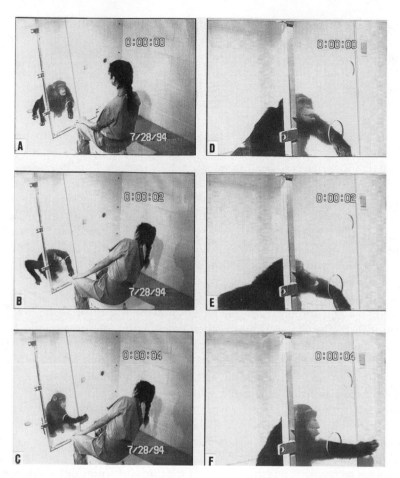

FIGURE 3.8. Two views of a chimpanzee responding to gaze and body motion in the direction of an obscured target (glance-to-partition, see Figure 3.7). In (A)–(C), the experimenter looks at a target on an opaque partition so that her line of sight (if projected straight ahead) strikes the back wall of the test unit. In response, the subject leans forward and sideways, apparently attempting to look at the surface of the partition facing the experimenter. In (D)–(E), the same trial is shown but from a videotape view shown to naive raters. Raters were told to note whether the subject attempted to look at the square target. The chimpanzees demonstrated an ability to appreciate the interaction between an experimenter's line of regard and an opaque surface (see Figure 3.7 for control treatments).

Do Chimpanzees Understand Seeing-as-Attention?

Over the past several years, our laboratory has conducted dozens of studies to examine what, if anything, chimpanzees know about seeing-as-attention. We initially began by focusing on their natural begging gesture described earlier (Povinelli & Eddy, 1996a; Figure 3.9B). The gesture is used in several communicative contexts, including situations in which one ape is seeking reassurance from another, or in cases where one ape is attempting to acquire food from another. Every day, our apes spontaneously use this gesture to request treats such as bananas, apples, sweet potatoes, onions, or carrots from their caretakers and trainers. It occurred to us that we might be able to use this gesture to investigate whether or not chimpanzees understand the attentional aspect of seeing. We first trained them to enter the test unit and look to see whether an experimenter was positioned on the right or the left (see Figure 3.9). With food just out of reach, the natural response of the apes was to look at the experimenter and beg through

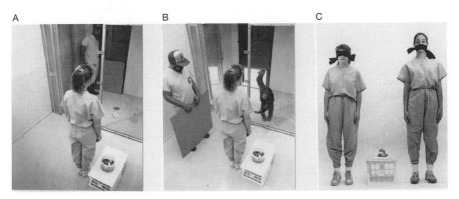

FIGURE 3.9. Although young chimpanzees are very sensitive to the eyes of others, they do not appear to understand their attentional significance in the manner that 2½-year-old humans do. Chimpanzees can be trained (A and B) to use their natural begging gesture to request food from an experimenter. If an experimenter stands on the left (A and B), the chimpanzee will gesture through the hole on the left side of the partition; if the experimenter stands to the right, the subject will respond to the right. However, this is no guarantee that the subjects realize that this person is subjectively linked to them via the mental state of attention. When two experimenters are used, one who can see the chimpanzees and one who cannot (C), the chimpanzees typically respond randomly. The apes apparently fail to understand that only one of the experimenters is visually and mentally connected to them, and thus able to provide food. Perhaps this is because they simply cannot represent mental states of attention. Alternatively, chimpanzees might simply fail to understand the specific role that eyes play in deploying attention. From Povinelli and Preuss (1995). Copyright 1995 by Elsevier Trends Journals. Reprinted by permission.

the holes in the Plexiglas toward the experimenter or food. We trained the apes to be selective about where they would gesture. If the experimenter was on the right, and the ape begged through the hole on the right, the experimenter handed him or her a piece of fruit or small cookie. If the ape gestured through the wrong hole, he or she was ushered back outside and another trial started. Once the apes were excellent at this, we were in a position to begin to ask them what they understood about seeing.

To accomplish this, we occasionally confronted the apes with not one experimenter, but two—one of whom could clearly see the ape, the other of whom could not. In considering the best way to go about this, we first selected objects and toys with which our apes had extensive experience. In particular, we selected objects that we had seen them using in ways that suggested they might understand something about seeing. For example, one of our apes' favorite games was to take large plastic buckets, pull them over their heads, and then walk around their enclosure bipedally. With one hand stretched out in front of them, and the other holding the bucket in place, they would amble about until they bumped into something. Indeed, in the context of these antics, they would even occasionally lift the bucket slightly: to peek, as it were. Another routine involved draping burlap sacks or other cloths over their heads and performing similar stunts. Some versions of these games did not rely on using objects at all. One, for instance, was quite simple: They would take the palms of their hands and cover their eyes. It is important to stress that none of these observations definitively established that our apes understood seeing (either in themselves or in others). Clearly, a sober dissection of these behaviors did not force us to accept this interpretation. But the subjective impression was unavoidable— surely, they at least understood seeing as an event that connects subject and object together via some kind of attentional glue (i.e., Level 1 perspective taking; see Figure 3.3a).

When testing began, the subjects entered the test unit and performed excellently on the trials involving a single experimenter. As they had previously learned, they approached the Plexiglas in front of the experimenter and requested a piece of food. In a preliminary control phase, the apes entered the test unit and encountered two familiar experimenters. One of them was holding out a desirable piece of food, and the other was holding out an undesirable block of wood. After taking a look, the chimpanzees responded by gesturing to the one offering the food. Thus, when the apes did not have to reason about who could see them, they had no trouble choosing between two people.

But what about the crucial trials—those on which the apes were confronted with two experimenters, one of whom could see them, the other of whom could not? Using the spontaneous games of the apes as a guide, we constructed several ways of posing this question. In one case, one experimenter held a bucket over her head, whereas the other experimenter held a bucket in such as way as to still allow her to clearly see the subject. In another condition, one experimenter covered her face with her hands, the other covered her ears. In still another condition, one tied a blindfold over her eyes, whereas the other tied it over her mouth. Much to our astonishment, the intuitive interpretation faltered, and the view counseled by sobriety prevailed. In all of the conditions just described, the apes acted exactly as if it did not matter that only one of the experimenters could see them (see Figure 3.9C). Despite their extensive and recent experience with the objects and scenarios used, the subjects gestured with equal frequency to the trainer who was visually connected to the scene and the one who was not (Povinelli & Eddy, 1996a, Experiment 1). Nor did their performance improve across the four trials they received in each condition. There was, however, a single, notable exception. One of the phases involved probe trials in which one trainer faced forward and the other faced backward. In direct contrast to all of the other conditions, here the apes gestured to the trainer facing forward from their very first trial onward. With this puzzle on our hands, we embarked on an additional 13 experiments to determine what factors the chimpanzees were using to make their selections.

Initially, we focused on determining why the apes had such an easy time with the back-versus-front trials, but such a difficult time with all of the others. The first idea that occurred to us was that back-versus-front trials were just a more natural or obvious instance of seeing versus not seeing. But we knew there was an alternative, more mundane explanation lurking nearby: Perhaps the apes were selecting the experimenter who was facing forward because that was precisely what we had initially trained them to do. From their perspective, in this condition one of the two people was configured in exactly the same manner as in training. In contrast, the other person was not. Thus, they may have simply been rehearsing a learned procedure to gesture to the frontal stimulus of a person. We realized that one way of subjecting these alternative explanations to a test would be to confront the apes with a new condition in which both of the experimenters faced away, but one looked back over her shoulder in the direction of the subjects (Figure 3.10). If the mentalistic interpretation were correct, the chimpanzees could be expected to gesture to the person looking over her shoulder. In contrast, if the apes' success on the original back-versus-front condition was based on a rule about a frontal stimulus, they ought to fail to discriminate between the two. Much to our surprise, this latter outcome is exactly the result we obtained (Povinelli & Eddy, 1996a, Experiment 3).

Despite the fact that the lower-level framework had generated some correct and impressively unexpected results, we did not abandon the idea that our chimpanzees might still understand that one of the experimenters was attentionally connected to them. Indeed, we conducted 11 additional

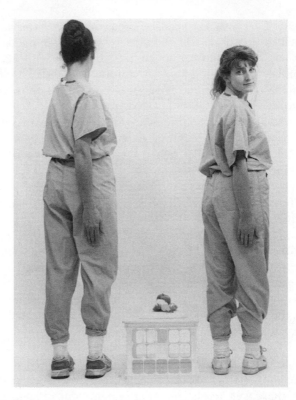

FIGURE 3.10. Looking-over-the-shoulder condition used to determine if our apes were relying on face and head visibility or if they were relying primarily upon front-ward- or backward-facing bodies. In contrast to their performance on back-versus-front trials, the chimpanzees responded randomly to experimenters in this configuration. See text for details. From Povinelli and Eddy (1996a). Copyright 1996 by University of Chicago Press. Reprinted by permission.

studies in which we varied the postures, locations, and eye direction of the experimenters in a systematic effort to uncover some hidden understanding on the chimpanzees part. Across these studies, it became clear that the subjects' performances were improving—at least in some conditions. However, further research revealed the probable reason for this improvement: The subjects were learning a rule that could be verbally summarized as, "*select the person whose face is visible.*" Thus, the apes began to perform well in those cases (such as the one involving buckets) where one of the experimenters face was completely absent (in this case, obscured by the bucket) and the other was clearly visible. Yet they continued to perform poorly when equal amounts of the faces were visible (e.g., blindfolds over the eyes versus the mouth)—despite the fact that only one of the experimenters could see them (see especially, Povinelli & Eddy, 1996a, Experiment 13). Indeed, after additional experience the apes even began to learn a hierarchy of rules: "*Choose the person whose face is visible, but if an equal amount of faces are visible, select the person whose eyes are visible*" (see especially, Experiment 14).

At this point, one might wonder what the difference might be between these chimpanzees and young children. After all, simply because it took the apes a while to learn these rules, if they were ultimately able to use them properly, who is to say that they did not now understand seeing as attention? In fact, there were at least two variations of this possibility (Povinelli & Eddy, 1996a). First, our chimpanzees may have possessed a sensorially undifferentiated theory of attention from the outset and simply required experience to learn how to apply it in the context of these tests. Second, perhaps something about participating in these tests actually forced the subjects to construct an understanding of attention that was not originally present. But how were we to distinguish between these two possibilities on the one hand, and the altogether leaner interpretation that they had simply learned a series of procedural rules, with no concomitant appreciation of attention at all?

Although our laboratory has been systematically chipping away at this problem from several directions, perhaps some of the most striking data on this point inadvertently arose just over a year after we had completed the prior studies. As part of a somewhat unrelated study, we wanted to demonstrate that our animals would respond appropriately to an eyes-open-versus-closed condition that we had used in one of the earlier investigations. By the end of the Povinelli and Eddy (1996a) studies, our subjects had displayed some evidence of responding selectively to faces in which the eyes were visible (Experiment 14). But 13 months later, as we readministered the eyes-open-versus-closed trials to our subjects, an outcome we had not anticipated emerged—the animals performed at chance levels. Even after receiving 48 trials of this type, they continued to respond randomly. Indeed, using another, less subtle condition (screen-over-the-face) to which they had previously learned to respond excellently, and with which they had far greater experience, the subjects' performances only gradually improved, barely bobbing above chance after 12 trials (Povinelli, 1996b). Their failure to show stable evidence of retention cast grave doubt on the idea that the chimpanzees had acquired a theory of attention during their participation in the initial round of research.

After pondering the contrast between the quite sophisticated gaze-following abilities of our subjects, and their relatively poor performance on the tests just described, we embarked on another set of studies to look for evidence that our apes might understand attention in other contexts

(Povinelli et al., in press). It occurred to us that perhaps the apes had trouble reasoning about the differing visual perspectives of two persons simultaneously, or perhaps they had difficulty understanding themselves as objects of the visual perspective of others. Given their excellent gaze-following abilities, we reasoned that they might show better evidence of understanding the "aboutness" of visual perception if we created the following situation. First, we trained the apes to search under two cups for a hidden treat. In the initial stages, the experimenter was equidistant between the two cups and pointed to one of the cups while staring at a target positioned midway between the two. After considerable trial and error, the subjects learned to use the experimenter's pointing behavior to select the correct cup.[4] This set the stage to test the chimpanzees by occasionally having them enter the test lab and discover that the experimenter was no longer pointing, but instead was gazing in one of three ways: at the correct cup (At Target), above the correct cup (Above Target), or at the correct cup but just using the eyes (Eyes Only) (see Figure 3.11). Our reasoning was simple. If the chimpanzees understood the referential significance of the gaze of the experimenter, they ought to select the correct cup on the At Target trials and possibly the Eyes Only trials, but should choose *randomly* between the two cups on the Above Target trials. The latter prediction is the key one, because we reasoned that organisms with a theory of seeing-as-attention (e.g., 3-year-old children) would interpret the distracted experimenter as not conveying any information about the location of the reward.

In order to evaluate our task, we first tested whether 3-year-old children met the central prediction of the mentalistic framework. As we predicted, these children selected the cup that the experimenter was looking at on the At Target trials, but chose randomly between the two cups on the Above Target trials (Povinelli, et al., in press, Experiment 3). This result is crucial, because it demonstrates that our theory of the task was supported. In direct contrast, the chimpanzees did not discriminate between the At Target and Above Target trials. Rather, they entered the test unit, moved to the side of the apparatus in front of the experimenter's face, and then chose the cup that was closest to them. Did the apes simply not notice the exact direction of the experimenter's gaze on the Above Target trials, thereby confusing these with the At Target trials? To the contrary, the results indicated that the apes looked above and behind themselves on over 71% of all Above Target trials, a level dramatically higher than on either the At Target or the Eyes Only trials (16 and 7% of the trials, respectively; Povinelli et al., in press, Experiment 1). Thus, although the chimpanzees clearly noticed and responded to the gaze direction of the experimenter in the Above Target trials, unlike 3-year-old children, they provided us with no unique reason to believe that they understood how this posture was connected to an internal state of attention.

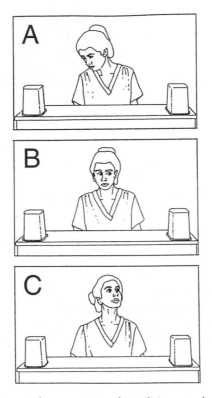

FIGURE 3.11. Experimental apparatus and conditions used with chimpanzees and children to assess chimpanzees' understanding of the "aboutness" property of visual perception in the context of gaze following. In the At Target condition (A), the experimenter directs her face and gaze at the correct cup. In the Eyes Only condition (B), the experimenter directs only her gaze at the correct cup. In the Above Target condition (C), the experimenter directs her face and gaze above the correct cup. See text for details of results and interpretation.

Finally, consider an additional study we conducted in which our chimpanzees were asked to reason about one person's attentional state. This study built on some previous work by Gómez (1996), who constructed a situation in which chimpanzees needed an experimenter's assistance in obtaining a food reward. Gómez varied the state of the experimenter's attention on each trial (e.g., eyes open vs. closed). He reasoned that if chimpanzees understood the differential attentional states of the experimenter in these situations, they ought to differentially deploy their spontaneous attention-getting behaviors. Indeed, Gómez reported that his 4-year-old chimpanzees did engage in attention-getting behaviors when an experimenter was inattentive. However, from his brief report of these findings, there is no evidence that the deployment of these behaviors was specific to these inattentive conditions, as opposed to other, attentive conditions. Thus, the data he has presented show that chimpanzees possess and deploy attention-getting behaviors. They in no way uniquely demonstrate that they understand the connection between these behaviors and the internal mental state of attention in the other person.

In order to address this question directly, we created several conditions in which we manipulated the state and location of the experimenter's attention. On most trials, the chimpanzees simply entered the test unit, gestured to the experimenter, and were immediately handed a food reward. On the crucial trials, however, the experimenter activated a 20-second timer as soon as the subject gestured to them. During these 20 seconds, the experimenter engaged in one of several behaviors. He or she either (1) stared directly at the subject while attempting to make direct eye contact, (2) made direct eye contact with the subject while engaging in slight back and forth movements of the head (a signal of attention in chimpanzees), (3) closed their eyes and waited, or (4) looked above and behind the chimpanzee. In the first two of these conditions, the experimenter maintained a state of visual attention to the subjects throughout the 20-second waiting period, whereas in the latter two, the experimenter was visually inattentive for the waiting period. We predicted that if the subjects appreciated the different state of the experimenter's attention in these conditions, they ought to deploy more attention-getting behaviors (touching/slapping at the experimenter or vocalizing) in the inattentive conditions than in the attentive ones. In fact, although the subjects readily engaged in such attention-getting behaviors, they did not exhibit more or longer episodes of them in the inattentive conditions as compared to the attentive ones (Povinelli, Davidson, & Theall, 1998).

Do Chimpanzees Understand Seeing-as-Knowing?

The results just described cast doubt on whether chimpanzees understand seeing-as-attention. It is curious, therefore, that earlier reports suggested that chimpanzees might understand even some of the more sophisticated aspects of visual perception. For example, can chimpanzees understand that someone who sees an event possesses knowledge different from someone who does not? Premack (1988) and Povinelli, Nelson, and Boysen (1990) both reported studies that attempted to address this question. However, both studies possessed serious methodological limitations that prevent a strong inference that chimpanzees were reasoning about knowledge states per se. In the Povinelli et al. (1990) study, chimpanzees were allowed to choose between the pointing "advice" of two experimenters, a "knower" who had previously seen where food was hidden, and a "guesser" who had been out of the room during the hiding procedure. Although 3 of the 4 subjects showed some evidence of understanding the task, this understanding did not emerge until after they had experienced many trials. Indeed, even when the subjects were confronted with a novel version of the task, evidence for comprehension did not emerge until after they had received a number of trials (Povinelli, 1994b). Thus, there is ample room to suppose that their performance was the result of trial-and-error learning. In contrast, and consistent with the findings of other developmental psychologists using different procedures, 4-year-old children performed excellently on this task from Trial 1 forward (Povinelli & deBlois, 1992b). (Other critiques of these studies are provided by Povinelli & Eddy, 1996a, and Tomasello, 1996.)

Other Mental States, Other Tasks, Important Caveats

Clearly, there is more to our understanding of the mind than our reasoning about the connection between visual perception on the one hand, and attention, knowledge, and belief on the other. Indeed, as we have seen, developmental psychologists have been exploring infant and young children's understanding of a wide range of mental states and activities including intentions, desires, emotions, inference, stream of consciousness, and false belief, just to name a few. Yet even though the theory of mind concept originated in the context of research with chimpanzees, current research with nonhuman primates has lagged behind. However, there have been a few attempts to investigate related topics such as their understanding of the distinction between accidental and intentional behaviors (Povinelli, Perilloux, Reaux, & Bierschwale, in press). In general, there is little evidence that nonhuman primates reason about other mental states any more than they do about attention (see Tomasello & Call, 1997, for a thorough review of the topic).

An important caveat concerning development is necessary in attempting to draw any firm conclusions about chimpanzees' understanding of mental states. In most of the studies reviewed here, and indeed in virtually all of the research to date, the age of the subjects has not been an independent variable. Thus, the majority of the tests we have described took place when our subjects were between 5 and 8 years of age. Admittedly, the abilities we were testing for are well consolidated in human children by 2 or 3 years of age. However, there is no *a priori* reason to suppose that even if there is extensive overlap in the cognitive developmental pathways of humans and chimpanzees, the rate of development is the same. Indeed, there are some data to suggest that certain behavioral capacities (including self-recognition in mirrors) that emerge at about 18–24 months in humans are

not acquired by chimpanzees until about 5–8 years of age (for a partial review, see Povinelli, 1996a). Thus, it is still possible that adult apes
might be reasoning about emotions, desires, and goals, for example, in a
manner comparable to 18- to 24-month-old human infants (e.g., Kagan,
1981; Baldwin, 1991; Baldwin & Moses, 1994; Brownell & Carriger,
1990; Lewis et al., 1989; Zahn-Waxler, Radke-Yarrow, Wagner, & Chapman, 1992; Repacholi & Gopnik, 1997; Bischof-Köhler, 1988; Meltzoff,
1995).

Summary: Chimpanzees' Understanding of Mental States

We conclude that at present there is little compelling evidence to suggest
that chimpanzees understand seeing-as-attention. Indeed, although we
have not provided as thorough a review as we would like, there is little evidence that chimpanzees understand anything at all about mental states.
These statements, however, must not be interpreted to mean that chimpanzees are open-ended, Skinnerian learning machines. Quite to the contrary. Chimpanzees, like other nonhuman primates, appear to be emotionally grounded to the dynamic social interactions that cascade around them.
They are intelligent, alert, and cognizing organisms, able to respond appropriately and dynamically to changing social circumstances. They possess
communicative signals with high emotional valence, signals that mediate
approaches, withdrawals, and a wide range of social behaviors (Goodall,
1986; for experimental data, see Povinelli & Eddy, 1996c). They evince evidence of the intelligent use of social information, such as following the
gaze of conspecifics, and engage in flexible patterns of social interactions
that allow them to manipulate and deceive each other. What they may *not*
possess is an elaborated theory of these social dynamics—a coherent, organized body of knowledge that recruits concepts such as attention, desire,
knowledge, and belief to explain the behavioral landscape that is constantly unfolding around them.

THE SOCIAL INTELLIGENCE MODEL REVISITED

Thus far, we have examined several lines of evidence concerning what nonhuman primates, and especially chimpanzees, know about self (as revealed
through self-recognition studies) and other (as revealed through theory-
of-mind studies). The data we have just reviewed suggest a complicated
mosaic of similarity and difference in the self and social intelligences of
chimpanzees and humans. It is always possible, of course, to react to such
findings as incoherent, and conclude that there must be something flawed

about the nature of the tests that have yielded this pattern. Elsewhere, we have assessed the extent to which such methodological objections have merit (Povinelli, 1996a). Rather than reexamining those arguments here, we note that as irritating as these findings may be to our preconceived notions of what a coherent account ought to look like, they are consistent with a growing body of data in related areas. For example, chimpanzees do not appear to possess genuine imitative skills (see Tomasello, Kruger, & Ratner, 1993). Although we recognize the immature nature of our current understanding of chimpanzee cognitive development, we nonetheless believe that the current findings demand some kind of evolutionary account. Thus, we ask whether there are any available theoretical frameworks that could explain the existence of such a radical incongruity between the similarity of human–chimpanzee behavioral patterns, on the one hand, and dissimilarity in theory-of-mind abilities on the other.

One possibility, of course, is the social intelligence model. As we have seen, this proposal points to the social arena as the sharpening stone against which self and social knowledge were honed. But a moment's reflection will reveal that the social intelligence framework is unable to explain the emerging pattern of data on self and social intelligence. For example, it cannot explain why the kind of self-representational system that chimpanzees may possess would have evolved in the common ancestor of the great apes and humans, but not earlier (see Povinelli & Cant, 1995). There are many ways of measuring social complexity, but however we choose to measure it, great ape societies do not stand out as being exceptionally sophisticated. Indeed, by certain measures, such as group size, baboons stand out as facing more social challenges per unit time than chimpanzees (Dunbar, 1988). For example, baboons display patterns of social intrigue every bit as captivating as chimpanzees (e.g., Smuts, 1985; Strum, 1987; Whiten & Byrne, 1988). Thus, if there are phylogenetic differences in self-representational abilities among nonhuman primate species, then attempting to map them onto differences in social complexity will immediately run into trouble.

The social intelligence framework fares even less well when it comes to explaining the evolution of theory-of-mind abilities. The first difficulty is the mounting evidence (some of which we have reviewed) that not even chimpanzees possess much in the way of an ability to represent the mental states of others. If humans are largely unique with respect to reasoning about the mental states of self and other, how can sociality—widespread and ancient in the primate order—be the adequate causal variable in explaining the emergence of theory-of-mind skills? Clearly, there must be some other factors or conditions that explain the restriction of this kind of cognitive system to the human species.

EVOLUTION OF THE SELF–OTHER FUSION: ALTERNATIVE HYPOTHESES

We have seen that humans appear to develop an understanding of the minds of others that parallels their understanding of their own. However, the exact connection between these conceptual knowledge bases is unclear. As we have seen, the theory-theorists argue for a kind of general theoretical system that is applied uniformly to self and other in developing ideas about intentional states (e.g., Gopnik, 1993). In contrast, many simulation theorists believe that information about the self's mental states is primary, whereas knowledge of comparable states of others is derived from such knowledge. Nonetheless, both theoretical positions acknowledge the intimate connection between self understanding and an understanding of others. To date, however, no one has seriously examined the possible evolutionary history of this fusion of self–other understanding in humans. Although we fully recognize that there are numerous possibilities, we now offer two broad hypotheses that we believe raise the central questions connected with this problem.

The Asynchrony Hypothesis

In principle, it is possible that despite the intimate connections between self–other knowledge in our species, the evolutionary history of these knowledge systems was separate. Although either sequence is logically possible, we wish to explore the possibility that knowledge of the self's intentional states evolved prior to knowledge of similar states in others.

In light of the failure of the social intelligence framework to account for the phyletic differences in self-recognition, Povinelli and Cant (1995) offered a hypothesis to explain why these abilities are restricted to the great apes and humans. Although we do not have space to review this proposal in detail here, they hypothesize that a large evolutionary increase in body weight, coupled with a commitment to an arboreal lifestyle, created unusually severe locomotor problems for the Miocene ancestor of great apes and humans. They speculate that the solution to this problem was the adoption of a nonstereotyped form of locomotion called orthograde clambering. This form of locomotion may have required an explicit representation of the self's actions (i.e., an explicit understanding of the self's agency). As we described earlier, this may be exactly the kind of self-representational system that allows for self-recognition in mirrors (see Povinelli, 1995). Thus, Povinelli and Cant's "clambering hypothesis" explains, from an evolutionary perspective, why self-recognition appears limited to organisms within the great ape–human group.

Their model also raises the possibility that this primitive representational system is not uniformly applied to self and other. In others words, although the extent of self-knowledge present in chimpanzees is debatable, if something like Povinelli and Cant's (1995) model turns out to be correct, it raises the possibility that selection may have favored a qualitatively new mechanism for representing the self's actions, without a concomitant alteration in their representation of others. If true, and if great apes have not undergone extensive psychological evolution in this regard since that point, the emergence of an integrated self–other representational system may have occurred at some point during the period of rapid brain size evolution during the course of the last 2 million years of human evolution (see Preuss & Kaas, in press). If these claims turn out to be correct, then the intimate psychological relation between self and other may turn out be one of the key psychological distinctions between humans and their closest living relatives. On this view, knowledge of the mental states of self and other evolved in asynchrony.

The Synchrony Hypothesis

Alternatives to the asynchrony hypothesis can be thought of as a family of hypotheses, which all share the common assumption that an interpretive stance toward self–other knowledge evolved in lockstep at each evolutionary juncture. For example, one possibility is that the presence of mirror self-recognition in chimpanzees indicates the presence of an explicit (albeit limited) self-concept, as well as a parallel understanding of others. Thus, if the evolution of such a self-concept evolved in the ancestor of the great apes and humans (e.g., the clambering hypothesis), this may have resulted in an immediate, correlated understanding of the actions of others. This particular hypothesis commits one to the view that chimpanzees and other great apes that display evidence of self-recognition possess an explicit self-concept. However, an alternative version of the general synchrony hypothesis is that great apes possess only implicit representations of self and other. In other words, Povinelli's (1995) model that explicit self-representations of one's actions are necessary for self-recognition in mirrors may be wrong. Instead, great apes may develop only implicit self-representations (although these representations must somehow differ from those possessed by other nonhuman primates in order to yield a system capable of engaging in self-exploratory actions in front of mirrors). Thus, in it broadest construction, the synchrony hypothesis concerns the uniformity of self–other representation, not the implicit–explicit nature of the representations.

In an era in which dedicated (if hypothetical) brain modules have somehow become a favored account of cognitive evolution, some may have trouble believing that such an understanding would evolve uniformly. However, if selection favored general representational structures, then regardless of whether the local reason for these evolutionary changes had to do with the need to explicitly represent the self's actions in relation to environmental effects of the self's actions (the clambering hypothesis), the consequences would be systemwide. Although current data have failed to produce compelling evidence for theory-of-mind abilities in chimpanzees, if the synchrony hypothesis is correct, and if the common ancestor of the great apes and humans evolved a limited understanding of self and other as mental agents (perhaps somewhat akin to the ways in which 18-month-old humans understand self and other), then Gallup's (1982) proposal could still be rescued. For example, if chimpanzees could be shown to understand attention in the ways in which 18-month-old human infants appear to, it would still be possible that self-recognition in mirrors might be a relatively easy-to-detect marker of a strong inferential connection between the psychology of self and other.[5]

REINTERPRETING BEHAVIOR

There remain several nagging ambiguities in our account. If, as we have suggested, chimpanzees possess a limited self-conceptual system, and at best possess only very circumscribed knowledge of other minds, then how can we account for the remarkably similar behaviors between our species? Perhaps the point is best put baldly: If we admit that our species reasons about mental states when we engage in a given behavior, how can we deny other species a similar understanding when they engage in similar behaviors? Conversely, in this sense our critique of Romanes' (1882) method can be stood on its head: If the complicated social intrigues practiced by chimpanzees and other nonhuman primates can be accounted for without granting them an understanding of mental states, why do we insist on this account of our own?

The mistake, we believe, lies in our understandable tendency to start with human behavior and psychology and work our way to other species. As an alternative, envision our planet long before humans or the modern great apes evolved. Envision a wide variety of language-less primate species negotiating their way through a complicated social milieu. Imagine untold generations of descendants evolving ever more sophisticated social rules and procedures—some tightly constrained developmentally, others more open-ended. And let us be especially careful not to oversimplify those abilities. Grant these species rich and diversified behavioral systems—systems composed of many behaviors still present in us. But now let us imagine that these species were devoid of any understanding of themselves or others as mental agents. Clever brains, in Humphrey's (1982) turn of phrase, but blank minds. Next, imagine the first spark of self-awareness emerging in common ancestor of the great apes and humans—perhaps a limited understanding of the self as causal agent (Povinelli & Cant, 1995). Finally, imagine that only one of that ancestor's living descendants—the genus *Homo*—evolved an additional cognitive specialization that produced a uniform, mentalistic understanding of self and other. What we are suggesting is that at some point during the emergence of hominids as a distinct group of species, a new conceptual understanding of others may have been woven into existing, ancestral developmental pathways controlling the expression of myrad ancient (but sophisticated) social behaviors.

In short, our hypothesis is that the terminal addition model of general psychological evolution proposed by Parker and Gibson (1979) cannot be fruitfully applied in the domain of psychological systems that we call theory of mind. Consider the self–other psychological systems that we are suggesting could be unique to humans. Rather than having tacked these systems onto the end of the cognitive developmental pathways that were present in the last common ancestor of humans and chimpanzees, we speculate that these skills were woven into those pathways at some point much earlier in development. If correct, human psychological evolution could be characterized less by the addition of new behaviors than by the weaving in of new cognitive systems alongside (and probably into) the old. Like a tapestry into which new colors were added, the old tapestry was not discarded or merely lengthened. Rather, the ancient neural systems may have served as both a substrate and/or constraint for the new systems. And, as we elaborate later, these new systems, in turn, may have had the effect of functionally reorganizing the older ones.

At this juncture, it may be instructive to consider a similar problem faced by researchers interested in cognitive development. During the course of development, young children often display the following enigmatic behavior. On the one hand, they will perform a given task immediately, effortlessly, and correctly; on the other hand, when queried about their actions, they will often offer a lucid, quite precise, but utterly irrelevant explanation for these actions. Indeed, we have discussed one example of this phenomenon in the context of children using obvious information to locate hidden objects, but failing to understand how this information assisted them in the first place. These kinds of mismatches between production and comprehension have led Annette Karmiloff-Smith (1992) to propose a process of "representational redescription" as a general feature of cognitive development. As she explains:

> [Representational redescription] involves a cyclical process by which information already present in the organism's independent functioning, special purpose representations, is made progressively available, via re-

descriptive processes, to other parts of the cognitive system. In other words, representational redescription is a process by which implicit information *in* the mind subsequently becomes explicit information *to* the mind, first within a domain and then sometimes across domains. (pp. 17–18)

In this way, cognitive development can be thought of as the process of

recoding information that is stored in one representational format or code into a different one. Thus, a spatial representation might be recoded into a linguistic format, or a proprioceptive representation into spatial format. Each redescription, or re-representation, is a more condensed or compressed version of the previous level.... [the representational redescription] model postulates at least four hierarchically organized levels at which the process of representational redescription occurs. (p. 23)

It is significant for our framework that the first two of these postulated levels are not necessarily available to conscious access. Thus, Karmiloff-Smith's (1992) account offers a particularly thoughtful explanation of how perfectly articulate children can possess and process knowledge that they cannot express—in other words, how information can be present in a system but reside at different levels of explicitness. It is our contention that the information controlling many of the common behaviors of humans and other primates inhabits these less explicit representational levels. To use Karmiloff-Smith's terminology, one interpretation of our hypothesis is that humans have uniquely evolved mechanisms allowing for the very process of representational redescription—the process of transforming implicit representations into forms available to consciousness.[6]

The significance of this view is that it dramatically alters how one interprets the behavioral and cognitive similarities and differences among humans, chimpanzees, and other nonhuman primates (Povinelli, Zebouni, & Prince, 1996; Povinelli, 1996b). Rather than interpreting such behavioral similarities as *prima facie* evidence in favor of the view that chimpanzees must possess the same psychological systems that attend our execution of these behaviors, it is possible to consider an alternative. Many behaviors that our species naively interprets through a mentalistic framework may have evolved and been in full operation millions of years before we appeared on the scene. We are not suggesting an extreme form of dualism (such as epiphenomenalism) in which these metacognitive states attend, but play no causative role in behavior. No, we suppose that representations of the mental states of self and other evolved because of their useful, causal connection to the behavior of the organisms possessing them. But until now, we may have been thinking in the wrong way about how the evolution of a novel psychological system may have affected the behavioral fabric of human ancestors. To date, many researchers (including ourselves) have been looking—largely unsuccessfully—for a simple connection between the emergence of a novel psychological system related to theory of mind and a class of novel behaviors onto which this novel system could be mapped. But this may turn out to have been a fool's quest. The evolution of a new psychological system may have functioned to support, optimize, or otherwise reorganize existing behavioral patterns, without necessarily leading to any individual elements that look definitively novel. Thus, although it may have had a profound effect on the complexity, rapidity, and informational density that a behavioral network could achieve, it may not have sprouted many "new" or "novel" behaviors per se. We shall return to this point later.

Lessons from Gaze Following

In order to illustrate our argument, we return to the case of gaze following—although we intend to show how the same logic may apply with equal force to other behavioral systems as well. First, when we follow the gaze of someone else, it is not clear when, exactly, we begin to entertain notions about the other person's internal mental state. Indeed, there are surely many cases in which we respond by looking where they are looking and then back again, without ever even registering the fact that we have done so. Thus, in principle, it is possible that the attribution of attention to the other person might occur after the behavioral act of gaze following, thus eliminating it as a possible causal explanation of the behavior itself. Some may object here, noting that the case of adults may have little to do with infants. For example, adults may have routinized the gaze-following response after considerable experience. But this already grants a large portion of our argument by demonstrating that the two elements—the response and the attribution—are, at least in principle, dissociable. In other words, imagine that we could selectively cripple the ability to represent the mental state of another's attention. Would the gaze-following response still occur? We suspect so.

Our suggestion is that gaze following evolved not as part of a theory-of-mind system, but because of its utility in the social ecology of our primate ancestors. For example, Povinelli and Eddy (1996a, 1996b) have proposed that gaze following may be a fairly ancient behavioral mechanism common to many social primates (and perhaps other social mammals) and may have evolved independently of an understanding of the mental state of attention. A psychological system that could rapidly process information about other group members' line of sight could be very advantageous in several naturalistic contexts such as the early detection of predators and anticipating the likely targets of the social behaviors of other group members (see Chance, 1967; van Schaik, van Noordwijk, Warsono, & Sutriono, 1983). To be sure, this system *might* mediate the perceptual input (the visual signal of seeing a group member turn and look) and the motor output (turning and looking in that same direction) with representations of the other's mental state (i.e., attention). However, there are other, more direct linkages that might mediate the input–output relation in question—ones that do not depend on representations of mental states per se. Thus, we propose that it is quite possible that in the context of the evolution of a broader psychological system, metacognitive representational systems were stitched into the broader psychological system of the 18-month-old, a system that already possessed the neural mechanisms supporting gaze following (see Figure 3.12). Thus, it may be accurate to say that the infant's representation of the other person's attention accompanies the act of gaze following; but we offer the hypothesis that this representation is not causally related to the triggering of the action.

At this point it may seem as if we are dangerously close to invoking a spectator-consciousness effect: The presence of an awareness of a certain action without this awareness having any causal effect in the real world. Quite to the contrary, we suppose that infants' growing abilities to represent and reason about attention (and other mental states) in those around them offers them a coherent explanatory account of what is occurring (see Gopnik & Meltzoff, 1997). This explanation in turn allows them to learn about relations among people, objects, and events in a faster and more direct manner. In the context of gaze following, we imagine that during its earliest evolutionary manifestation, the response may have functioned as a fairly automatic response to sudden shifts in the head movements of other conspecifics and may have then operated as described in Figure 3.6A. However, a first evolutionary step toward understanding gaze as attention may have involved understanding shifts in others' gaze as behaviors that signal important impending relations between the organism and other events. Indeed, it is quite possible that even chimpanzees acquire this kind of understanding of gaze, as opposed to a mentalistic understanding of gaze as a psychological spotlight emanating from the other organism.

A glaring weakness in our account is that, at present, we cannot precisely specify the nature or scope of the neural systems mediating the gaze-following response, or the attribution of attention. For example, until now, we have used metaphors related to sewing and weaving to help us convey the idea of the interleaving of new systems alongside older ones. Yet this metaphor fails in at least one crucial respect. It suggests that the newer structures could be easily pulled out, leaving the old ones intact. However, the emergence of new representational systems may often involve the direct use of ancestral neural architecture, thereby resulting in a functional interdependency between the old and the new. Some cases of human psy-

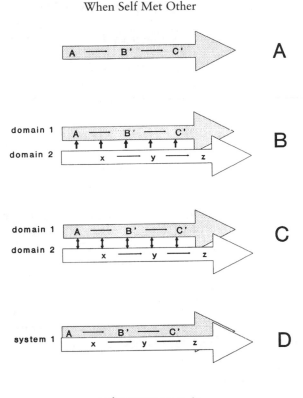

ontogeny ———▶

FIGURE 3.12. A hypothetical evolutionary history of an ancestral cognitive-behavioral pathway (A) becoming altered through the interleaving of a novel pathway (B–D). Note that the ancestral behavioral markers at various points in ontogeny are conserved, although their interpretation or possible uses may be altered by the weaving in of the new pathway. Despite information flow between them, dissociations of the two pathways may still be somewhat possible in the hypothetical cases (B) and (C). However, in the hypothetical extreme illustrated in (D), the two pathways are essentially fused (no longer dissociable), despite their separate evolutionary history.

chopathology may help to illustrate this point. Autism, for example, has received considerable attention recently as being characterized by a severe deficit in theory-of-mind and other metacognitive abilities (e.g., Baron-Cohen, 1995). Autistic individuals have been shown to be profoundly impaired on tasks relating to an understanding of mental states such as desires, knowledge and beliefs. Interestingly, there is evidence that the inability of an 18-month-old infant to follow an adult's gaze (accompanied by a pointing gesture) predicts a diagnosis of autism at 3 years of age (Baron-Cohen et al., 1996). Given that the dysfunction of gaze following in autistic infants and children predicts a later dysfunction in their understanding of attention, one interpretation of this finding is that gaze following is, in fact, causally related to an understanding of the mental state of attention.[7]

However, this conclusion need not follow. It is possible that during the course of recent human psychological evolution, new and old structures have combined into related functional systems in such a manner that it is not easily possible to selectively impair the new elements in this system. Some researchers, for example, have recently shown how the suite of behavioral and cognitive impairments associated with autism may in fact be the result of damage very early in development—perhaps as early as the fourth week of gestation (Rodier, Ingram, Tisdale, Nelson, & Romano, 1996). Thus, many of the cognitive dysfunctions of autism may be the result of abnormal inputs from the brain stem to a relatively normal forebrain. Conversely, it is possible that there are some as-of-yet undetected abnormalities in the forebrain that are secondary, cascading consequences of such early injury. In either case, the implication is clear: The evolution of new psychological systems may neither replace nor sit insulated alongside ancestral systems. Rather, new systems or subsystems may be created by building inside ancestral templates in such a manner that most of the useful behavioral propensities of ancestral organisms are conserved.

Broadening the Explanatory Framework

We suspect that a wide class of social behaviors can be usefully thought of in this same manner: deception, reconciliation, selective retaliation against less dominant allies of one's aggressors, certain forms of social learning—indeed, most of the core elements of the fabric of human (and ape) behavior. Consider Frans de Waal's (1982, 1996) marvelous expositions of the striking commonalities in chimpanzee and human politics and morality. He interprets these behavioral similarities as indicating not just basic psychological similarity between us and them, but also similarity in the kind of metacognitive processes that attend those behaviors in our own species. In contrast, we are suggesting that there are both real, and "really real" explanations for our behaviors. Certainly our folk psychological explanations count for something; indeed, they may reflect the discovery of an important shorthand for reasoning about enormously complicated neural processes. Rather than having to specify that so-and-so has such-and-such a configuration of neural activation in the prefrontal cortex, we can simply say, "She thinks that I want to trick her." But although accurate in a sense, it may not be at this level of explanation that the utility of these representations is to be found. Indeed, many (if not most) of the behavioral propensities now associated with our species' higher-level metacognitive representations were present long before these representations were even possible.

But if the view that we are exploring here is correct, then the striking *behavioral* similarities between humans and chimpanzees, for example, are only a guide to superficial psychological similarities. Similarity at the level of interpretation and meaning—folk psychology—is another matter altogether.

There is another manner in which our folk narratives constitute a real explanation for why we behave in the manner we do. To the extent that we form beliefs about the causative role of mental states in behavior (i.e., to the extent that we ponder why so-and-so has done such-and-such), we quickly construct elaborate narrative accounts of why events have taken the shape they have. These narrative accounts may serve as a database in their own right, a storehouse of information not so much different from information received and translated through the primary senses. Thus, in considering what we may do next, we appear to take into account not just what has happened, but why (at least from the perspective of our narrative) it has happened. Such narratives, while not strictly "accurate," may provide human beings with a powerful adaptive device, a means of rapidly reorganizing existing fundamental behavioral units into novel cultural configurations. In this sense, the diversity of human cultural beliefs and non-material traditions may be the construction of narratives well suited for the current and historical ecological challenges faced by individual cultures. If folk narratives function in this fashion, this may help to explain why human populations have expanded into virtually every conceivable ecogeographic zone on the planet, whereas chimpanzees and other great apes have remained restricted to the tropics and neotropics, and even there inhabit very few niches (tropical rain forest, open woodland savannas). Indeed, part of the consequence (and perhaps function) of such causal narratives is that they open up the possibility of mental time travel, the ability to understand how the past, present, and future are all part of a flow of connected events (see Suddendorf, 1994; Povinelli, 1995). Nonetheless, despite the diversity of human habitat and culture, and despite the enormity of possibilities opened up by these cognitive specializations, we suspect that with but rare exceptions the fundamental behavioral building blocks that humans use to generate novel future actions have remained largely unaltered from those present millions of years ago in the common ancestor of great apes and humans (and, in many cases, even earlier).

Our approach can therefore be interpreted as an explicit alternative to Romanes' (1882) hope of using the spontaneous behavior of animals as portals into their minds. Although this approach may have fit the bill charged by the complete absence of information, it now seems doomed to miss the historical complexities of the evolutionary process. Just as the morphological structures and systems of modern organisms reflect the baggage of developmental constraints laid down in the Cambrian, so too do psychological structures carry the ancient alongside the new. The error in Romanes' reasoning was that he assumed that the behavioral similarities between humans and other animals must reflect underlying psychological similarities. First, he assumed that an accurate introspective assessment of the mental states that accompany our own behavior could yield an accurate inference as to their cause. Worse still, he assumed it could yield an equally accurate inference for species other than our own. And although it may be true that humans form internal representations of mental states such as desires, knowledge, and belief that provide us with a useful means of anticipating what others will do, the basic blueprint of the behaviors we can respond with evolved long before those meta-level representations were possible.

As a final point, it is worth considering how this view articulates with the problem of human uniqueness. Faced with the apparent distinctiveness of the human species, philosophers, anthropologists and psychologists have offered numerous honorary titles for our species, each emphasizing some would-be unique characteristic. But from tool-making, to culture, to language, the animal kingdom has relentlessly refused to unambiguously concede any of these traits as exclusively human. The view we have offered here suggests a way out of this impasse by recognizing that we may never locate any truly unique and universal human behavioral traits. Even with respect to those phenomena where we seem to differ most from other species—cultural learning, pedagogy, ethics, and language (Premack, 1984; Tomasello et al., 1993; Povinelli & Godfrey, 1993)—the individual elements of the behaviors involved may have similar ancient precursors, revealed by observing a diverse array of other species. And in each case, the similarities may be great enough that there will always be adequate grist for the mill for those who wish forever to deny the existence of qualitative differences between humans and other species. Yet the real differences between humans and even our nearest living primate relatives may be much deeper, rooted in our *interpretation* of behavior. Human uniqueness may stem first and foremost from underlying cognitive—not behavioral—specializations. These cognitive differences, in turn, may then translate into a radical reorganization and redeployment of existing ancestral behavioral abilities. And it is this generativity, manifested through ancient behavioral patterns, that at a proximate level has yielded the remarkable diversity and flexibility of human culture.

EPILOGUE: AN APE'S EYE VIEW OF ROMANES' PROBLEM

If the alternative view we have sketched here is correct, we might do well to close by imagining the reaction of our apes to the questions we continue to pose to them: "Yes," they reply obliquely through their responses to our still-clumsy tasks, "we share with you an enormous range of useful, complicated, and sophisticated social behaviors. Yes, we share with you a psychological system that is able to knit these behaviors into novel and productive strategies that serve to fulfill our goals and desires. And it is true that our emotions, mannerisms, and reactions are much like your own. We even possess a self-concept that offers us an objective perspective on our own behavior. But what ever gave you the idea that we have a theory of mind? Why do you want to believe so desperately that we are able to construct a self–other narrative like you? After all, it was your lineage, not ours, that tripled the size of its brain during the past 5 million years. It was your species, not ours, that constructed the idea that there are unobservable mental states that mediate behavior. And thus it is you," they conclude, "not us, who are in the position of reinterpreting ancient behavioral patterns in terms of mentalistic notions—notions that never even occurred to us."

ACKNOWLEDGMENTS

This work was supported by National Institutes of Health Grant No. RR-03583-05 to the University of Southwestern Louisiana New Iberia Research Center and National Science Foundation Young Investigator Award SBR-8458111 to Daniel J. Povinelli. Photographs and original drawings are by Donna T. Bierschwale. We thank Todd M. Preuss for offering valuable comments on the manuscript, and Karen Wright for thought-provoking discussions about these issues.

NOTES

1. This assumption may seem naive to some cultural anthropologists in that it appears to ignore a wealth of data on the unique ways in which concepts such as self are constructed across different societies (e.g., Geertz, 1973; La Fontaine, 1984; Lienhardt, 1984; Duranti, 1988). However, we follow those who recognize universal aspects of human social understanding, while simultaneously exploring the ways in which this understanding is shaped by cultural frames of reference (Hallowell, 1971; White, 1980; Heelas, 1981; Lock, 1981). Recently, there have been several direct examinations of cross-cultural understanding of mental states such as false belief (Avis & Harris, 1991; Vinden, 1996; see Lillard, 1998, for a comprehensive review).

2. A number of careful studies have failed to demonstrate self-recognition in gorillas (Lethmate & Dücker, 1973; Suarez & Gallup, 1981; Ledbetter & Basen, 1982). These results stand in contrast to convincing evidence for this ability in a female gorilla who was reared by Francine Patterson as part of an intensive attempt to teach her sign language (Patterson & Cohn, 1994). Povinelli (1994a) has re-

viewed evidence suggesting that, compared to the other great apes, gorillas may have undergone a secondary reversal in key aspects of their schedule of physical maturation. These heterochronic processes may have had cascading consequences for aspects of general cognitive development, including their capacity for self-recognition in mirrors. However, secondary losses of derived traits generally are not the result of the loss of specific genetic instructions, but rather result from the shutdown of their expression due to changes in the rate and timing of other developmental events (Frazetta, 1975; Gould, 1977; Albrech, Gould, Oster, & Wake, 1979; for examples see Hampé, 1960; Kollar & Fisher, 1980). One consequence of this is that ancestral characteristics—traits that have been "lost" for tens of millions of years—can be reinstated by slight, abnormal changes in the developmental pathways involved. Given that gorillas presumably descended from a common ancestor of the great apes and human beings that was capable of self-recognition (Povinelli, 1987), it seems quite possible that the genetic instructions for the construction of the neural structures necessary for self-recognition still exist in modern gorillas, and can be reinstated through abnormal rearing environments such as that received by Patterson's gorilla (see Povinelli, 1994a).

3. Montgomery, Moran, and Bach (in press) have challenged the depth of 3- and even 4-year-old's understanding of gaze direction as attention. They have provided evidence suggesting that when the orientation of someone's body contrasts with the direction of their gaze, preschoolers are generally unclear as to the person's goal.

4. This is not to say that the chimpanzees necessarily grasped the referential significance of the pointing gesture. Indeed, in a separate series of studies, we explored what our chimpanzees really understood about this gesture (Povinelli, Reaux, Bierschwale, Allain, & Simon, 1997). These results indicated that the subjects used the gesture as a landmark cue, adopting the following rule: "*Pick the cup that is closest to the experimenter's hand/finger.*" In contrast, even the youngest children we tested (26-month-olds) had no difficulty using the gesture in a referential manner.

5. As an example, consider the pointing gesture. In human infants, understanding of the attentional significance (the referential aspect) of the pointing gesture is consolidated between 12 and 15 months (see Murphy & Messer, 1977; Lempers, 1979; Butterworth & Grover, 1988; Morissette, Ricard, & Décarie, 1995). Thus, this would seem to qualify as an excellent arena to search for similarities. However, current experimental evidence suggests that although chimpanzees can easily be taught to learn to respond appropriately to pointing, they appear to interpret it in a different manner from humans. Unlike even very young human children who interpret pointing as indicating the attentional focus of the person making the gesture, chimpanzees appear to interpret the gesture in a nonmentalistic manner (see note 4). In addition, a majority of children do not display self-recognition in mirrors until 18 months (see earlier references).

6. Indeed, Karmiloff-Smith (1992) speculates that the "pervasiveness of representational redescription in human cognition is, I maintain, what makes human cognition specifically human" (p. 192). She further speculates that if the process of representational redescription is available to other species (such as chimpanzees), "the higher-level codes into which representations are translated during redescrip-

tion are very impoverished" (p. 192). As a related aside, if our account of the nature of the self-representation underwriting the chimpanzee's capacity for self-recognition in mirrors is correct, it would suggest that in this domain, chimpanzees may at least translate implicit representations of their own proprioceptive states into representations at the first level of explicitness (or E1) in Karmiloff-Smith's (1992) current model—a translation that does not occur in species outside the great ape/human group.

      7. The situation cannot be quite so simple, however. Recent studies have shown that although gaze following is severely impaired in autistic people, their ability to engage in simple Level 1 perspective-taking tasks (e.g., "What toy am I looking at?") is not (Hobson, 1984; Dawson & Fernald, 1987; Baron-Cohen, 1989; Tan & Harris, 1991; Leekam, Baron-Cohen, Perrett, Milders, & Brown, 1995). One interpretation of these findings is that Level 1 perspective-taking tasks can be solved geometrically, without understanding attention (Hobson, 1980, 1982; Baron-Cohen, 1994; Leekam et al., 1995). However, another possibility is that autistic individuals do retain an understanding of attention, but other systems that automatically process information about the eyes and head direction are crippled or impaired. In any event, if our research with apes is replicated, the opposite situation would seem to be possible as well: Organisms may possess systems for automatically processing information about gaze direction without an understanding of the mentalistic aspect of gaze.

REFERENCES

Adams, D. K. (1928). The inference of mind. *Psychological Review, 35*, 235–252.

Albrech, P., Gould, S. J., Oster, G. F., & Wake, D. B. (1979). Size and shape in ontogeny and phylogeny. *Paleobiology, 5*, 296–317.

Amsterdam, B. (1972). Mirror self-image reactions before age two. *Developmental Psychobiology, 5*, 297–305.

Anderson, J. R. (1983). Responses to mirror image stimulation and assessment of self-recognition in mirror- and peer-reared stumptail macaques. *Quarterly Journal of Experimental Psychology, 35B*, 201–212.

Anderson, J. R., & Roeder, J. -J. (1989). Responses of capuchins monkeys (*Cebus apella*) to different conditions of mirror-image stimulation. *Primates, 30*, 581–587.

Andrew, R. J. (1962). Evolution of intelligence and vocal mimicking. *Science, 137*, 585–589.

Argyle, M., & Cook, M. (1976). *Gaze and mutual gaze.* Cambridge, England: Cambridge University Press.

Asendorpf, J. B., & Baudonniere, P.-M. (1993). Self-awareness and other-awareness: Mirror self-recognition and synchronic imitation among unfamiliar peers. *Developmental Psychology, 29*, 88–95.

Avis, J., & Harris, P. L. (1991). Belief-desire reasoning among Baka children: Evidence for a universal conception of mind. *Child Development, 62*, 460–467.

Baldwin, D. A. (1991). Infants' contribution to the achievement of joint reference. *Child Development*, *63*, 875–890.

Baldwin, D. A. (1993a). Early referential understanding: Infants' ability to recognize referential acts for what they are. *Developmental Psychology*, *29*, 832–843.

Baldwin, D. A. (1993b). Infants' ability to consult the speaker for clues to word reference. *Journal of Child Language*, *20*, 395–418.

Baldwin, D. A., & Moses, L. J. (1994). Early understanding of referential intent and attentional focus: Evidence from language and emotion. In C. Lewis & P. Mitchell (Eds.), *Children's early understanding of mind* (pp. 133–156). Hillsdale, NJ: Erlbaum.

Baron-Cohen, S. (1989). Perceptual role-taking and protodeclarative pointing in autism. *British Journal of Developmental Psychology*, *7*, 113–127.

Baron-Cohen, S. (1991). Precursors to a theory of mind: Understanding attention in others. In A. Whiten (Ed.), *Natural theories of mind: Evolution, development and simulation of everyday mindreading* (pp. 233–251). Oxford, England: Blackwell.

Baron-Cohen, S. (1994). How to build a baby that can read minds: Cognitive mechanisms in mindreading. *Current Psychology of Cognition*, *13*, 513–552.

Baron-Cohen, S. (1995). *Mindblindness: An essay on autism and theory of mind.* Cambridge, MA: MIT Press.

Baron-Cohen, S., Campbell, R., Karmiloff-Smith, A., Grant, J., & Walker, J. (1995). Are children with autism blind to the mentalistic significance of the eyes? *British Journal of Developmental Psychology*, *13*, 379–398.

Baron-Cohen, S., Cox, A., Baird, G., Swettenham, J., Nightingale, N., Morgan, K., Drew, A., & Charman, T. (1996). Psychological markers in the detection of autism in infancy in a large population. *British Journal of Psychiatry*, *168*, 158–163.

Bayart, F., & Anderson, J. R. (1985). Mirror-image reactions in a tool-using, adult male *Macaca tonkeana*. *Behavioural Processes*, *10*, 219–227.

Benhar, E. E., Carlton, P. L., & Samuel, D. (1975). A search for mirror-image reinforcement and self-recognition in the baboon. In S. Kondo, M. Kawai, & S. Ehara (Eds.), *Contemporary primatology: Proceedings of the 5th International Congress of Primatology* (pp. 202–208). New York: Karger.

Bernstein, I. S. (1988). Metaphor, cognitive belief, and science. *Behavioral and Brain Sciences*, *11*, 247–248.

Bertenthal, B. I., & Fischer, K. W. (1978). Development of self-recognition in the infant. *Developmental Psychology*, *14*, 44–50.

Bischof-Köhler, D. (1988). Uber der zusammenhang von empathie und der fahigkeit, sich im spiegel zu erkennen [On the association of empathy and the ability to recognize oneself in a mirror]. *Schweizerische Zeitschrift für Psychologie*, *47*, 147–159.

Bischof-Köhler, D. (1991). The development of empathy in infants. In M. E. Lamb & H. Keller (Eds.), *Infant development: Perspectives from German speaking countries* (pp. 245–273). Hillsdale, NJ: Erlbaum.

Blest, A. D. (1957). The function of eyespot patterns in the Lepidoptera. *Behaviour*, *11*, 209–255.

Boccia, M. L. (1994). Mirror behavior in macaques. In S. T. Parker, R. W. Mitchell, & M. L. Boccia (Eds.), *Self-awareness in animals and humans* (pp. 350–360). New York: Cambridge University Press.

Brentano, F. (1960). The distinction between mental and physical phenomena. In R. Chisholm (Ed.), *Realism and the background of phenomenology*. New York: Free Press. (Original work published 1874 in German)

Brooks-Gunn, J., & Lewis, M. (1984). Development of early visual self-recognition. *Developmental Review, 4,* 215–239.

Brownell, C. A., & Carriger, M. S. (1990). Changes in cooperation and self–other distinction during the second year. *Child Development, 61,* 1164–1174.

Burger, J., Gochfeld, M., & Murray, B. G., Jr. (1991). Role of a predator's eye size in risk perception by basking black iguana, *Ctenosaura similis. Animal Behaviour, 42,* 471–476.

Burghardt, G. M. (1988). Anecdotes and critical anthropomorphism. *Behavioral and Brain Sciences, 11,* 248–249.

Burghardt, G. M., & Greene, H. W. (1988). Predator simulation and duration of death feigning in neonate hognose snakes. *Animal Behaviour, 36,* 1842–1844.

Butterworth, G., & Cochran, E. (1980). Towards a mechanism of joint visual attention in human infancy. *International Journal of Behavioral Development, 3,* 253–272.

Butterworth, G., & Grover, L. (1988). The origins of referential communication in human infancy. In L. Weiskrantz (Ed.), *Thought without language* (pp. 5–24). Oxford, England: Clarendon.

Butterworth, G., & Jarrett, N. (1991). What minds have in common is space: Spatial mechanisms serving joint visual attention in infancy. *British Journal of Developmental Psychology, 9,* 55–72.

Byrne, R., & Whiten, A. (1985). Tactical deception of familiar individuals in baboons (*Papio ursinus*). *Animal Behaviour, 33,* 669–673.

Calhoun, S., & Thompson, R. L. (1988). Long-term retention of self-recognition by chimpanzees. *American Journal of Primatology, 15,* 361–365.

Chance, M. R. A. (1967). Attention structure as the basis of primate rank orders. *Man, 2,* 503–518.

Corkum, V., & Moore, C. (1994). Development of joint visual attention in infants. In C. Moore & P. Dunham (Eds.), *Joint attention: Its origins and role in development* (pp. 61–83). Hillsdale, NJ: Erlbaum.

Darwin, C. (1871). *The descent of man.* (Reprinted, New York: Modern Library, 1982).

Davis, L. H. (1989). Self-consciousness in chimps and pigeons. *Philosophical Psychology, 2,* 249–259.

Dawson, G., & Fernald, M. (1987). Perspective-taking ability and its relationship to the social behaviour of autistic children. *Journal of Autism and Developmental Disorders, 17,* 487–498.

de Waal, F. (1982). *Chimpanzee politics: Power and sex among apes.* New York: Harper & Row.

de Waal, F. (1986). Deception in the natural communication of chimpanzees. In R. W. Mitchell & N. S. Thompson (Eds.), *Deception: Perspectives on human and nonhuman deceit* (pp. 221–244). Albany, NY: SUNY Press.

de Waal, F. (1996). *Good natured*. Cambridge, MA: Harvard University Press.

Dunbar, R. I. M. (1988). *Primate social systems*. London: Croom Helm.

Duranti, A. (1988). Intentions, language, and social action in a Samoan context. *Journal of Pragmatics, 12*, 13–33.

Eddy, T. J., Gallup, G. G., Jr., & Povinelli, D. J. (1996). Age differences in the ability of chimpanzees to distinguish mirror-images of self from video-images of others. *Journal of Comparative Psychology, 110*, 38–44.

Epstein, R., Lanza, R. P., & Skinner, B. F. (1981). "Self-awareness" in the pigeon. *Science, 212*, 695–696.

Fehr, B. J., & Exline, R. B. (1987). Social visual interaction: A conceptual and literature review. In A. W. Siegman & S. Feldstein (Eds.), *Nonverbal behavior and communication* (2nd ed., pp. 225–325). Hillsdale, NJ: Erlbaum.

Flavell, J. H., Everett, B. A., Croft, K., & Flavell, E. R. (1981). Young children's knowledge about visual perception: Further evidence for the level 1–level 2 distinction. *Developmental Psychology, 17*, 99–103.

Flavell, J. H., Flavell, E. R., Green, F. L., & Wilcox, S. A. (1980). Young children's knowledge about visual perception: Effect of observer's distance from target on perceptual clarity of target. *Developmental Psychology, 16*, 10–12.

Flavell, J. H., Shipstead, S. G., & Croft, K. (1978). What young children think you see when their eyes are closed. *Cognition, 8*, 369–387.

Flavell, J. H., Green, F. L., & Flavell, E. R. (1989). Young children's ability to differentiate appearance-reality and level 2 perspectives in the tactile modality. *Child Development, 60*, 201–213.

Fodor, J. (1983). *Modularity of the mind*. Cambridge, MA: MIT Press.

Fodor, J. (1987). *Psychosemantics: The problem of meaning in the philosophy of mind*. Cambridge, MA: Bradford Books/MIT Press.

Fodor, J. (1992). A theory of the child's theory of mind. *Cognition, 44*, 283–296.

Fornasieri, I., Roeder, J. -J., & Anderson, J. A. (1991). Les reactions au miroir chez trois especes de lemuriens (*Lemur fulvus, L. macaco, L. catta*) [Responses to mirror-image stimulation in three species of lemurs (*Lemur fulvus, L. macaco, L. catta*)]. *C. R. Acad. Sci. Paris, 312*, 349–354.

Franco, F. (in press). The development of meaning in infancy: Early communication and social understanding. In S. Hala (Ed.), *The development of social cognition*. London: University College London Press.

Frazetta, T. H. (1975). *Complex adaptations in evolving populations*. Sunderland, MA: Sinauer.

Frye, D., Zelazo, P. D., & Palfai, T. (1995). Theory of mind and rule-based reasoning. *Cognitive Development, 10*, 483–527.

Gallup, G. G., Jr. (1968). Mirror-image stimulation. *Psychological Bulletin, 70*, 782–793.

Gallup, G. G., Jr. (1970). Chimpanzees: Self-recognition. *Science, 167*, 86–87.

Gallup, G. G., Jr. (1975). Toward an operational definition of self-awareness. In R. H. Tuttle (Ed.), *Socio-ecology and psychology of primates* (pp. 309–341). The Hague, The Netherlands: Mouton.

Gallup, G. G., Jr. (1977). Absence of self-recognition in a monkey (*Macaca fascicularis*) following prolonged exposure to a mirror. *Developmental Psychobiology, 10*, 281–284.

Gallup, G. G., Jr. (1979). Self-awareness in primates. *American Scientist, 67,* 417–421.

Gallup, G. G., Jr. (1982). Self-awareness and the emergence of mind in primates. *American Journal of Primatology, 2,* 237–248.

Gallup, G. G., Jr. (1983). Toward a comparative psychology of mind. In R. E. Mellgren (Ed.), *Animal cognition and behavior* (pp. 473–510). New York: North-Holland.

Gallup, G. G., Jr. (1985). Do minds exist in species other than our own? *Neuroscience and Biobehavioral Reviews, 9,* 631–641.

Gallup, G. G., Jr. (1991). Toward a comparative psychology of self-awareness: Species limitations and cognitive consequences. In G. R. Goethals & J. Strauss (Eds.), *The self: An interdisciplinary approach* (pp. 121–135). New York: Springer-Verlag.

Gallup, G. G., Jr. (1994). Self-recognition: Research strategies and experimental design. In S. Parker, R. Mitchell, & M. Boccia (Eds.), *Self-awareness in animals and humans* (pp. 35–50). Cambridge, England: Cambridge University Press.

Gallup, G. G., Jr., Nash, R. F., & Ellison, A. L., Jr. (1971). Tonic immobility as a reaction to predation: Artificial eyes as a fear stimulus for chickens. *Psychonomic Science, 23,* 79–80.

Gallup, G. G., Jr., Povinelli, D. J., Suarez, S. D., Anderson, J. R., Lethmate, J., & Menzel, E. W. (1995). Further reflections on self-recognition in primates. *Animal Behaviour, 50,* 1525–1532.

Gallup, G. G., Jr., & Suarez, S. D. (1986). Self-awareness and the emergence of mind in humans and other primates. In J. Suls & A. G. Greenwald (Eds.), *Psychological perspectives on the self* (Vol. 3, pp. 3–26). Hillsdale, NJ: Erlbaum.

Gallup, G. G., Jr., & Suarez, S. D. (1991). Social responding to mirrors in rhesus monkeys (*Macaca mulatta*): Effects of temporary mirror removal. *Journal of Comparative Psychology, 105,* 376–379.

Gallup, G. G., Jr., Wallnau, L. B., & Suarez, S. D. (1980). Failure to find self-recognition in mother–infant and infant–infant rhesus monkey pairs. *Folia Primatologica, 33,* 210–219.

Geertz, C. (1973). *The interpretation of cultures.* New York: Basic Books.

Gómez, J.-C. (1996). Non-human primate theories of (non-human primate) minds: Some issues concerning the origins of mind-reading. In P. Carruthers & P. K. Smith (Eds.), *Theories of theories of mind* (pp. 330–343). New York: Cambridge University Press.

Goodall, J. (1971). *In the shadow of man.* Boston: Houghton Mifflin.

Goodall, J. (1986). *The chimpanzees of Gombe: Patterns of behavior.* Cambridge, MA: Belknap.

Gopnik, A. (1993). How we know our minds: The illusion of first-person knowledge of intentionality. *Behavioral and Brain Sciences, 16,* 1–14.

Gopnik, A., & Graf, P. (1988). Knowing how you know: Young children's ability to identify and remember the sources of their beliefs. *Child Development, 59,* 1366–1371.

Gopnik, A., & Meltzoff, A. (1997). *Words, thoughts and theories.* Cambridge, MA: MIT Press.

Gopnik, A., Meltzoff, A. N., & Esterly, J. (1995, February). *Young children's understanding of visual perspective-taking*. Poster presented at the 1st annual Theory of Mind Conference, Eugene, Oregon.

Gopnik, A., & Wellman, H. M. (1994). The theory theory. In L. A. Hirschfeld & S. A. Gelman (Eds.), *Mapping the mind: Domain specificity in cognition and culture* (pp. 257–293). New York: Cambridge University Press.

Gordon, R. M. (1986). Folk psychology as simulation. *Mind and Language, 1,* 158–171.

Gould, S. J. (1977). *Ontogeny and phylogeny*. Cambridge, MA: Harvard University Press.

Hallowell, A. I. (1971). *Culture and experience*. Philadelphia: University of Philadelphia Press.

Hampé, A. (1960). La competition entre les elements osseaux du zeugopode de poulet [Competition between the elements of the bones of the chicken embryo]. *Journal of Embryology and Experimental Morphology, 8,* 241–245.

Harlow, H. F. (1965). Total social isolation: Effects on macaque monkeys. *Science, 148,* 666.

Harlow, H. F., Schlitz, K. A., & Harlow, M. K. (1968). Effects of social isolation on the learning performance of rhesus monkeys. In *Proceedings of the 2nd International Congress in Primatology* (Vol. 1, pp. 178–185). New York: Karger.

Harris, P. L. (1991). The work of the imagination. In A. Whiten (Ed.), *Natural theories of mind: Evolution, development and simulation of everyday mindreading* (pp. 283–304). Cambridge, England: Basil Blackwell.

Hauser, M. D., Kralik, J., Botto-Mahan, C., Garrett, M., & Oser, J. (1995). Self-recognition in primates: Phylogeny and the salience of species-typical features. *Proceedings of the National Academy of Sciences, 92,* 10811–10814.

Heelas, P. (1981). The model applied: Anthropology and indigenous psychologies. In P. Heelas & A. Lock (Eds.), *Indigenous psychologies: The anthropology of the self* (pp. 39–63). London: Academic Press.

Heyes, C. (1994). Reflections on self-recognition in primates. *Animal Behaviour, 47,* 909–919.

Heyes, C. (1995). Self-recognition in mirrors: Further reflections create a hall of mirrors. *Animal Behaviour, 50,* 1533–1542.

Hobson, P. (1980). The question of egocentrism: The young child's competence in the coordination of perspectives. *Journal of Child Psychology and Psychiatry, 21,* 325–331.

Hobson, P. (1982). The question of childhood egocentrism: The coordination of perspectives in relation to operational thinking. *Journal of Child Psychology and Psychiatry, 23,* 43–60.

Hobson, P. (1984). Early childhood autism and the question of egocentrism. *Journal of Autism and Developmental Disorders, 14,* 85–104.

Hobson, R. P. (1993). *Autism and the development of mind*. Hillsdale, NJ: Erlbaum.

Hume, D. (1888). *Treatise of human nature*. Oxford, England: L. A. Selby-Bigge. (Original work published 1739)

Humphrey, N. K. (1976). The social function of intellect. In P. P. G. Bateson & R. A. Hinde (Eds.), *Growing points in ethology* (pp. 303–317). Cambridge, England: Cambridge University Press.

Humphrey, N. K. (1980). Nature's psychologists. In B. D. Josephson & V. S. Ramachandran (Eds.), *Consciousness and the physical world* (pp. 57–75). New York: Pergamon Press.

Humphrey, N. K. (1982, August 19). Consciousness: A just-so story. *New Scientist*, 474–477.

Itakura, S. (1987). Use of a mirror to direct their responses in Japanese monkeys (*Macaca fuscata fuscata*). *Primates*, 28, 343–352.

James, W. (1950). *The principles of psychology*. New York: Dover. (Original work published 1890)

Johnson, D. B. (1982). Altruistic behavior and the development of the self in infants. *Merrill–Palmer Quarterly*, 28, 379–388.

Johnson, D. B. (1983). Self-recognition in infants. *Infant Behavior and Development*, 6, 211–222.

Johnson, M. H., & Vicera, S. P. (1993). Cortical parcellation and the development of face processing. In B. de Boysson-Bardies, S. de Schonen, P. Jusczyk, P. McNeilage, & J. Morton (Eds.), *Developmental neurocognition: Speech and face processing in the first year of life* (pp. 135–148). Dordrecht, The Netherlands: Kluwer.

Jolly, A. (1964a). Prosimians' manipulation of simple object problems. *Animal Behaviour*, 12, 560–570.

Jolly, A. (1964b). Choice of cue in prosimian learning. *Animal Behaviour*, 12, 571–577.

Jolly, A. (1966). Lemur social intelligence and primate intelligence. *Science*, 153, 501–506.

Kagan, J. (1981). *The second year: The emergence of self-awareness*. Cambridge, MA: Harvard University Press.

Karmiloff-Smith, A. (1992). *Beyond modularity: A developmental perspective on cognitive science*. Cambridge, MA: MIT Press.

Kollar, E. J., & Fisher, C. (1980). Tooth induction in chick epithelium: Expression of quiescent genes for enamel synthesis. *Science*, 207, 993–995.

Kummer, H. (1967). Tripartite relations in hamadryas baboons. In S. A. Altmann (Ed.), *Social communication among primates* (pp. 67–77). Chicago: University of Chicago Press.

La Fontaine, J. S. (1984). Person and individual: Some anthropological reflections. In M. Carrithers, S. Collins, & S. Lukes (Eds.), *The category of the person: Anthropology, philosophy, history* (pp. 123–140). Cambridge, England: Cambridge University Press.

Lasky, R. E., & Klein, R. E. (1979). The reactions of five-month-olds to eye contact of the mother and stranger. *Merrill–Palmer Quarterly*, 24, 163–170.

Ledbetter, D. H., & Basen, J. D. (1982). Failure to demonstrate self-recognition in gorillas. *American Journal of Primatology*, 2, 307–310.

Leekam, S., Baron-Cohen, S., Perrett, D., Milders, M., & Brown, S. (1995). *Eye-direction detection: A dissociation between geometric and joint attention skills in autism*. Unpublished manuscript.

Lempers, J. D. (1979). Young children's production and comprehension of nonverbal diectic behaviors. *Journal of Genetic Psychology, 135*, 93–102.

Lempers, J. D., Flavell, E. R., & Flavell, J. H. (1977). The development in very young children of tacit knowledge concerning visual perception. *Genetic Psychology Monographs, 95*, 3–53.

Leslie, A. M. (1994). ToMM, ToBY, and agency: Core architecture and domain specificity. In L. A. Hirschfeld & S. A. Gelman (Eds.), *Mapping the mind: Domain specificity in cognition and culture* (pp. 119–148). New York: Cambridge University Press.

Lethmate, J., & Dücker, G. (1973). Untersuchungen am sebsterkennen im spiegel bei orangutans einigen anderen affenarten [Self-recognition by orangutans and some other primates]. *Zietschrift fur Tierpsychologie, 33*, 248–269.

Lewis, M., & Brooks-Gunn, J. (1979). *Social cognition and the acquisition of self.* New York: Plenum.

Lewis, M., Sullivan, M. W., Stanger, C., & Weiss, M. (1989). Self development and self-conscious emotions. *Child Development 60*, 146–156.

Lienhardt, G. (1984). Self: Public, private. Some African representations. In M. Carrithers, S. Collins, & S. Lukes (Eds.), *The category of the person: Anthropology, philosophy, history* (pp. 141–155). Cambridge, England: Cambridge University Press.

Lillard, A. (in press). Ethnopsychologies: Cultural variations in theories of mind. *Psychological Bulletin, 123*, 3–32.

Lock, A. (1981). Universals in human conception. In P. Heelas & A. Lock (Eds.), *Indigenous psychologies: The anthropology of the self* (pp. 19–36). London: Academic Press.

Lyon, T. D. (1993). *Young children's understanding of desire and knowledge.* Unpublished doctoral dissertation, Stanford University, Palo Alto, CA.

Marino, L., Reiss, D., & Gallup, G. G., Jr. (1994). Mirror–self-recognition in bottle-nosed dolphins: Implications for comparative investigations of highly dissimilar species. In S. Parker, R. Mitchell, & M. Boccia (Eds.), *Self-awareness in animals and humans* (pp. 380–391). Cambridge, England: Cambridge University Press.

Marten, K., & Psarakos, S. (1994). Evidence of self-awareness in the bottle-nosed dolphin (*Tursiops truncatus*). In S. Parker, R. Mitchell, & M. Boccia (Eds.), *Self-awareness in animals and humans* (pp. 361–379). Cambridge, England: Cambridge University Press.

Masangkay, Z. S., McKluskey, K. A., McIntyre, C. W., Sims-Knight, J., Vaughn, B. E., & Flavell, J. H. (1974). The early development of inferences about the visual precepts of others. *Child Development, 45*, 357–366.

Meltzoff, A. N. (1995). Understanding the intentions of others: Re-enactment of intended acts by 18–month-old children. *Developmental Psychology, 31*, 838–850.

Menzel, E. W., Jr. (1974). A group of young chimpanzees in a one-acre field. In A. Schrier & F. Stollnitz (Eds.), *Behavior of non-human primates: Modern research trends* (pp. 83–153). New York: Academic Press.

Mitchell, R. B. (1993). Mental models of mirror–self-recognition: Two theories. *New Ideas in Psychology, 11*, 295–325.

Montgomery, D. E., Moran, C., & Bach, L. M. (in press). The influence of nonverbal cues associated with looking behavior on young children's mentalistic attributions. *Journal of Nonverbal Behavior.*

Moore, C. P. (1994). Intentionality and self–other equivalence in early mindreading: The eyes do not have it. *Current Psychology of Cognition, 13,* 661–668.

Morissette, P., Ricard, M., & Décarie, T.G. (1995). Joint visual attention and pointing in infancy: A longitudinal study of comprehension. *British Journal of Developmental Psychology, 13,* 163–175.

Mumme, D. L. (1993). *Rethinking social referencing: The influence of facial and vocal affect on infant behavior.* Unpublished doctoral dissertation, Stanford University, Stanford, CA.

Murphy, C., & Messer, D. (1977). Mothers, infants and the pointing gesture. In H. R. Schaffer (Ed.), *Studies in mother–infant interaction* (pp. 325–353). London: Academic Press.

Nelson, K. (1992). Emergence of autobiographical memory at age 4. *Human Development, 35,* 172–177.

Nelson, K. (1993). The psychological and social origins of autobiographical memory. *Psychological Science, 4,* 7–14.

Olson, D., & Campbell, R. (1993). Constructing representations. In C. Pratt & A. F. Garton (Eds.), *Systems of representation in children: Development and use* (pp. 11–26). New York: Wiley.

Olson, D. R. (1993). The development of representations: The origins of mental life. *Canadian Psychology, 34,* 1–14.

O'Neill, D. K. (1996). Two-year-old children's sensitivity to a parent's knowledge state when making requests. *Child Development, 67,* 659–677.

O'Neill, D. K., Astington, J. W., & Flavell, J. H. (1992). Young children's understanding of the role that sensory experiences play in knowledge acquisition. *Child Development, 63,* 474–490.

O'Neill, D. K., & Gopnik, A. (1991). Young children's ability to identify the sources of their beliefs. *Developmental Psychology, 27,* 390–397.

Parker, S. T. (1991). A developmental approach to the origins of self-recognition in great apes. *Human Evolution, 2,* 435–449.

Parker, S. T., & Gibson, K.R. (1979). A developmental model for the evolution of language and intelligence in early hominids. *Behavioral and Brain Sciences, 2,* 367–408.

Patterson, F. G. P., & Cohn, R.H. (1994). Self-recognition and self-awareness in lowland gorillas. In S. T. Parker, R. W. Mitchell, & M. L. Boccia (Eds.), *Self-awareness in animals and humans* (pp. 273–290). New York: Cambridge University Press.

Perner, J. (1991). *Understanding the representational mind.* Cambridge, MA: MIT Press.

Perner, J., & Ogden, J. (1988). Knowledge for hunger: Children's problems with representation in imputing mental states. *Cognition, 29,* 47–61.

Perrett, D., Harries, M., Mistlin, A., Hietanen, J., Benson, P., Bevan, R., Thomas, S., Oram, M., Ortega, J., & Brierly, K. (1990). Social signals analyzed at the single cell level: Someone is looking at me, something touched me, something moved! *International Journal of Comparative Psychology*, 4, 25–55.

Pillow, B. H. (1989). Early understanding of perception as a source of knowledge. *Journal of Experimental Child Psychology*, 47, 116–129.

Povinelli, D. J. (1987). Monkeys, apes, mirrors and minds: The evolution of self-awareness in primates. *Human Evolution*, 2, 493–509.

Povinelli, D. J. (1989). Failure to find self-recognition in Asian elephants (*Elephas maximus*) in contrast to their use of mirrors cues to discover hidden food. *Journal of Comparative Psychology*, 102, 122–131.

Povinelli, D. J. (1991). *Social intelligence in monkeys and apes*. PhD dissertation, Yale University, New Haven, CT.

Povinelli, D. J. (1994a). How to create self-recognizing gorillas (but don't try it on macaques). In S. Parker, R. Mitchell, & M. Boccia (Eds.), *Self-awareness in animals and humans* (pp. 291–294). Cambridge, England: Cambridge University Press.

Povinelli, D. J. (1994b). Comparative studies of mental state attribution: A reply to Heyes. *Animal Behaviour*, 48, 239–241.

Povinelli, D. J. (1995). The unduplicated self. In P. Rochat (Ed.), *The self in early infancy* (pp. 161–192). Amsterdam: North-Holland/Elsevier.

Povinelli, D. J. (1996a). Chimpanzee theory of mind? The long road to strong inference. In P. Carruthers & P. Smith (Eds.), *Theories of theories of mind* (pp. 293–329). Cambridge, England: Cambridge University Press.

Povinelli, D. J. (1996b). Growing up ape. *Monographs of the Society for Research in Child Development*, 61 (2, Serial No. 247), 174–189.

Povinelli, D. J., Bierschwale, D. T., & Čech, C. G. (in press). Comprehension of seeing as a referential act in young children, but not juvenile chimpanzees. *British Journal of Developmental Psychology*.

Povinelli, D. J., & Cant, J. G. H. (1995). Arboreal clambering and the evolution of self-conception. *Quarterly Review of Biology*, 70, 393–421.

Povinelli, D. J., & deBlois, S. (1992a). On (not) attributing mental states to monkeys: First, know thyself. *Behavioral and Brain Sciences*, 15, 164–166.

Povinelli, D. J., & deBlois, S. (1992b). Young children's (*Homo sapiens*) understanding of knowledge formation in themselves and others. *Journal of Comparative Psychology*, 106, 228–238.

Povinelli, D. J. , Davidson, C. A. & Theall, L. A. (1998). *Chimpanzees deploy attention-getting behaviors independent of the attentional states of others*. Unpublished manuscript.

Povinelli, D. J., & Eddy, T. J. (1994). The eyes as a window: What young chimpanzees see on the other side. *Current Psychology of Cognition*, 13, 695–705.

Povinelli, D. J., & Eddy, T. J. (1996a). What young chimpanzees know about seeing. *Monographs of the Society for Research in Child Development*, 61 (2, Serial No. 247).

Povinelli, D. J., & Eddy, T. J. (1996b). Chimpanzees: Joint visual attention. *Psychological Science*, 7, 129–135.

Povinelli, D. J., & Eddy, T. J. (1996c). Factors influencing young chimpanzees' (*Pan troglodytes*) recognition of attention. *Journal of Comparative Psychology*, 110, 336–345.

Povinelli, D. J., & Eddy, T. J. (1997). Specificity of gaze-following in young chim-panzees. *British Journal of Developmental Psychology, 15,* 213–222.

Povinelli, D. J., Gallup, G. G., Jr., Eddy, T. J., Bierschwale, D. T., Engstrom, M. C., Perilloux, H. K., & Toxopeus, I. B. (1997). Chimpanzees recognize themselves in mirrors. *Animal Behaviour, 53,* 1083–1088.

Povinelli, D. J., & Godfrey, L. R. (1993). The chimpanzee's mind: How noble in reason? How absent of ethics? In M. Nitecki & D. Nitecki (Eds.), *Evolution-ary ethics* (pp. 277–324). Albany, NY: SUNY Press.

Povinelli, D. J., Landau, K. R., & Perilloux, H. K. (1996). Self-recognition in young children using delayed versus live feedback: Evidence of a developmental asyn-chrony. *Child Development, 67,* 1540–1554.

Povinelli, D. J., Nelson, K. E., & Boysen, S. T. (1990). Inferences about guessing and knowing by chimpanzees *(Pan troglodytes). Journal of Comparative Psy-chology, 104,* 203–210.

Povinelli, D. J., Perilloux, H. K., Reaux, J. E., & Bierschwale, D. T. (in press). Young and juvenile chimpanzees' *(Pan troglodytes)* reactions to intentional versus accidental and inadvertent actions. *Behavioral Processes.*

Povinelli, D. J., & Preuss, T. M. (1995). Theory of mind: Evolutionary history of a cognitive specialization. *Trends in Neurosciences, 18,* 418–424.

Povinelli, D. J., Reaux, J. E., Bierschwale, D. T., Allain, A. D., & Simon, B. B. (1997). Exploitation of pointing as a referential gesture in young children, but not adolescent chimpanzees. *Cognitive Development, 12,* 327–365.

Povinelli, D. J., Rulf, A. R., Landau, K. R., & Bierschwale, D. T. (1993). Self-recognition in chimpanzees *(Pan troglodytes):* Distribution, ontogeny, and patterns of emergence. *Journal of Comparative Psychology, 107,* 347–372.

Povinelli, D. J., & Simon, B. B. (in press). Young children's reactions to briefly ver-sus extremely delayed visual feedback of the self: Evidence for the emergence of the autobiographical stance. *Developmental Psychology.*

Povinelli, D. J., Zebouni, M. C., & Prince, C. G. (1996). Ontogeny, evolution and folk psychology. *Behavioral and Brain Sciences, 19,* 137–138.

Pratt, C., & Bryant, P. (1990). Young children understand that looking leads to knowing (so long as they are looking into a single barrel). *Child Development, 61,* 973–982.

Premack, D. (1984). Pedagogy and aesthetics as sources of culture. In M. S. Gaz-zaniga (Ed.), *Handbook of cognitive neuroscience* (pp. 15–35). New York: Plenum.

Premack, D. (1988). "Does the chimpanzee have a theory of mind" revisited. In R. Byrne & A. Whiten (Eds.), *Machiavellian intelligence* (pp. 160–179). New York: Oxford University Press.

Premack, D., & Dasser, V. (1991). Perceptual origins and conceptual evidence for theory of mind in apes and children. In A. Whiten (Ed.), *Natural theories of mind* (pp. 46–65). Oxford: Blackwell.

Premack, D., & Woodruff, G. (1978). Does the chimpanzee have a theory of mind? *Behavioral and Brain Sciences, 1,* 515–526.

Preuss, T. M., & Kaas, J. H. (in press). Human brain evolution. In F. E. Bloom, S. C. Landis, J. L. Roberts, L. R. Squire, & M. J. Zigmond (Eds.), *Fundamental neuroscience.* San Diego: Academic Press.

Repacholi, B. M., & Gopnik, A. (1997). Early reasoning about desires: Evidence from 14– and 18–month-olds. *Developmental Psychology, 33,* 12–21.

Ristau, C. A. (1991). Before mindreading: Attention, purposes and deception in birds? In A. Whiten (Ed.), *Natural theories of mind: Evolution, development and simulation of everyday mindreading* (pp. 209–222). Cambridge, England: Blackwell.

Rodier, P. M, Ingram, J. L., Tisdale, B., Nelson, S., & Romano, J. (1996). Embryological origin for autism: Developmental anomalies of the cranial nerve motor nuclei. *Journal of Comparative Neurology, 370,* 247–261.

Romanes, G. J. (1882). *Animal intelligence.* London: Kegan Paul.

Romanes, G. J. (1883). *Mental evolution in animals.* New York: Appleton.

Ruffman, T. K., & Olson, D. R. (1989). Children's ascriptions of knowledge to others. *Developmental Psychology, 25,* 601–606.

Russell, B. (1948). *Human knowledge: Its scope and limits.* London: Unwin Hyman.

Sackett, G. P. (1966). Monkeys reared in isolation with pictures as visual input: Evidence for an innate releasing mechanism. *Science, 154,* 1468–1473.

Scaife, M., & Bruner, J. (1975). The capacity for joint visual attention in the infant. *Nature, 253,* 265–266.

Schulman, A. H., & Kaplowitz, C. (1977). Mirror image response during the first two years of life. *Developmental Psychobiology, 10,* 133–142.

Schutz, A. (1962). *Collected papers* (Vol. 1). The Hague, The Netherlands: Nijhoff.

Smith, A. (1759). *A theory of moral sentiments.* (Reprinted, Oxford, England: Clarendon, 1961)

Smuts, B. (1985). *Sex and friendship in baboons.* New York: Aldine.

Stewart, D. (1956). *Preface to empathy.* New York: Philosophical Library.

Strum, S. (1987). *Almost human: A journey into the world of baboons.* New York: Norton.

Suarez, S. D., & Gallup, G. G., Jr. (1981). Self-recognition in chimpanzees and orangutans, but not gorillas. *Journal of Human Evolution, 10,* 175–188.

Suarez, S. D., & Gallup, G. G., Jr. (1986). Face-touching in primates: A closer look. *Neuropsychologia, 24,* 597–600.

Suddendorf, T. (1994). *Discovery of the fourth dimension: Human evolution and mental time travel.* Unpublished master's thesis, University of Waikato, Hamilton, New Zealand.

Swartz, K. B., & Evans, S. (1991). Not all chimpanzees show self-recognition. *Primates, 32,* 483–496.

Tan, J., & Harris, P. (1991). Autistic children understand seeing and wanting. *Development and Psychopathology, 3,* 163–174.

Taylor, M., Esbensen, B. M., Bonnie, M., & Bennett, R. T. (1994). Children's understanding of knowledge acquisition: The tendency for children to report that they have always known what they have just learned. *Child Development, 65,* 1581–1604.

Thompson, R. L., & Boatright-Horowitz, S. L. (1994). The question of mirror-mediated self-recognition in apes and monkeys: Some new results and reservations. In S. T. Parker, R. W. Mitchell, & M. L. Boccia (Eds.), *Self-awareness in animals and humans* (pp. 330–349). New York: Cambridge University Press.

Tomasello, M. (1995). The power of culture: Evidence from apes. *Human Development, 38,* 46–52.

Tomasello, M. (1996). Chimpanzee social cognition. *Monographs of the Society for Research in Child Development, 61*(2, Serial No. 247), 161–173.

Tomasello, M., & Call, J. (1994). The social cognition of monkeys and apes. *Yearbook of Physical Anthropology, 37,* 273–305.

Tomasello, M., & Call, J. (1997). *Primate cognition.* Oxford University Press.

Tomasello, M., Kruger, A. C., & Ratner, H. H. (1993). Cultural learning. *Behavioral and Brain Sciences, 16,* 495–552.

van Schaik, C. P., van Noordwijk, M. A., Warsono, B., & Sutriono, E. (1983). Party size and early detection of predators in Sumatran forest primates. *Primates, 24,* 211–221.

Vinden, P. G. (1996). Junín Quechua children's understanding of mind. *Child Development, 67,* 1707–1716.

Welch-Ross, M. K. (1995). An integrative model of the development of autobiographical memory. *Developmental Review, 15,* 338–365.

Wellman, H. M. (1990). *The child's theory of mind.* Cambridge, MA: Bradford.

White, G. M. (1980). Conceptual universals in interpersonal language. *American Anthropologist, 82,* 759–781.

Whiten, A., & Byrne, R.W. (1988). Tactical deception in primates. *Behavioral and Brain Sciences, 11,* 233–244.

Wimmer, H., Hogrefe, G-J., & Perner, J. (1988). Children's understanding of informational access as a source of knowledge. *Child Development, 59,* 386–396.

Wooley, J. D., & Wellman, H. M. (1993). Origin and truth: Young children's understanding of imaginary mental representations. *Child Development, 64,* 1–17.

Zahn-Waxler, C., Radke-Yarrow, M., Wagner, E., & Chapman, M. (1992). Development of concern for others. *Developmental Psychology, 28,* 126–136.

Zelazo, P. D., & Frye, D. (in press). Cognitive complexity and control: A theory of the development of deliberate reasoning and intentional control. In M. Stamenov (Ed.), *Language: Structure, discourse and the access to consciousness.* Amsterdam and Philadelphia: John Benjamins.

CHAPTER FOUR

A Social Selection Model for the Evolution and Adaptive Significance of Self-Conscious Emotions

✧

SUE TAYLOR PARKER

> I find it impossible to see, if we assume the
> Darwinian theory of the origin of emotional
> attitudes and expressions, why the class of
> emotions which we cover by the term "shame"
> should be cut in two, and those which are simply
> social should be said to have grown up in race-
> history in union with their expression, while the
> other half, those which are called ethical, although
> showing the same organic reactions, should be
> supposed to have acquired their connection with
> the organism in some extra evolutionary way. This
> agreement, in fact, in the expressions of the ethical
> and social, taken with the social rise of the ethical
> emotions in the child, furnishes, to my mind, a
> twofold and irresistible proof of the evolution of
> the ethical sentiments in race history. No other
> theory seems to explain *the blush of moral shame.*
> —BALDWIN (1897, p. 209; emphasis in original)

How do self-conscious emotions (SCEs) such as shame and guilt differ
from other emotions? Are they unique to humans or do other primates display them? What is their evolutionary history and adaptive significance?
How are they tied to cognition and self-awareness? How are they related
to altruism? When do they develop? What are their functions? The social

108

selection model I propose to answer these questions produces answers strikingly similar to those James Mark Baldwin gave to similar questions a century ago (see epigraph). Baldwin approached these questions by comparing the stages of development of emotions in human children and other animals and using ethnographic data to speculate on the origins of those that are uniquely human. A century later, the same approach supported by more powerful evolutionary models and more comprehensive data on the emotional development of children and great apes yields similar but more sharply defined answers.

DEFINING SELF-CONSCIOUS EMOTIONS

The seventh edition of the *Merriam–Webster Dictionary* defines emotion as "a departure from the normal calm state of an organism of such nature as to include strong feeling, an impulse toward open action, and certain internal physical reactions; any of the states designated as fear, anger, disgust, grief, surprise, yearning, etc." In other words, all emotions have components of physiological response, motivation, feeling, cognition, and communication (McNaughton, 1989).

Human emotions can be broadly categorized into two classes: the self-conscious emotions (SCEs) and the non-self-conscious emotions (NSCEs). The NSCEs include those emotions that are shared by most mammalian species: fear, surprise, anger, sadness, and joy. Of these, all but fear and surprise are social emotions, that is, emotions aroused primarily through interactions with conspecifics. Most of these emotions develop early in life in the context of mother–infant and peer interactions. They continue to operate throughout life in various social contexts, most notably in mating competition, courtship, and parenting.

In humans, there are apparently seven pancultural NSCEs: happiness, anger, surprise, fear, disgust, sadness, and interest (Ekman, 1972). These are communicated by species-typical facial expression and bodily states (Darwin, 1965; Eibl-Eibesfeldt, 1989; Ekman, Friesen, & Ellsworth, 1972). Many of these species-typical expressions are present in certain contexts from the first few months or even days of life (Darwin, 1965; Izard, 1984).

Similar expressions of some of these emotions are found in humans and great apes, especially in our closest relatives, the African apes. Both humans and great apes, for example, communicate joy through the relaxed open-mouthed (play) face and staccato grunts or laughter (Goodall, 1968; van Hoof, 1972). They both communicate anger through glaring and compressed lips, and frustrated desire by whimpering and pouting (Goodall, 1968).

In humans, NSCEs can be triggered indirectly by mental associations, for example, by "mental tickling," as well as directly by physical stimulation, such as by tickling in the case of joy (Darwin, 1965). This pattern of dual elicitation suggests dual pathways of neurological control. Indirect elicitation pathways allow greater voluntary control over the expression of emotions than direct elicitation pathways (Ekman, 1972).

Several features suggest that these NSCEs have been phylogenetically ritualized for their signal value (Eibl-Eibesfeldt, 1979). These features include the species-typical forms of facial expressions of emotions, their associated morphological displays of lips, cheeks and eyebrows, their early development, and their homologues in closely related species. The variety of expressions and their contexts is increased by the human capacity to blend and modify these species-typical displays through such "management techniques" as intensification, deintensification, neutralization, and masking. These management techniques, or "display rules," vary from society to society, and between age classes and genders in the same society (Ekman, 1972).[1]

The SCEs include embarrassment, shame, guilt, pride, and envy, and according to some, jealousy and empathy (Barrett, 1995) and hubris (Lewis, Sullivan, Stanger, & Weiss, 1989). These SCEs differ from the other class of emotions in several ways. First, they generally lack specific facial expressions (e.g., see Barrett, 1995). Second, SCEs do seem to be triggered only by mental reflection—not by direct physical stimulation. Third, they develop in the second or third year of life rather than early in infancy.

Fourth, and related to their indirect stimulation, SCEs are self-reflexive. They depend upon self-awareness and on the ability to evaluate the self or its actions relative to a social standard. Fifth, they are associated with concepts of good and bad, competence and incompetence. As such, they depend upon the development of self-awareness and awareness of social standards. In humans, awareness of social standards, in turn, depends upon language and certain cognitive capacities that develop in the second year of life and later.

These features suggest that unlike NSCEs, SCEs have not been strongly phylogenetically ritualized for social communication (Eibl-Eibesfeldt, 1979). The one possible exception to this is blushing, which Darwin (1965) saw as a specific display of shame. He noted that it was generally accompanied by attempts to hide the face or at least the gaze.[2]

Comparative data suggest that SCEs and the higher cognitive capacities on which they depend are uncommon among mammals. There is some evidence though, that great apes develop some of the cognitive prerequisites for self-conscious emotions, and may display embarrassment and empathy (e.g., Patterson & Cohn, 1994). If so, great apes display only incipient forms of SCEs. The unique occurrence of SCEs in hominoids raises the question of how and why our ancestors evolved these emotions. Analysis of the contexts and outcomes of SCEs may suggest some answers to these questions.

A FUNCTIONAL MODEL FOR
SELF-CONSCIOUS EMOTIONS AND THEIR
SOCIAL CONTEXTS AND OUTCOMES

According to Izard's discrete theory of emotion, emotions are primary motivators of behavior, which have been shaped by past selection. As such, they entail typical neural substrates, forms of expression, and behavioral outcomes (Zahn-Waxler & Kochanska, 1988). According to the functionalist approach to SCEs, which elaborates the discrete theory, each SCE has associated with it several functions. These include (1) behavioral regulatory functions, (2) social regulatory functions, (3) internal regulatory functions, (4) a goal, (5) appreciation of self, (6) appreciation of other, (7) an action tendency, and (8) a focus of attention (Barrett, 1995). Please see Table 4.1 for a summary of these functions for various SCEs that are elaborated in the following paragraphs.

Guilt entails feelings of anxiety and restless tension that impel its experiencers to repair intimate relationships. Indeed, guilt induction is a strategy intimates may use to stimulate communication and increase closeness in intimate relationships. Guilt does this by inducing the target to make reparations for specific sins of omission or commission (Baumeister, Stillwell, & Hetherton, 1995). It thereby impels experiencers to change their guilt-provoking behaviors. Guilty persons appraise their actions as bad, but think they can expiate their sins through appropriate reparations. In other words, they experience themselves as being competent, having control over their behaviors (Ferguson & Stegge, 1995). Guilt is closely allied with shame and is sometimes difficult to distinguish. Unlike shame, however, guilt seems to develop in association with empathy (Hoffman, 1984).

In contrast to guilt, shame engenders feelings of sadness and depression. Unlike guilt, shame is associated with feelings of helplessness owing to fundamental inability or defectiveness or incompetence (Ferguson & Stegge, 1995). It seems to impel its experiencers to avoid those who have witnessed their shameful appearance or deed, and thereby to reduce contact with them. It may also impel its experiencers to change their behavior by avoiding the appearance and actions that elicited their shame. The worst aspect of shame comes from experiencers' feeling that they are irreparably, overwhelmingly, or globally inadequate and therefore unable to remedy the situation. In extreme cases, shame may lead individuals to exile

TABLE 4.1. Characteristics of Some Social Emotion Families

Family	Behavioral regulatory functions	Social regulatory functions	Internal regulatory functions	Goal for self	Appreciation re: self	Appreciation re: other	Action tendency	Focus of attention	Vocalic pattern	Physiological reaction
Shame	Distance oneself from evaluating agent; reduce "exposure"	Communicate deference/submission; communicate self as "small" or inadequate	Highlight standards and importance of standards; aid in acquisition of knowledge of self as object; reduce arousal	Maintenance of others' respect and/or affection; preservation of positive self-regard	"I am bad." (Self-regard is perceived to be impaired.)	"Someone thinks I am bad. Everyone is looking at me."	Withdrawal: avoidance of others; hiding of self	Self as object	"Narrow," moderately lax, thin	Low heart rate; blushing
Guilt	Repair damage	Communicate awareness of proper behavior; communicate contrition/good intentions	Highlight standards and importance of standards; aid in acquisition of knowledge of self as agent	Meeting known standards	"I have done something contrary to my standards."	"Someone has been injured by my act."	Outward movement; inclination to make reparation, tell others, and punish oneself	The wrongdoing; consequences of one's act; self as agent and experiencer	"Narrow," tense, moderately full voice	High heart rate and skin conductance; irregular respiration
Envy	Protect or obtain possession or access to loved one	Inform others re: whom/what one cares about; prevent others from taking one's possessions	Highlight what one cares about/values	Obtaining an object/person that someone else possesses	"I cannot obtain the object."	"Someone has what I want."	Withdrawal and outward movement; inclination to avoid and/or hurt the one who possesses the desired object/person/quality	The possession; the possessor	"Narrow," moderately lax, thin	Irregular respiration; slightly elevated heart rate
Pride	Decrease distance from evaluating agents	Show others one has achieved standard; show dominance/superiority	Highlight standards and importance of standards; aid in acquisition of knowledge of self as object and agent	Maintenance of good feelings about oneself	"I am good."	"Someone/everyone thinks (or will think) I am good."	Outward movement; inclination to show/tell others	Self as agent and as object	"Wide," moderately tense, full voice	Flushed face; high heart rate

Note. From Barrett (1995, pp. 42–43). Copyright 1995 by The Guilford Press. Reprinted by permission.

or even suicide. (Embarrassment seems to be a simpler and less painful form of shame.)

Pride entails positive feelings that impel experiencers to show off or display their abilities. It also impels them to repeat successful actions in order to reexperience the feeling of success and the good opinion of others. People who are proud appraise themselves as superior performers or achievers and want people to know it. Their pride may impel them to lead or dominate others in his arena of special competence. Parents and teachers may induce pride in an effort to reinforce and increase the frequency of behaviors they desire or value in their youngsters.

Envy is a negative feeling that impels its experiencers to resent and avoid those who induce that feeling. It may also motivate its experiencers to strive to get the object or status they envy. They appraise the other as unlike themselves and believe that that person is less deserving than they are. They may use envy to justify their own hostile behaviors toward a rival. Or they may induce envy to stimulate an associate to competitive behaviors.

Hubris is an overweening, global sense of pride that impels its experiencers to see themselves as being intrinsically better than others, and hence motivates them to dominate others globally. Experiencers appraise themselves himself as intrinsically good, superior in a global way that does not depend upon performing specific good or successful acts. In fact, they tend to experience themselves as perfect and therefore above criticism or correction (Lewis, 1992).

Empathy is a vicarious feeling based on an appraisal that the self is like the other. It impels its experiencers to approach and succor those who induce that feeling. Experiencers feel a strong connection with the other and want to help the other escape his or her painful dilemma through actions or through blaming others. Inducing empathy may motivate approach and succoring behaviors or prevent hostile behaviors toward others.

Research in the functionalist paradigm suggests that guilt—anxiety—is aroused by moral transgressions, but that shame—helplessness—is aroused by such nonmoral transgressions as breaches of propriety or loss of face. These latter breaches imply social defectiveness or incompetence rather than the carelessness or lack of effort that engenders guilt. Studies suggest that guilt is inculcated through parental discipline, that is, by emphasizing responsibility and explaining behavioral outcomes to the child. Shame, in contrast, is inculcated by absence of discipline, and by expressions of hostility and/or disappointment to the child (Ferguson & Stegge, 1995).

The functionalist approach to socialization of SCEs identifies four ways in which parents and other socialization agents influence emotional development in children. First, they affect children by directly modeling affective styles. Second, they provide directives about how the child should feel in particular situations. Third, they provide feedback about how they perceive the child's behavior, helping the child construct attributions. Finally, they attribute emotional reactions to the child, especially in situations in which rules have been violated (Ferguson & Stegge, 1995). In other words, this approach suggests how socialization provides children with a theory of emotion.

As these brief sketches indicate, appraisal of the self by the self and others are key elements of SCEs (Barrett, 1995). These elements infuse SCEs with an aura of goodness or badness and/or adequacy or inadequacy. Self-appraisal and moralizing judgments underlie the self-regulatory aspects of SCEs, which make them powerful aids to socialization and enculturation. Before addressing the evolution of SCEs, it is necessary to introduce some evolutionary concepts.

A SOCIAL SELECTION MODEL FOR THE FUNCTIONS OF SELF-CONSCIOUS EMOTIONS

Evolutionary biologists have identified four major mechanisms by which organisms in social species can increase their direct individual fitness and/or their indirect inclusive fitness through others. These four mechanisms explain why social behavior has evolved. They can all be subsumed under the rubric of social selection, because they all involve selection by members of the same species, and usually, the same social group (Trivers, 1985).

These forms of social selection are (1) sexual selection through male competition and female choice (Darwin, 1930); (2) kin selection for nepotism or altruistic behaviors toward relatives according to their degree of relatedness (Wilson, 1975); (3) parental manipulation, or differential investment in various offspring according to their reproductive potential (Alexander, 1974; Trivers, 1974); and (4) reciprocal altruism, or altruism toward nonkin according to the likelihood of reciprocation (Trivers, 1971). A brief summary of four these mechanisms provides the context for my social selection model for the evolution of SCEs.

According to these social selection models, social organisms have evolved the tendency to interact strategically in their own genetic interests. Nepotistic strategies entail behavioral choices that are contingent on assessment of the relationship between altruists and potential beneficiaries, as well as the reproductive potential of possible beneficiaries.

Parental strategies depend on the parent's ability to assess the relative reproductive potential of offspring. Reproductive strategies depend upon the male's ability to assess the relative abilities of the self and competitors, and on the female's ability to assess the quality and relatedness of potential mates (Dawkins & Guilford, 1991). Altruistic strategies depend upon the altruist's ability to assess the beneficiary's probability of reciprocating or cheating. Of these four forms of social selection, parental manipulation and reciprocal altruism are probably the most relevant to the evolution of SCEs.

Sexual selection, for example, favors a male's capacity to "decide" whether to compete with a potential rival strategically on the basis of his relative strength, status, or alliances. In the arena of mate choice, for example, selection favors a female's ability to assess her genetic similarity to a potential mates (as a means for avoiding deleterious inbreeding or outbreeding).

Similarly, parental selection favors a parent's capacity to assess the relative reproductive potential of its own offspring and to strategically invest in those with the greatest reproductive value (Alexander, 1974; Trivers, 1974). In polygynous species, for example, male offspring of high-ranking females have greater reproductive potential than female offspring. In these circumstances, mothers can increase their own reproductive success by preferentially investing in male offspring (Trivers & Williard, 1973).

Likewise, selection for reciprocal altruism favors an individual's capacity to assess the likelihood that a potential beneficiary will reciprocate altruistic actions at some future date, either to the self or to kin. This likelihood depends first on individual recognition and the stability of social relations (Trivers, 1971). It also depends upon the donor's ability to coerce reciprocity and to detect and punish lack of reciprocity. Conversely, selection favors a beneficiary's ability to communicate the impression of his or her own integrity, and hence to attract altruistic behaviors.

Application of social selection and assessment theory suggests that various forms of self-knowledge have evolved in social species as adaptations for assessing the self's characteristics relative to those of conspecifics (Parker, 1998). In the case of humans, I suggest that SCEs were favored by parental manipulation and reciprocal altruism as strategies for socialization and enculturation by parents, kin, peers, and others. SCEs were favored because they increased the ability of parents, kin, peers, and others to socially manipulate their offspring, kindred, peers, and so on. Self-regulation functions of SCEs were favored because they allowed parents and others to inculcate and enforce values and behavioral codes in others that increased their own inclusive fitness.

First, I suggest that humans evolved the capacity to experience SCEs through parental manipulation. This occurred because parents who had offspring who could be manipulated directly through induction of SCEs, and indirectly through self-induced SCEs had greater inclusive fitness than those who could not. Self-induction of SCEs in their offspring saves parental energy, which can be used to invest in other offspring.

This energy-saving innovation is analogous to the energy-saving innovation involved in communication as opposed to direct physical action
(Dawkins & Krebs, 1978). Direct and indirect SCE induction allows parents to minimize energy expenditure in direct control of their offspring by
hitchhiking on the self-socializing and self-enculturating functions of SCEs.
In other words, it allows parents and other kin to stimulate children to act
as their own parents by responding to internalized values and standards.

Likewise, I suggest that SCEs were favored by reciprocal altruism as
means for manipulating beneficiaries of altruistic acts into reciprocating.
First, SCEs allow individuals to assess potential beneficiaries and altruists
among nonkin. Once they have identified these potential altruists (e.g.,
friends, allies, mentors, mentees), they must attract and woo them. Conversely, once they have conferred benefits on a beneficiary, they must maintain contact and encourage reciprocity. SCEs, especially inducing guilt in
the beneficiary, play an important role in motivating reciprocal relations.
Finally, if all else fails, they may use shame and ostracism to punish lack of
reciprocity.

I also suggest that sexual selection has favored SCEs. First, it has favored guilt and shame as means for males to control female sexuality,
specifically, to enforce fidelity in wives and virginity in daughters (Dickemann, 1981). Second, sexual selection has favored guilt and shame as
means for females to enforce exclusive paternal investment in their offspring. Likewise, I suggest that reciprocal altruism has favored guilt between mates as means for enforcing mutual sharing.

In other words, social selection has favored SCEs because they facilitate increased inclusive fitness through social manipulation. Pride allows
individuals to assess their likelihood of being successful leaders and en-
forcers of bargains. Empathy allows individuals to assess the feelings and
thoughts of others, to display concern, and hence to indicate that they are
likely to reciprocate. Guilt impels them to reciprocate or to make restitu-
tion for failures of reciprocity. Guilt induction is a powerful means for motivating reciprocity. Shame induction by others is a powerful means for directly and indirectly punishing cheaters by damaging their reputation within the group. Shame and guilt are induced by moral indignation, which, in
turn, is one of the means for enforcing social reciprocity (de Waal, 1996).
In contrast to empathy, pride, guilt, and shame, envy induction seems to be
used to incite competition and hostility.

The social selection model of SCEs implies that susceptibility to SCEs
should be context- and relationship-specific rather than generalized; that
is, individuals will be more likely to respond with SCEs to certain classes of
individuals in certain contexts as opposed to others in other contexts.

Individuals should be more likely to respond with empathy to members of their kin group, their mates, and their friends and benefactors who

have experienced strong emotions than they are to others who have experienced the same emotions. They should be more likely to respond with guilt and shame to kin, mates, friends, and benefactors when they have violated some norm or failed in reciprocity than they are to others who are not kin, mates, friends, or benefactors. They should be more likely to respond with empathy and envy to others whom they perceive as being like themselves (in age, status, group membership) than to others whom they perceive as unlike themselves. Finally, individuals should be more likely to respond with shame and guilt to reprimands by those who have power over them than to others who lack such power. It is noteworthy that just these sets of relationships are those that are central to socialization and enculturation: parents, kin, peers, and external authority figures.

THE ROLE OF SELF-CONSCIOUS EMOTIONS IN SOCIALIZATION AND ENCULTURATION

Taken together, the social selection and functionalist models for shame, guilt, and pride suggest that SCEs are powerful tools for socialization and enculturation, both directly by others and indirectly by self through internalized self-regulation according to learned values.

Socialization is a concept sociologists developed to describe how young individuals in social species are shaped and trained by other individuals in their society into performing specific behaviors and avoiding others. These processes may be mediated by explicit or implicit instruction from parents and other adult members of the group (vertical transmission) or by peers (horizontal transmission) (Zahn-Wexler & Kochanska, 1988).

Enculturation, in contrast, is a concept social/cultural anthropologists developed to describe how young humans acquire the culture of their own particular society or subgroup, that is, its language, values, rituals, recipes, and procedures (Goodenough, 1981). Although some ethologists use the term "culture" to describe social learning, culture in the anthropological sense is unique to humans (Fox, 1971; Parker & Russon, 1996). Social transmission is also both vertical and horizontal, both explicit and implicit. Therefore, both socialization and enculturation entail conscious and unconscious modes of transmission.

According to Margaret Mead (1963), socialization refers broadly to social learning on a specieswide level, whereas enculturation refers to learning on a specific cultural level. (Although this definition is an oversimplification, it makes a useful point.) This definition also implies that socialization occurs in all social mammals, whereas enculturation occurs only in cultural animals.

Although socialization depends primarily on social learning through operant conditioning, enculturation depends upon higher cognitive processes involving self-awareness and symbolic thought. Socialization is much more widespread among mammals, whereas enculturation is limited to humans and possibly great apes. In humans who are both social and cultural animals, both phenomena occur in an interpenetrated complex that is virtually impossible to separate.

Using Mead's distinction, I would like to suggest that although all emotions, including NSCEs, serve socializing functions, only SCEs serve both socializing and enculturating functions. Fear, for example, usually motivates flight and avoidance. Anger usually motivates attack; joy motivates approach and attachment, and so on. Responses to these emotions allow external (social and nonsocial) forces to condition young animals to avoid certain situations and conspecifics, and to approach and engage with others.

In contrast, the SCEs impel experiencers to consciously assess or appraise their behavior relative both to internalized and external standards. Consequently, SCEs socialize and enculturate their experiencers on two interrelated levels, through direct self-regulation and through self-regulation induced by others (Lewis, 1992). On the first level, SCEs directly motivate the self to behave in socially prescribed ways, for example, to treat parents, siblings, and kin as they wish to be treated and to avoid selfish actions they want him or her to avoid. On the second level, SCEs can be used indirectly by others to manipulate experiencers into regulating themselves in the direction of certain kinds of altruistic behaviors. The self-regulating functions of SCEs must arise in part from the internalization of cultural forms that Vygotsky describes as "the distinguishing feature of human psychology, the basis of the qualitative leap from animal to human psychology" (Vygotsky, 1978, p. 57).

These self-regulating elements operate strongly, for example, in the development of behaviors and attitudes associated with eating, sexual behavior, and gender roles. It is interesting that—both from the perspective of both the self and the other—SCEs entail a strong sense of right and wrong, superiority and inferiority. As a consequence, socialization and enculturation through SCEs carry moral overtones. SCEs thereby serve as means for enforcing moral codes.

DEVELOPMENT OF SELF-CONSCIOUS EMOTIONS AND SELF-UNDERSTANDING IN HUMAN CHILDREN

There are a variety of approaches to the development of self-conscious emotions. The development of guilt, for example, has been studied from the perspective of discrete emotion, learning, psychodynamic, and cognitive theories (for a review, see Zahn-Waxler & Kochanska, 1990). In this chapter, I focus on cognitive-developmental approaches that identify the sequential emergence of various aspects of self, including SCEs.

Even within the cognitive-developmental school, there is no single, comprehensive scheme that describes development of the self from infancy through adolescence. There are, however, several complementary schemes that address various aspects of the developing self, including SCEs.

Michael Lewis and his colleagues, and other developmental psychologists, have identified the following sequences in the development of self-awareness in human infants and toddlers:

1. Mirror self-recognition, coyness, and verbal self-labeling emerge between 15 and 24 months of age during the latter part of the Sensorimotor Period (Lewis & Brooks-Gunn, 1979).
2. Pictorial self-recognition and use of personal pronouns emerges around 24 months of age, at the end of the Sensorimotor Period (Lewis & Brooks-Gunn, 1979).
3. Self-evaluative emotions of guilt, shame, pride, and hubris emerge at about 36 months of age (Lewis et al., 1989; Cameron & Gallup, 1988) have discovered that shadow self-recognition also begins at this time.
4. Autobiographical self begins to emerge at about 48 months of age (Snow, 1990).

According to Mascolo and Fischer (1995), children's self-conscious or self-evaluative emotions are stimulated by appraisals of their own behavior and its effects on others. Each self-conscious emotion has its own developmental trajectory. In the development of guilt, for example, at each step the child becomes capable of appraising a new, more complex dimension of wrongdoing that can cause harm to another.

It is interesting to note that the nature of children's appraisals changes as they develop new levels of cognitive skill in pretend play (Fischer, 1980; Watson, 1984; Watson & Fischer, 1980). Before the age of 2 years, in the sensorimotor steps, children seem to appraise responses based on contingencies. Between ages 2 and 4, in the single representation steps, they understand that feelings affect the behavior of others and therefore appraise responses in these terms. After age 4 years, at the level of representational mappings, they understand that the failure to reciprocate hurts others.[3]

Fischer and Watson's scheme confirms Mead's (1970) theory that pretend play, particularly role play, is a major arena for the development of understanding of the emotions and thoughts of self and other. Pretend play, which preoccupies children between the ages of 2 and 5 years, is the first expression of role playing (e.g., Bretherton, 1984). Role playing presupposes and reinforces a distinction between self and other. More crucially, it allows children to experience actions and the emotions of others by playing their roles.

Taking the perspectives of others during pretend play is also implicated in the development of theory of mind. Each level in the development of theory of mind is characterized by a different level of understanding of emotions (Wellman, 1990).

> Stage 1, "Simple desire psychology," is displayed by 2½-year-old children who understand that feelings may affect behavior. Children in this stage can correctly predict how others will act based on observation of the their emotions.
> Stage 2, "Desire-belief psychology," is displayed by 3-year-old children who understand that feelings and beliefs may affect behavior but cannot conceive that beliefs could be false. Children in this stage can correctly predict how others will act based on their stated beliefs.
> Stage 3, "False belief psychology," is displayed by 4-year-old children who understand that false beliefs can affect behavior. Children in this stage can correctly predict that others will act on the basis of their incorrect knowledge of a situation.

It is interesting to note that the steps in the development of appraisals correspond in age and apparent cognitive complexity to the developmental sequence in theory of mind in 2- to 4-year-old children. Development in both domains involves an increasing understanding of the motivational role of emotions, as well as the cognitive prerequisites for that understanding.

Understanding of the social role of emotions continues to develop after early childhood. Related research indicates that 4-year-old children believe that they cannot experience two emotions simultaneously. Six-year-old children believe that they can experience two emotions in close temporal sequence. Only at 8 years do children understand that two emotions can coexist. Interestingly, this is the first age at which children can verbally distinguish between shame and guilt (Griffin, 1995).

These later developmental stages in the understanding of emotions are consistent with Watson and Fischer's (1980) description of development of social roles in 7- and 8-year-old children at the level of representational systems. It is only at this level that children understand that the same person can play two or three roles simultaneously (e.g., being a doctor and husband, and a father).

These and other studies reveal that self-awareness and self-conscious emotions undergo a long and complex developmental trajectory beginning in infancy and continuing through later childhood and into adolescence. They demonstrate that the focus and complexity of children's self-conscious emotions change as they develop more powerful cognitive processes. As part of this change, children display increasingly integrated and abstract foundations for their appraisals of self and others.

Damon and Hart (1988), for example, show how children's appraisals of the self-as-object (the physical, active, social, and psychological aspects) and the self-as-subject (agency, continuity, and distinctness) develop in complexity from early childhood through adolescence. These appraisals are expressed in increasingly complex verbal judgments about the nature of the physical, social, and psychological self.

These studies also reveal that appraisal or judgment of self and others is intrinsic to the development of SCEs. Appraisals address the competence or incompetence, the appropriateness or inappropriateness, and/or the goodness or badness of the actions of the self and other. The fact that SCEs depend upon appraisals clearly implies that these emotions are implicated in the development of moral judgment. Before addressing the evolution of SCEs and moral judgment, we must consider the comparative evidence for the occurrence of self-awareness, SCEs, and the sense of justice in living monkeys and apes.

TAXONOMIC DISTRIBUTION OF SELF-AWARENESS AND SELF-CONSCIOUS EMOTIONS IN PRIMATES

Recently, in responses to reciprocal altruism theory, primatologists have returned to Darwin's (1930) focus on the evolution of morality. Investigators have discovered that great apes display empathy, generosity, and revenge for lack of reciprocity (de Waal, 1996). They may display rudimentary understanding of role reversals (Povinelli, Nelson, & Boysen, 1992; but see Povinelli, Chapter 3, this volume). They display the capacity to learn symbols and to use these symbols productively to label themselves and others, and to express feelings (Gardner, Gardner, & van Cantford, 1989). They also display the capacity to recognize their images in mirrors and photographs (Parker, Mitchell, & Boccia, 1994b). These and other behaviors strongly suggest self-awareness and even an incipient sense of justice and injustice (de Waal, 1996). But how are we to identify degrees of self-awareness and SCEs in monkeys and great apes?

The preceding stages of development of self-awareness and SCEs in human children provide a framework for comparing these phenomena in monkeys, great apes, and humans. They are useful for comparative purposes because their development is tied to cognitive development. Cognitive development is epigenetic in the sense that each succeeding stage is constructed from raw materials of the preceding stage. As Baldwin (1897) noted, the epigenetic nature of cognitive development implies an evolutionary sequence (Parker, 1990, 1996a).[4]

Comparative data suggest that all great apes display signs of self-awareness that are similar to those seen in young children between the ages of 1½ and 3 years of age (mirror self-recognition, self-conscious behaviors, and verbal self-labeling, pictorial self-recognition, and judgmental labeling of self and others). It is significant that great apes display the same range of self-aware phenomena as young humans. Moreover, great apes apparently traverse the same early stages of development of self-awareness as young human children (Parker, Mitchell, & Boccia, 1994a). They differ from human children in their slower developmental rate as well as their truncated level of maximum achievement. In sharp contrast to the great apes, monkeys fail to display these signs of self-awareness (e.g., Parker, Mitchell, & Boccia, 1994a).

Beginning at about 3 or 4 years of age—as compared to about 1½ to 2 years in human children—great apes begin to display mirror self-recognition (MSR; Linn, Bard, & Anderson, 1992; Miles, 1994; Povinelli, Rulf, Landau, & Bierschwale, 1993; Patterson & Cohn, 1994). As in human children, MSR is sometimes accompanied by signs of embarrassment or self-consciousness (Patterson & Cohn, 1994). Self-consciousness is also suggested by anecdotal reports of strategic hiding of facial expressions from conspecifics (de Waal, 1983; Tanner & Bryne, 1993).

By the time they are 5 or 6 years old—as compared to about 3 years of age in human children—apes are able to recognize photographs of themselves (Itakura, 1994) and to recognize their own shadows (Boysen, Bryan, & Shreyer, 1994). Language-trained great apes also begin to use personal pronouns at about the same age they show MSR (Miles, 1994; Patterson & Cohn, 1994). Somewhat later, they begin to use words such as "good," "bad," and "dirty" to characterize disapproved behaviors in themselves and others (Miles, 1994; Patterson & Cohn, 1994). Although it is difficult to judge the exact meaning of these words, their use suggests an incipient sense of meeting or failing to meet standards.

Developmentally, as indicated, MSR may be based on Piaget's fifth- and sixth-stage sensorimotor abilities to imitate novel actions, whereas subsequent use of feeling and judgment terms seems to be based on symbolic-level abilities characteristic of early preoperations. Unlike human children, great apes apparently develop no further than early symbolic substage of preoperations (Parker, 1990; Russon, Bard, & Parker, 1996). There is no evidence that great apes have elaborated the self-conscious emotions beyond incipient forms equivalent to those of 2- or 3-year-old children.

THE COMPARATIVE METHOD FOR RECONSTRUCTING THE ANCESTRY OF SELF-CONSCIOUS EMOTIONS

When the taxonomic distribution of characteristics of interest is known, tracing the evolutionary origins of characteristics is a straightforward procedure. We can do this by using the comparative method of evolutionary reconstruction. This method entails mapping the occurrence of the characteristics in monkeys and apes onto a previously derived phylogeny of living primates (Brooks & McLennan, 1991). See Figure 4.1 for a generally accepted family tree of living primates (Maddison & Maddison, 1992).

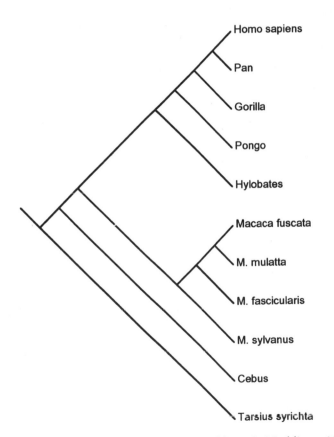

FIGURE 4.1. Phylogeny of the primates (Maddison & Maddison, 1992).

In this case, the characteristics of interest are *MSR*, *incipient SCEs*, and *elaborated SCEs*. If we map these characteristics according to their distribution in living primates, we see several things. First, we see that the capacity for MSR and incipient SCEs first arose in the common ancestor of the great apes and humans (B). We also see that *MSR* arose in conjunction with the capacity for Piaget's fifth- and sixth-stage imitation and sensorimotor intelligence (which develops at about 1 and 1½ years, respectively, in human children). Second, *incipient SCEs* arose in conjunction with the capacity for symbolic intelligence (which develops at about 2 years in human children) (Mitchell, 1993; Parker, Mitchell, & Boccia, 1994b).[5]

Figure 4.1 shows that *elaborated SCEs* first arose in an unspecified *Homo* species (A) sometime after the divergence of hominids from the common ancestor of hominids and chimpanzees. These emotions arose in conjunction with, and apparently depend upon, the capacity for Piaget's intuitive thought, which develops at about 4½ years. Fuller development of elaborated SCEs arose in conjunction with, and apparently depends upon, the capacity for conceptual thought, which develops after 6 or 7 years of age (Piaget & Inhelder, 1969). See Figure 4.2 for a depiction of this reconstruction.

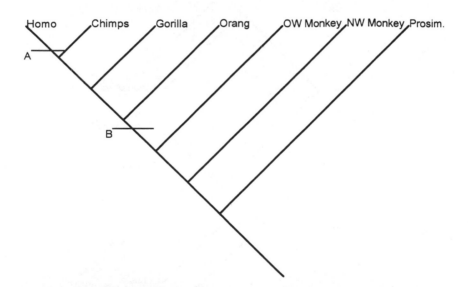

FIGURE 4.2. A, Common ancestor of *Homo* species; B, common ancestor of Great Apes and *Homo*. B was the first species to display the symbolic intelligence, mirror self-recognition, and incipient self-conscious emotions. A was the first species to display conceptual intelligence and elaborated self-conscious emotions. Lines are below the common ancestors who originated various aspects of self-awareness.

THE POSSIBLE ORIGINS OF GUILT, SHAME, AND PRIDE

It is important to note that new adaptations are constructed from preexisting adaptations. Therefore, the SCEs must have originated from interplay of preexisting emotions and emerging capacities for cognitive evaluations. Comparative data on the form and function of various emotional displays in monkeys, apes, and humans offer some clues as to the various amalgams that may underlie SCEs.

Like shame, guilt is associated with shrinking of the body and aversion of the gaze (Darwin, 1965). It is interesting to note that in monkeys and apes, these responses are associated with fear and subordination. Fear can be expressed in freezing, flight, or in appeasement gestures. The withdrawal behaviors associated with shame suggest that it may be derived from the freezing and flight reactions to fear, whereas the reparative behaviors associated with guilt suggest that it may be derived, in part, from the appeasement reaction to fear. Guilt may also be derived from empathetic feelings for the victim (Hoffman, 1984).

The close relationship between shame and guilt is indicated by the fact that different individuals may respond to the same event with either shame or guilt, whereas some individuals are prone to one or the other response (Ferguson & Stegge, 1995). The primary cognitive factor differentiating the response seems to be the actor's perception of his or her own competence or incompetence, responsibility or lack of responsibility for the violation. This differentiation is consistent with the notion that shame is akin to overwhelming, helpless fear leading to flight or freezing, whereas guilt is akin to moderate but manageable fear leading to appeasement gestures.

In contrast to shame and guilt, pride is associated with holding out the chest and enlargement of the body and engagement of gaze—Darwin (1965) allies pride with disdain. In monkeys and apes, these responses are associated with dominance. In the case of African apes, dominance displays can include either bluffing-over movements or reassurance gestures, depending upon the behavior of the subordinate (de Waal, 1983). Dominance is apparently experienced as a positive emotion, which may be akin to pleasure. When positive, cognitively mediated appraisals of one's performance may engender a mixture of dominance and pleasure that we experience as pride. Again, the experience of pride may relate to the characteristics of the audience and the relationship between the praiser and the praisee.

Whatever the raw materials from which SCEs were fashioned, we would like to understand the adaptive significance of these emotions. In order to do so, we must look at the peculiar nature of hominid adaptations. These adaptations can be understood best in terms of various forms of social selection, described at the beginning of this chapter.

SOCIAL SELECTION AND THE ADAPTIVE
SIGNIFICANCE OF SELF-CONSCIOUS EMOTIONS
IN HOMINID EVOLUTION

Humans are preeminent reciprocal altruists and parental manipulators. The relationship between husbands and wives and their families is a prototypical case of reciprocal altruism (Irons, 1983). This pattern is associated with prolonged care of offspring by mates and/or their families. Reciprocity between mates is based on and necessitated by sexual dimorphism in subsistence behavior.

Among most foraging peoples, human females are specialists in gathering and preparing vegetable foods, whereas males are specialists in hunting, butchering, and distributing meat. The degree of differentiation differs among groups and may differ within groups according to the season of the year (Friedl, 1975). Sexual dimorphism in hunting-and-gathering specializations favored reciprocal altruism in the form of food sharing between mates, which is the hallmark of human adaptation (Isaac, 1978; McGrew & Fiestner, 1992).

Like humans, our closest living relatives, chimpanzees, display a tendency toward sexual dimorphism in foraging and feeding that was probably present in our common ancestor. Females engage more in foraging and feeding on termites and ants, whereas males engage more in hunting and meat consumption. Males may share scraps of meat from their kills with others, especially with estrus females, and mothers may share difficult-to-process foods with their young. Male chimpanzees do not provide food for females and their young on a regular or systematic basis (Goodall, 1986; McGrew, 1992; McGrew & Fiestner, 1992).

Paleontological and archeological evidence suggest that sexually dimorphic hunting-and-gathering adaptations must have arisen by about 1.6 million years ago, the time of the emergence of *Homo erectus*. *Homo erectus* was the first hominid to hunt and or scavenge large game animals whose bulk exceeded the demands of immediate consumption and hence encouraged delayed sharing (McGrew & Feistner, 1992). Hunting, butchering, and transporting large game, and excavating, transporting, and processing vegetable foods with tools, in turn, entailed increased intelligence and brain size, which forced birth of immature infants as well as more prolonged infant and juvenile dependency. The more prolonged development of *Homo erectines* as compared to australopithecines has been documented through comparative studies of molar eruption (Smith, 1992). Greater and more prolonged dependency favored increased parental investment, including prolonged food sharing with offspring.

Prolonged dependency of offspring, in turn, favored parental manipulation for "helpers in the nest." Parental manipulation of offspring to help in feeding, transporting, and protecting younger offspring is common among birds and mammals (Brown, 1975). The willingness to share food with a sibling is a necessary aspect of sibling help in many species. Among modern humans, offspring and siblings are socialized through pride, shame, and guilt to share food (Schieffelin, 1990).

It is interesting to note that human infants show an innate tendency to offer and share food and other objects with adults and other children (Eibl-Eibesfeldt, 1989). The showing and/or giving gesture of the outstretched palm develops at about 13 months of age and is a precursor of the first word (Bates, 1979). Parker and Gibson (1979) speculated that this gesture occurred as part of a food-sharing complex in early hominids. This innate tendency to offer objects suggests strong selection for reciprocal altruism in ritualized food sharing.

Evolutionarily early forms of reciprocal altruism in food sharing and parental manipulation of offspring for helpers in the nest were subsequently expanded and augmented by new contractual forms. These included familial control of mating by "arranged marriages" between their offspring and, in more elaborated forms, exchange of wives among kin groups (Lévi-Strauss, 1969; Fox, 1967). Exchange of mates between families was probably part of an adaptation for buffering local resources shortages through long-term exchanges with kin groups living in different habitats (Gamble, 1976).

Exchanges within and among human groups are another characteristic feature of human societies. Enforcement of such contractual exchanges between mates, families, and larger groups depends upon social ideology and social sanctions. Family honor is an ideology maintained through pride, guilt, and shame (e.g., Friedl, 1975).

It is interesting to note that ritualized food sharing plays a key role in motivation and enforcement of all of these forms of social reciprocity. Older siblings feed younger siblings, lovers and husbands and wives feed each other, families and friends feed each other at weddings, political leaders feed each other (Lévi-Strauss, 1969; Eibl-Eibesfeldt, 1989). Cross culturally, generosity, and especially generosity in sharing food, seems to be a source of pride and political status in most human societies (see, e.g., Lévi-Strauss, 1969; Lee, 1979). In many societies, an ideology of sharing is inculcated through guilt and shame. These many convergent elements suggest that shame, guilt, and empathy may have arisen in *Homo* as means for socialization for food sharing. (In the absence of cross-cultural surveys of food-sharing practices and ideology, this interpretation must remain speculative.)

Like failures of reciprocity in food sharing, failure to control sexuality is another common cause of anger and hurt between spouses, and between parents and offspring. It is also the occasion for shame and guilt in perpetrators. The capacity to suffer guilt and shame about sex emerges early in human development (Freud, 1962). Concern with sexuality makes sense given that a male's ability to prevent his mate from engaging in such liaisons increases his confidence in paternity and hence his reproductive success (e.g., Daly & Wilson, 1983). Given the adaptive significance of control of female sexuality, it seems likely that sexual selection favored females who could be socialized by shame and guilt to control their sexuality.

In later periods of human evolution, under circumstances of accumulated wealth, parents (especially fathers, who had made substantial investment in their offspring) displayed more elaborated forms of parental manipulation. These included use of SCEs to indoctrinate daughters into the ideology of virginity and fidelity, which rendered them more desirable mates for wealthy suitors who sought confidence in paternity (Dickemann, 1981). Under these circumstances, control of female sexuality became a familial responsibility. This control is accomplished through a variety of direct and indirect means, ranging from claustration to ideology (Dickemann, 1981).

Use of ideology to control behavior depends upon guilt and shame, which are powerful means for socializing offspring. Ideology depends upon the capacity for declarative planning, which in turn depends upon language. Social contracts also depend upon these capacities. Social manipulation through symbolic communication and SCEs may have been one of the major selective forces on the evolution of language and intelligence in anatomically modern humans (Parker, 1985; Parker & Milbrath, 1993).

Finally, it is important to note that the experience of guilt and shame are not the only penalty for failure to act in accord with familial and/or larger social contracts. Refusing to honor thy parents and or to honor social contracts can lead to ostracism and, in the case of families, to blood feuds, as indicated in the following quotation from Mauss's (1967) famous treatise, *The Gift.*

> It is groups, and not individuals, which carry on exchange, make contracts, and are bound by obligations; the persons represented in the contracts are moral persons—clans, tribes, and families; the groups, or the chiefs as intermediaries for the groups, confront and oppose each other. Further, what they exchange is not exclusively goods and wealth, real and personal property, and things of economic value. They exchange rather courtesies, entertainments, ritual, military assistance, women, children, dances, and feasts; . . . Finally, although the prestations and counter-prestations take place under a voluntary guise they are in essence strictly obligatory, and their sanction is private or even open warfare. (p. 3)

SUMMARY AND CONCLUSIONS

I have presented a social selection model for the evolution and adaptive functions of such SCEs as shame and guilt in humans. This model incorporates recent evolutionary concepts into a comparative developmental approach that harkens back to James Mark Baldwin. According to the social selection model, SCEs were favored by sexual selection, parental manipulation, kin selection, and reciprocal altruism as adaptations for social manipulation. These emotions were favored because they facilitated direct and indirect socialization and enculturation into values that serve the genetic interests of parents, mates, kindred, and authority figures. Comparative data on great apes suggest that incipient SCEs arose in the common ancestor of the great apes and humans. They may have evolved through parental manipulation as an adaptation of imitative learning and demonstration teaching of tool-mediated extractive foraging. Paleonotological and archeological data suggest that elaborated SCEs probably originated in the *Homo* lineage, probably in *Homo erectines*. They may have evolved through parental manipulation and reciprocal altruism as adaptations for hunting and gathering and food sharing. The main conclusion of this analysis is that elaborated SCEs are apparently unique hominid adaptations for social manipulation.

ACKNOWLEDGMENTS

I would like to thank the following individuals for their helpful comments, which in no way implies their agreement with the ideas presented in this chapter: Drs. Michel Ferrari, Sulamith Hines Potter, and Andrew P. Wilson. I also want to thank the Department of Anthropology and Sonoma State University for time to work on this project during my sabbatical leave.

NOTES

1. It is important to distinguish facial expressions of emotions from language-related facial gestures that are culturally variable. These include regulators and illustrators used to supplement speech, as well as emblems, which are equivalent to words and phrases (Ekman & Friesen, 1969).

2. Given the dire consequences that shame can have for its experiencer, selection may have favored communication of the affect for the purpose of eliciting protective or reassuring behaviors. Alternatively, sexual selection may have favored the display of shame in females because it communicates their ability to be controlled by males.

3. Fischer and Watson and have proposed a neo-Piagetian model for the de-

velopment of social role playing in pretend play. According to their model, the child traverses 14 steps through four "skill levels": sensorimotor systems, single representations, representational mappings, and representational systems. (Fischer's levels correspond roughly to Piaget's early and late sensorimotor period, early and late preoperations period, and concrete operations period.) During the period from 2 to 4 years of age, children become capable of single representations. Early in this period, for example, they can make a doll perform several actions. Later, at about 2½ years, they can pretend that an object is another object. Finally, at about 3 years, they can perform several behaviors befitting a social role. At about 4 years, they become capable of representational mappings, in which two or more representations are related. It is at this level that children first become capable of understanding social roles, that is, the complementarity of roles.

4. Comparative developmental studies raise the specter of recapitulation theory. Early attempts to use recapitulation theory created a backlash against evolutionary models that result in recapitulation. This controversy is beyond the scope of this chapter (see Parker & McKinney, in press, for discussion).

5. Elsewhere, I have argued that the capacity for mirror self-recognition arose in the common ancestor of the great apes as an adaptation for apprenticeship (imitative learning by young and demonstration teaching by mothers) in tool-mediated extractive foraging on a variety of embedded foods (Parker, 1996a). See Povinelli, Chapter 3, this volume for an alternative scenario.

REFERENCES

Alexander, R. (1974). The evolution of social behaviour. *Annual Review of Ecology and Systematics, 5*, 325–383.

Baldwin, J. M. (1897). *Social and ethical interpretations of mental development.* New York: Macmillan.

Barrett, K. C. (1995). A functionalist approach to shame and guilt. In J. P. Tangney & K. W. Fischer (Eds.), *Self-conscious emotions: The psychology of shame, guilt, embarrassment, and pride* (pp. 25–63). New York: Guilford Press.

Bates, E. (1979). *The emergence of symbols.* New York: Academic Press.

Baumeister, R., Stillwell, A. M., & Heatherton, T. F. (1995). Interpersonal aspects of guilt: Evidence from narrative studies. In J. P. Tangney & K. W. Fischer (Eds.), *Self-conscious emotions: The psychology of shame, guilt, embarrassment, and pride* (pp. 255–273). New York: Guilford Press.

Boysen, S. T., Bryan, K. M., & Shreyer, T. A. (1994). Shadows and mirrors: Alternative avenues to the development of self-recognition in chimpanzees. In S. T. Parker, R. W. Mitchell, & M. L. Boccia (Eds.), *Self-awareness in animals and humans* (pp. 227–240). New York: Cambridge University Press.

Bretherton, I. (1984). Representing the social world in symbolic play: Reality and fantasy. In. I. Bretherton (Ed.), *Symbolic play: The development of social understanding* (pp. 32–41). New York: Academic Press.

Brooks, D., & McLennan, D. (1991). *Phylogeny, ecology, and behavior.* Chicago: University of Chicago Press.

Brown, J. L. (1975). *The evolution of behavior*. New York: Norton.

Cameron, P. A., & Gallup, G. G., Jr. (1988). Shadow self-recognition in human infants. *Infant Behavior and Development, 11*, 465–471.

Daly, M., & Wilson, M. (1983). *Sex, evolution, and behavior* (2nd ed.). Boston: PWS Publishers.

Damon, W., & Hart, D. (1988). *Self understanding in childhood and adolescence*. New York: Cambridge University Press.

Darwin, C. (1930). *The descent of man*. New York: Appleton.

Darwin, C. (1965). *The expression of the emotions in man and animals*. Chicago: University of Chicago Press.

Dawkins, M. S., & Guilford, T. (1991). The corruption of honest signaling. *Animal Behaviour, 41*, 865–873.

Dawkins, R., & Krebs, J. (1978). Animal signals: Information or manipulation? In J. Krebs & N. B. Davies (Eds.), *Behavioural ecology* (pp. 282–309). London: Blackwell.

de Waal, F. (1983). *Chimpanzee politics*. New York: Harper & Row.

de Waal, F. (1996). *Good natured: The origins of right and wrong in humans and other animals*. Cambridge, MA: Harvard University Press.

Dickemann, M. (1981). Paternal confidence and dowry competition: Biocultural analysis of purdah. In R. D. Alexander & D. W. Tinkle (Eds.), *Natural selection and social behavior: Recent research and new theory* (pp. 417–438). New York: Chiron Press.

Eibl-Eibesfeldt, I. (1989). *Human ethology*. New York: Aldine de Gruyter.

Ekman, P. (1972). Universals and cultural differences in facial expressions of emotion. In J. Cole (Ed.), *Nebraska Symposium on Motivation* (Vol. 19). Lincoln: University of Nebraska Press.

Ekman, P., & Friesen, W. (1969). The repertoire of nonverbal behavior—Categories, origins, usage, and coding. *Semiotica, 1*, 49–98.

Ekman, P., Friesen, W., & Ellsworth, P. (1972). *Emotion in the human face*. New York: Pergamon Press.

Ferguson, T. J., & Stegge, H. (1995). Emotional states and traits in children: The case of guilt and shame. In J. P. Tangney & K. W. Fischer (Eds.), *Self-conscious emotions: The psychology of shame, guilt, embarrassment, and pride* (pp. 174–197). New York: Guilford Press.

Fischer, K. W. (1980). A theory of cognitive development: The control and construction of hierarchies of skills. *Psychological Review, 87*(6), 477–531.

Fox, R. (1967). *Kinship and marriage*. Middlesex, England: Penguin Books.

Fox, R. (1971). The cultural animal. In J. Eisenberg (Ed.), *Man and beast* (pp. 275–292). Washington, DC: Smithsonian Press.

Freud, S. (1962). Three contributions to the theory of sex. New York: Basic Books.

Friedl, E. (1975). *Women and men: An anthropologist's view*. Prospect Heights, IL: Waveland Press.

Gamble, C. (1976). Interaction and alliance in Palaeolithic society. *Man, 17*, 92–107.

Gardner, R. A., Gardner, B. T., & van Cantford, T. E. (Eds.). (1989). *Teaching sign language to chimpanzees*. Albany: SUNY Press.

Goodall, J. (1968). Expressive movements and communication in free-ranging chimpanzees: A preliminary report. In P. C. Jay (Ed.), *Primates: Studies in adaptation and variability* (pp. 313–374). New York: Holt, Rinehart & Winston.

Goodall, J. (1986). *Chimpanzees of the Gombe*. Cambridge, MA: Harvard University Press.

Goodenough, W. (1981). *Language, culture, and society* (2nd ed.). Menlo Park, CA: Benjamin Cummings.

Griffin, S. (1995). A cognitive-developmental analysis of pride, shame, and embarrassment in middle childhood. In J. P. Tangney & K. W. Fischer (Eds.), *Self-conscious emotions: The psychology of shame, guilt, embarrassment, and pride* (pp. 219–236). New York: Guilford Press.

Hoffman, M. L. (1984). Interaction of affect and cognition in empathy. In C. E. Izard, J. Kagan, & R. Zajonc (Eds.), *Emotions, cognition, and behavior* (pp. 103–131). New York: Cambridge University Press.

Irons, W. (1983). Human female reproductive strategies. In S. Wasser (Ed.), *Social behavior of female vertebrates* (pp. 169–213). New York: Academic Press.

Isaac, G. (1978). The food-sharing behavior of protohuman hominids. *Scientific American, 238*(4), 90–108.

Itakura, S. (1994). Symbolic representation of possession in a chimpanzee. In S. T. Parker, R. W. Mitchell, & M. L. Boccia (Eds.), *Self-awareness in animals and humans* (pp. 248–253). New York: Cambridge University Press.

Izard, C. E. (1984). Emotion–cognition relationships and human development. In C. E. Izard, J. Kagan, & R. B. Zajonc (Eds.), *Emotions, cognition, and behavior* (pp. 17–37). New York: Cambridge University Press.

Lee, R. B. (1979). *The !Kung San: Men, women and work in a foraging society*. New York: Cambridge University Press.

Lévi-Strauss, C. (1969). *The elementary structures of kinship* (Rev. ed.; J. H. Bell & J. R. von Sturmer, Trans.). Boston: Beacon Press.

Lewis, M. (1992). *Shame: The exposed self*. New York: Free Press.

Lewis, M., & Brooks-Gunn, J. (1979). *Social cognition and the acquisition of self*. New York: Plenum.

Lewis, M., Sullivan, M. W., Stanger, C., & Weiss, M. (1989). Self-development and self-conscious emotions. *Child Development, 60*, 146–156.

Linn, A., Bard, K., & Anderson, J. R. (1992). Development of self-recognition in chimpanzees. *Journal of Comparative Psychology, 106*, 120–127.

Maddison, D., & Maddison, W. (1992). *MacClade*. Sunderland, MA: Sinaur Associates.

Mascolo, M. F., & Fischer, K. W. (1995). Developmental transformations in appraisals for pride, shame, and guilt. In J. P. Tangney & K. W. Fischer (Eds.), *Self-conscious emotions: The psychology of shame, guilt, embarrassment, and pride* (pp. 64–113). New York: Guilford Press.

Mauss, M. (1967). *The gift* (I. Cunnison, Trans.). New York: Norton.

McGrew, W. (1992). *Chimpanzee material culture*. New York: Cambridge University Press.

McGrew, W., & Feistner, A. T. C. (1992). Two nonhuman primate models for the evolution of human food sharing: Chimpanzees and Callitrichids. In J. H.

Barkow, L. Cosmides, & J. Tooby (Eds.), *The adapted mind: Evolutionary psychology and the generation of culture* (pp. 229–243). New York: Oxford University Press.

McNaughton, N. (1989). *Biology and emotion.* Cambridge, England: Cambridge University Press.

Mead, M. (1963). Socialization and enculturation. *Current Anthropology, 4*(2), 184–188.

Mead, G. H. (1970). *Mind, self and society.* Chicago: University of Chicago Press.

Miles, H. L. (1994). Me Chantek: The development of self-awareness in a signing gorilla. In S. T. Parker, R. W. Mitchell, & M. L. Boccia (Eds.), *Self-awareness in animals and humans* (pp. 254–272). New York: Cambridge University Press.

Mitchell, R. W. (1993). Mental models of mirror-self-recognition: Two theories. *New Ideas in Psychology, 11,* 295–325.

Parker, S. T. (1985). A social technological model for the evolution of language. *Current Anthropology, 26*(5), 617–639.

Parker, S. T. (1990). The origins of comparative developmental evolutionary studies of primate mental abilities. In S. T. Parker & K. R. Gibson (Eds.), *"Language" and intelligence in monkeys and apes* (pp. 3–64). New York: Cambridge University Press.

Parker, S. T. (1996a). Using cladistic analysis of comparative data to reconstruct the evolution of cognitive development in hominids. In E. Martins (Ed.), *Phylogenies and the comparative method in animal behavior* (pp. 361–398). New York: Oxford University Press.

Parker, S. T. (in press). The evolution and development of self-knowledge: Adaptations for assessing the nature of self relative to others. In J. Langer & M. Killen (Eds.), *Piaget, evolution, and development.* Hillsdale, NJ: Erlbaum.

Parker, S. T., & Gibson, K. R. (1979). A developmental model for the evolution of language and intelligence in early hominids. *Behavioral and Brain Sciences, 2,* 367–408.

Parker, S. T., & McKinney, M. L, (in press). *Origins of intelligence in apes and humans.* Baltimore: Johns Hopkins University Press.

Parker, S. T., & Milbrath, C. (1993). Higher intelligence, propositional language, and culture as adaptations for planning. In K. R. Gibson & T. Ingold (Eds.), *Tools, language, and cognition in human evolution* (pp. 314–333). Cambridge, England: Cambridge University Press.

Parker, S. T., Mitchell, R. W., & Boccia, M. L. (1994a). Expanding dimensions of the self: Through the looking glass and beyond. In S. T. Parker, R. W. Mitchell, & M. L. Boccia (Eds.), *Self-awareness in animals and humans* (pp. 3–19). New York: Cambridge University Press.

Parker, S. T., Mitchell, R. W., & Boccia, M. L. (Eds.). (1994b). *Self-awareness in animals and humans.* New York: Cambridge University Press.

Parker, S. T., & Russon, A. (1996). On the wild side of culture and cognition in the great apes. In A. Russon, K. Bard, & S. T. Parker (Eds.), *Reaching into thought* (pp. 130–450). Cambridge, England: Cambridge University Press.

Patterson, F., & Cohn, R. (1994). Self-recognition and self-awareness in lowland gorillas. In S. T. Parker, R. W. Mitchell, & M. L. Boccia (Eds.), *Self-awareness in animals and humans* (pp. 273–290). New York: Cambridge University Press.

Piaget, J., & Inhelder, B. (1969). *The psychology of the child.* New York: Basic Books.

Povinelli, D., Rulf, A., Landau, K., & Bierschwale, D. (1993). Self-recognition in chimpanzees (Pan troglodytes): Distribution, ontogeny, and patterns of emergence. *Journal of Comparative Psychology, 107*(4), 347–372.

Povinelli, D. J., Nelson, K. E., & Boysen, S. T. (1992). Comprehension of role reversal in chimpanzees: Evidence of empathy? *Animal Behaviour, 43,* 633–640.

Russon, A., Bard, K., & Parker, S. T. (Eds.). (1996). *Reaching into thought: The minds of great apes.* Cambridge, England: Cambridge University Press.

Schieffelin, B. (1990). *The give and take of everyday life: Language socialization of Kaluli children* (Vol. 9). New York: Cambridge University Press.

Smith, B. H. (1992). Life history and the evolution of human maturation. *Evolutionary Antrhopology, 1,* 134–142.

Snow, C. E. (1990). Building memories: The ontogeny of autobiography. In D. Cicchetti & M. Beeghly (Eds.), *The self in transition: Infancy to childhood* (pp. 213–242). Chicago: University of Chicago Press.

Tanner, J., & Bryne, R. (1993). Concealing facial evidence of mood: Perspective-taking in a captive gorilla? *Primates, 34*(4), 451–457.

Trivers, R. (1971). The evolution of reciprocal altruism. *Quarterly Review of Biology, 46,* 35–57.

Trivers, R. (1974). Parent–offspring conflict. *American Zoologist, 14,* 249–264.

Trivers, R. (1985). *Social evolution.* Menlo Park: Benjamin Cummings.

Trivers, R., & Williard, D. E. (1973). Natural selection of parental ability to vary the sex ratio of offspring. *Science, 191,* 249–263.

van Hoof, J. A. R. A. M. (1972). A comparative approach to the phylogeny of laughter and smiling. In R. A. Hinde (Ed.), *Nonverbal communication* (pp. 209–237). Cambridge, England: Cambridge University Press.

Vygotsky, L. S. (1978). *Mind in society: The development of higher psychological processes.* Cambridge, MA: Harvard University Press.

Watson, M. W. (1984). Development of social role understanding. *Developmental Review, 4,* 192–213.

Watson, M. W., & Fischer, K. W. (1980). Development of social roles in elicited and spontaneous behavior during preschool years. *Developmental Psychology, 16*(5), 483–94.

Wellman, H. M. (1990). *The child's theory of mind.* Cambridge, MA: MIT Press.

Wilson, E. O. (1975). *Sociobiology.* Cambridge, MA: Harvard University Press.

Zahn-Waxler, C., & Kochanska, G. (1988). The origins of guilt. In R. A. Thompson (Ed.), *Socioemotional development* (Vol. 36, pp. 183–258). Lincoln: University of Nebraska Press.

PART THREE

DEVELOPMENT OF SELF-AWARENESS ACROSS THE LIFESPAN

✧

CHAPTER FIVE

Is There a Self in Infancy?

✧

JEROME KAGAN

The invention of scientific constructs follows one of two very different strategies. Investigators using an *a priori* model select theoretically useful ideas based on intuition, formal argument, or preliminary, but incomplete, empirical evidence. The contrasting *a posteriori* model, associated with Francis Bacon, invents constructs after a reliable set of observations, or functional relations, has been discovered. The Freudian concept of id is in the *a priori* mode; the concept of implicit memory is in the *a posteriori* mode.

A broadly based faith in the concept of self, which is in an *a priori* frame, has its foundation in each person's subjective awareness of his or her stable attributes—beliefs, moods, intentions, actions, and acute changes in feeling. These phenomena define the consensual, sense meaning of self among contemporary Western scholars. Unhappily, the referential meaning of self recruits less agreement. The most popular source of evidence is a questionnaire or interview that many scientists criticize on the grounds that a large proportion of individuals distort their descriptions of self, knowingly or unconsciously, when answering direct questions.

A second controversy derives from the fact that some scholars use the concept of self to refer to a set of qualities that might not be available to awareness; this idea is closer in meaning to the concept of identity in Erikson's (1963) writings. This view holds that each adult has acquired a set of symbolic concepts for personal attributes. Although some of these concepts are easily articulated, others are not available to consciousness. All men know they belong to the category *male;* but some self-aggrandizing, arrogant men do not know that they also belong to the psychologists' category *narcissistic.* Questionnaires and interviews will not be a valid source of information about the identity of these men. Thus, part of the tension surrounding the concept of self derives from the fact that some psychologists treat the concept in the frame of an impartial observer rather than the subjective frame of the agent. The previous discussion, albeit brief, represents a fair summary of current views on the concept of self in older children, adolescents, and adults, each of whom possesses symbolic categories for their personhood.

However, the application of these ideas to infants under 18 months old is questionable. It is not at all certain that 12-month-olds, who experience sensations, possess any concepts about their person, and it is dubious that they are consciously aware of their intentions, feelings, appearance, or actions. Mice also have sensations of pain, light, and touch, but few would suggest that these animals are aware of and reflect on these sensations.

Empirical evidence supporting a skeptical attitude toward a concept of self in infants is moderately convincing. For example, when children see in a mirror a reflection of themselves with a dot of rouge on their nose, they do not touch their nose until they are in the middle of the second year (Lewis & Brooks-Gunn, 1979). This robust fact has been used to argue that infants under age 18 months do not have a sense of self. It should be noted that some scientists question the validity of this procedure as a measure of self-recognition (Johnson, 1983; Robinson, Connell, McKenzie, & Day, 1990; Priel & deSchonen, 1986).

However, other behaviors that emerge by age 2 are more convincing. In the months before the second birthday, children first use their name, direct the behavior of others, and show anxiety following failure on a task. Few 12-month-old infants will show these behaviors (Kagan, 1981). The more mature responses imply that 2-year-old children are aware of their attributes, believe that their directions of others will be successful, and appreciate that they will be judged following some goal-related behavior. These psychological signs suggest that 2-year-olds have some limited awareness of their qualities, but there is no evidence that 1-year-olds possess this competence in even rudimentary form.

If this austere conclusion is correct, the meaning of self when applied to older children is not appropriate to children in the first year or year and one-half of life, a claim in accord with Piaget's (1954) views. If a concept of self in infancy is theoretically useful, it will have to rest on a different meaning. There are many qualities we attribute to adolescents that we would never attribute to infants. For example, adolescents can be loyal, sincere, dishonest, envious, jealous, and have a fear of death; infants are incapable of these ideas or emotions.

Thus, it is reasonable to ask if there is any evidence that would force psychologists to assume that infants under 1 year possess any structures that might require a concept of self as it is applied to older children. Some psychologists might reply that the appearance of fear to strangers or to separation at 8 to 12 months, smiles to familiar caretakers at 3 months, and frustration to loss of an object in the first month require a concept of self. But that claim may be too bold. Retreat or a cry of distress to novel objects is present in mice as a biologically prepared reaction to novelty. The bark of a dog to a stranger walking up the driveway does not require attributing a self to the animal. Similarly, the tail wagging of a dog upon seeing its owner, often regarded as analogous to an infant's smile to its mother, does not require a concept of self. Finally, the thrashing and crying to loss of a nipple in a 1-week-old infant is a biologically prepared response.

Meltzoff (1981) has argued that early imitation of an adult's response—for example, sticking out a tongue—requires the assumption that the infant knows the part of its body it should move in order to match the actions of the adult model. But this apparent imitation in a very young infant can be explained more parsimoniously as a biologically prepared reaction to slender objects slowly approaching the infant's mouth. Very young infants also stick out their tongue when an adult directs a small pencil at their mouth (Kagan, 1984). No one would claim that a biologically prepared motor act requires a concept of self. The fact that 5-month-olds can discriminate between a video display of their own leg movements and a display of themselves making other body movements might require positing a concept of self (Bahrick & Watson, 1985). However, this discrimination might require only the ability for cross-modal matching of visual and proprioceptive percepts. Vasta, Haith, and Miller (1992) suggest that the tendency of young infants to look in the same direction for an object they saw even after they have been rotated in the opposite direction might be evidence for self-awareness. Finally, some might argue that the temperamental qualities of irritability or relaxation, which have their partial roots in brain chemistry, must have derivatives in the child's conscious emotional experience. But, as noted earlier, rodents also differ in emotionality, and no theorist is prepared to assume that rats are aware of their feeling states.

The post-Freudian scholars had a stronger theoretical reason for assuming that an infant possessed a self. The ego was a biological given that had to exist early in life. Mahler (1968) assumed, therefore, that the infant's self was merged with that of the primary caretaker. However, because other parts of the analytic theoretical structure have fallen, this rationale is not sufficient to maintain a belief in an infant self.

If no infant behavior requires a concept of self in order to be understood, we must ask why some presume that this construct might be useful. I believe that one reason for its popularity is that Western scholars wish to believe in a connectedness between infancy and the subsequent stages of development (Kagan, 1984). There is a wish for an unbroken trail from the earliest days of life to senility. If a 5-year-old has a concept of self, the 5-month-old must possess some part of this structure, albeit in a primitive form. However, that argument resembles the idea of preformation, which held that inside every sperm was a very tiny person that grew larger with time. This notion reflects the Enlightenment scholars' dislike of dramatic transformations that are not aesthetic (Farley, 1974). I suspect that those who believe that an infant has a sense of self are bothered by the untidy idea of discontinuities in development.

THE EVIDENCE FOR SELF IN THE SECOND YEAR

There are several behaviors that occur just before the second birthday that imply the emergence of a self for the first time. For example, during the months before the second birthday, a child will cry when an examiner models a behavior with toys and then withdraws to a chair. This phenomenon was observed both in American and in Fijian children (Kagan, 1981, 1991). The most frequent distress reactions included clinging to the mother, cessation of play, and crying. These behaviors first appeared around 15 to 16 months, increased with age, and reached a peak just before the second birthday. The distress is not due to the number of prior exposures to the model. After 2-year-olds had been exposed to an adult model on six prior occasions, a majority still showed distress when the model displayed some acts and then withdrew. Very few children living in atolls in a Fiji chain showed distress prior to 20 months of age, and there was a steep increase between 20 and 27 months, followed by a decline in crying after the model finished her behaviors. Finally, 5 Vietnamese children who had recently arrived in the United States were observed at home every 2 to 3 weeks, beginning at 15 to 17 months through 26 to 28 months. None of the children showed distress during the initial visit to the home; every one displayed signs of distress to the model when they were between 17 and 24 months (Gellerman, 1981).

These phenomena imply processes that require positing a concept of self. It is suggested that the children feel an obligation to imitate the model, together with an awareness of their inability to implement the actions, either because the children forgot what the model did or because they are unsure of their ability to perform the acts. As a result, the children become uncertain and begin to cry. The distress implies that the children have some awareness of their inability to meet the standard represented by the model's prior behaviors. Faith in this interpretation requires elimination of the possibility that the distress was due simply to the adult's interruption of the children's play. Distress was rare when the examiner simply interrupted the children's play and did not model many acts.

Fewer children become distressed when the mother, rather than an unfamiliar woman, is the model. The familiarity of the relationship between children and adult monitors the likelihood that a state of uncertainty will be produced. Most 2-year-olds have seen their mothers display many behaviors on varied occasions. The children did not imitate the mother on most of these occasions and, therefore, had many opportunities to extinguish a feeling of obligation to imitate the parent every time she displayed a coherent action. The children have a less familiar relationship with the model in the laboratory and, therefore, feel a stronger motive to display the acts that were modeled. The uncertainty in the social relationship motivates children to be concerned with the stranger's reaction toward them. As a result, the children are vulnerable to increased apprehension after the model has displayed her responses because of a self-imposed obligation to perform behaviors that they are not sure they can implement.

Another sign of a sense of self that is seen late in the second year is a smile when the child has mastered a task. These smiles accompany the child's solitary activity with objects (Kagan, 1981). These mastery smiles begin to increase in frequency in the months before the second birthday. These smiles can be interpreted as signifying that the child generated a goal for an external behavioral sequence, persisted in his or her attempts to gain that goal, and smiled upon attainment.

A third sign of an emergence of self is seen in directives to adults, usually to a parent, which do not occur until the middle of the second year. Some common examples include putting a toy telephone to the mother's ear and gesturing or vocalizing in a way to indicate that the child wants the mother to talk on the telephone. The child may point to a place in the room to indicate that the child wants the mother to sit there or to move to that location. One 19-month-old girl made motions to the mother indicating she wanted the mother to pick up a telephone. The girl handed the mother a telephone; the mother gave it back to her, but the child said, "No, no," and pushed the phone toward the mother's face. A 23-month-old child handed a doll and a bottle to her mother, indicating she wanted her mother to feed the doll. Another child requested that the mother take her feet off the sofa and put them on the floor. The goal in all these requests seems to be to produce a specific reaction in the adult. Often, a child will have the mother move in one direction and several minutes later indicate that he or she wants her to move back to her original place. A second, related type of behavior involves asking the parent for assistance with a puzzle or opening a box.

A 2-year-old would not direct the behavior of an adult if he or she did not have an expectation that the request would be met. The appearance of this category of response suggests that the child expects to influence the behaviors of others. Although 8-month-olds point to desired objects and whine, suggesting they want the object, they have no conscious conception that the cry or gesture will change the adult's behavior. The pointing or whining by an 8-month-old may resemble, superficially, the 2-year-old's request for help with a puzzle but, as 19th-century observers argued, the two responses differ profoundly in underlying competencies.

A monkey, a 4-month-old infant, and a 2-year-old child can be conditioned to make an operant motor response when it is followed by a reinforcement. But that single element of similarity does not mean that the accompanying cognitive processes are the same in all three organisms. The request of the 2-year-old differs from that of the infant in at least two ways. First, the older child recognizes that the adult's reaction is necessary to gain the goal—in some cases, it is the goal. Second, the older child has an expectation that the adult will respond appropriately.

The speech of 2-year-old children also suggests they have an awareness of self. When speech first emerges, the vast majority of one-word utterances are names for objects in the child's visual field. The child looks at an object, or points to it and says its name. Another frequent class of utterance during this first stage of speech is the communication of a desire for an object or event by using the word "more," pointing to an object and saying its name with the high-pitched voice characteristic of a frustrated motive.

A third, less common class of utterance occurs when the child is engaged in an action and refers to that action—called self-descriptive utterances. For example, the 2-year-old child says "climb" as she is climbing up on a chair or says "up" as she tries to get up on a box. The 2- to 3-year-old also issues utterances containing the words "I," "my," "mine," or the child's name, together with an object or a predicate (e.g., "My book" or "I sit"). As the child begins to speak two- and three-word utterances, there is little difficulty deciding whether a phrase is descriptive of self. Self-descriptive utterances are absent during the first few months of the second year, but by age 2 years, they are observable, and by 27 to 28 months can be sophisticated and include phrases such as, "I step on my ankle" or "I do it myself" (Kagan, 1981). Bloom, Lightbown, and Hood (1975) collected a set of language protocols from 4 children during repeated visits to their homes. Their coding of self-descriptive utterances matched our data, despite the differences in the context of observation and the investigators' goals. Both studies revealed that about one-third of a 2-year-old's utterances are descriptive of what the child is doing. The increase in descriptions of the child's activity are not due simply to the increased use of verbs and modifiers. Most children showed a decline with age in the frequency of utterances describing objects. The age of sharpest decline was the time when the child displayed a sharp increase in self-descriptions. It is likely that when 2-year-olds become aware of his ability to gain goals through their actions, they feel pressed to comment upon these behaviors. These acts suddenly become a salient incentive for linguistic description, or at least more salient than the qualities of toys or other people. The children do not begin to talk about themselves because they can speak verbs, but because they are preoccupied with what they are doing.

There is also a major increase in recall memory for the spatial location of objects in the months before the second birthday. In addition, when the children are asked to draw a schematic face from a model, the likelihood that they will draw a circle and put some marks within the circle increases dramatically after 22 to 23 months of age. The ability to make a circle probably requires special motor coordination, but the children must believe that they can draw a face before attempting to do so. Thus, it is not an accident that the era between 19 and 23 months marks the time when the children put some marks within the circle while attempting to draw a face. The marks are placed inside the circle because the children reflect on what is required and have some expectation of meeting that standard (Kagan, 1981). In summary, during the last months of the second year, children display for the first time a preoccupation with adult standards, emotion appropriate to successful or unsuccessful mastery, directives to adults to change their behavior, and language that is descriptive of their actions. Although these phenomena do not reflect one, simple overarching function, they appear to be related. The similarity in age of onset and era of growth across children implies that these diverse phenomena are likely to be consequences of maturational changes in the central nervous systems of children growing up in any environment containing people and objects. Although there are many constructs that psychologists might use to label the process, or processes, that permit these behaviors, the word "self-awareness" seems not to be terrorist in connotation.

This phrase refers to the specific behaviors described earlier. Other observers have used different terms for the same period of growth. Tiedemann (1897) wrote that the child develops *eigenliebe*, meaning love of self. The Utku of Hudson Bay believe a child develops *ihuma*, best translated as reason (Briggs, 1970). Preyer (1888) suggested that *ichheit*, meaning selfhood, emerged during this period. Different premises are hidden by each of these different phrases. Although Tiedemann and Preyer assumed that the new behaviors reflected an appreciation of individuality, the Utku emphasized the child's ability to appreciate the difference between right and wrong. If these data had been discovered by a 17th-century Chinese observer, he would probably have used a different term. Weiss's (1968) comment on the use of the term "growth" is appropriate in this context. "Growth is not a simple and unitary phenomenon. Growth is a word, a term, a notion covering a variety of diverse and complex phenomena. It is . . . a popular label that varies with the accidental traditions, predilections, and purposes of the individual or school using it" (p. 241).

The suggestion that an important maturational advance occurs in the months before the second birthday finds support in data on histological changes in the brain. The length and degree of branching of dendrites in the human cortex do not approach adult magnitudes until 2 years of age. Indeed, at birth, the dendritic pattern resembles that of the rat. During the second year, there is enhancement in the rate of myelination in association areas. Most important, Rabinowicz (1979) suggested that the number of neurons per unit volume decreased very rapidly until birth. "From birth and between 3 and 6 months the decrease is slower and it ceases at about 15 months. . . . One has to deduce that a very important moment in cortical maturation appears to be the period between 15 and 24 months, a period when almost all of the layers reached for the first time a similar state of maturation" (p. 122). It is probably not a coincidence that this period corresponds to the interval when the child displays the behaviors regarded as indicative of an emergence of self.

SUMMARY

Why does psychology need a concept of self? As noted earlier, the compelling argument for the construct originates in the human quality of conscious awareness of one's name, ideas, feelings, and attributes. If a woman were asked, "Why did you help the old man across the street?", and answered, "Because I am a kind person," her reply invites us to assume that she holds a conscious belief about self. However, when psychologists say that an immigrant in Los Angeles has a self that is identified with Mexican culture, they usually mean that the person feels pride when a Mexican American does something praiseworthy, but shame when the Mexican American is disloyal to the standards the agent treats as definitive of desirable behavior. Thus, there is a good reason for positing at least two different conceptions of self, even though neither one is among the Big Five personality traits.

We live in an era of strong positivism due, in part, to the technical advances in the natural sciences. Accurate and reliable measurement is the *sine qua non* of acceptance by a prestigious journal. A person's concept of self is so private that few scientists have faith in its valid measurement. The current method of asking individuals to describe themselves is not sufficiently sensitive, despite the fact that many personality theorists use self-report questionnaires.

The most important critique of a concept of self is the assumption that it is unitary because, subjectively, it seems that way. However, phenomenological intuition is not always a valid guide. It is reasonable to suggest that reflection on one's feeling tone leads to one category of self, reflection on one's talents leads to another, and reflections on one's morality to a third. No average self is computed across this trio, just as no average of a row of poplar trees is computed by the perceptual apparatus. Social scientists like unitary concepts and, therefore, resist attempts to analyze abstract ideas. A large number of psychologists believe that there is one type of learning, one type of memory, one type of secure attachment, and one self. There are several reasons why psychologists are averse to analysis and attracted to synthetic ideas.

First, analysis requires powerful methodology that reveals new aspects of a phenomenon. Examples include electron microscopes, gene cloning, and positron emission tomography (PET) scans. Psychology has not had the privilege of such extraordinary advances in method, and without sources of new information, there is less motivation for analysis.

Second, analysis proceeds best when the existing level of understanding is firm and can serve as a secure base for analysis. Biologists could not differentiate between the Purkinje neurons in the cerebellum and the smaller neurons in the hippocampus until the concept of the neuron had been accepted as valid. Psychologists lack such a firm set of conceptual ideas; hence, analysis seems premature.

Finally, complexity is a factor. The phenomena of psychology—behaviors, cognitive processes, and emotions—are so context bound that social scientists are frustrated in their attempt to freeze-frame a broad generalization. Their data continually remind them that all conclusions are context dependent. But, unfortunately, there is no theory of contexts, and the number of different settings seems unlimited. The reaction to this frustration is to move away from the complexity of a concrete phenomenon to abstractions that deny, or at least repress, the extraordinary specificity of psychological evidence. The form in which an examiner asks the question—for example, "Would you buy a Ford or a BMW?" compared with "How much money are you prepared to spend on a new car?"—influences the respondent's answer in a nontrivial way. Data of this kind are extremely discouraging, for one can list dozens of different ways to ask a given question. If each method yields a different answer, intellectual anarchy results. Because it is difficult to construct a compelling argument that gives one method obvious priority over others, psychologists retreat to abstract ideas that ignore specific contexts. Although this defense is effective temporarily, it is, in the end, dangerous, for it denies nature's preferred forms.

When individuals reflect on their verbal ability, they arrive at one self-concept; when they think about their attractiveness, they have another evaluation. People have as many self-concepts as they have salient categories for their attributes. I am moderately competent at tennis, very poor at skiing, and of the opinion that I am quite good at lecturing to undergraduates. I do not average these qualities to create a unitary concept of self.

Finally, the reader should note that the term "self-concept" does not have the same meaning as the idea of self-esteem. The concept of self-esteem was invented to explain the degree of anxiety, or uncertainty, compared with feelings of pride or confidence, that follows reflection on one's attributes. Self-esteem has an emotional component that is based on the relation between people's standards and their evaluation of their attributes. A professor who is good at tennis but poor at lecturing will have a more fragile self-esteem than a professional tennis instructor who is a poor lecturer. Thus, the person's standards and the social context in which the person lives have important influences on self-esteem. Because no children under 5 years of age are cognitively mature enough to reflect on how closely their personal attributes match their standards and those of their community, self-esteem is an inappropriate construct to apply to them. When the phrase "low self-esteem" is applied to very young children, it is simply a way to claim that the child is anxious. But anxiety in children has many causes—for example, neglect, harsh punishment, temperamental vulnerability to challenge—which do not require a concept of self.

An awareness of feelings and attitudes about self emerges at the end of the second year as circuits that unite limbic structures with the frontal lobe reach an important stage of development. This suggestion is no different from the hypothesis that speech emerges in the second year or that reproductive maturity emerges at 13 to 14 years of age. Linguists do not claim that 3-month-olds possess a little language, and no biologist suggests that 2-year-olds have a tiny bit of fertility. There is no evidence requiring the positing of a concept of self before the middle of the second year. Scientists who believe that infants possess such a structure must either provide more convincing evidence or create a persuasive theoretical argument that renders this idea necessary.

ACKNOWLEDGMENT

Preparation of this chapter was supported in part by a grant from the W. T. Grant Foundation and the John D. and Catherine T. MacArthur Research Network on Psychopathology and Development.

REFERENCES

Bahrick, L. E., & Watson, J. S. (1985). Detection of intermodal proprioceptive–visual contingency as a potential basis for self-perception in infancy. *Developmental Psychology, 12,* 963–973.

Bloom, L., Lightbown, P., & Hood, L. (1975). Structure and variation in child language. *Monographs of the Society for Research in Child Development, 40*(2), 1–97.

Briggs, J. L. (1970). *Never in anger*. Cambridge, MA: Harvard University Press.

Erikson, E. H. (1963). *Childhood and society*. New York: Norton.

Farley, J. (1974). *The spontaneous generation controversy*. Baltimore: Johns Hopkins University Press.

Gellerman, R. L. (1981). *Psychological development of the Vietnamese child in the second year of life*. Unpublished doctoral dissertation, Harvard University, Cambridge, MA.

Johnson, D. B. (1983). Self-recognition in infants. *Infant Behavior and Development, 6*, 211–222.

Kagan, J. (1981). *The second year*. Cambridge, MA: Harvard University Press.

Kagan, J. (1984). *The nature of the child*. New York: Basic Books.

Kagan, J. (1991). The theoretical utility of constructs for self. *Developmental Review, 11*, 244–250.

Lewis, M., & Brooks-Gunn, J. (1979). *Social cognition and the acquisition of self*. New York: Plenum.

Mahler, M. S. (1968). *Human symbiosis and the viscissitudes of individuation*. New York: International Universities Press.

Meltzoff, A. N. (1981). Imitation, intermodal coordination, and representation in early infancy. In G. Butterworth (Ed.), *Infancy and epistemology* (pp. 85–114). Brighton, England: Harvester.

Piaget, J. (1954). *The construction of reality in the child* (M. Cook, Trans.). New York: Basic Books. (Original work published in French, 1936)

Preyer, W. (1888). *The mind of a child: Part 1*. New York: Appleton.

Priel, B., & deSchonen, S. (1986). Self-recognition. *Journal of Experimental Child Psychology, 41*, 237–250.

Rabinowicz, T. (1979). The differentiate maturation of human cerebral cortex. In F. Falkner & J. M. Tanner (Eds.), *Human growth* (Vol. 3, pp. 97–123). New York: Plenum.

Robinson, J. A., Connell, S., McKenzie, B. E., & Day, R. H. (1990). Do infants use their own images to locate objects reflected in a mirror? *Child Development, 61*, 1558–1568.

Tiedemann, D. (1897). *Beobachtungen Uber die Entwicklung der Seelenfahigkeiten*. Altenburg, Germany: Oskar Bonde. (First edition published 1787)

Vasta, R., Haith, M. M., & Miller, S. A. (1992). *Child psychology*. New York: Wiley.

Weiss, P. A. (1968). *Dynamics of development*. New York: Academic Press.

CHAPTER SIX

Inching toward a Mature Theory of Mind

✧

MICHAEL J. CHANDLER
JEREMY I. M. CARPENDALE

This chapter is all about young people's developing understanding of their own and others' mental lives. More particularly, it is meant as a summary and broad critique of the building program of research that has come to be known as the study of children's "developing theories of mind." Why some such account might deserve a place in a collection of essays about self-awareness is perhaps simultaneously self-evident and more than a little obscure. What makes it obscure is that, at least to date, the theories-of-mind literature has been dominated by an all but exclusive preoccupation with the topic of so-called "false-belief understanding" and its special role in preschool children's changing beliefs about beliefs. As a consequence of this singular focus, the large bulk of those contributing to this literature simply have not given a fig about self-awareness, or other closely related matters. Notwithstanding this fact, it nevertheless obviously remains the case that thoughts about one's own and others' thoughts necessarily occupy some natural place front-and-center on the stage of self-awareness; that is, high on any list of things that must be demonstrated by anyone hoping to qualify as even minimally self-aware has to be some understanding of how it is that they, like others, have ended up with the particular mental life that they have. Although not, then, the usual stuff that generally occupies contemporary theorists of mind, there is, living on the margin of this new literature, some need to know, and some evidence to be had about how and when it is that young persons initially come to some such self-understanding by seeing themselves reflected in their own intellectual efforts.

Any account of the developmental route by means of which young persons gradually—perhaps even fitfully—come to an increasingly mature understanding of themselves as repositories of signature beliefs is a story that is necessarily made up of at least two parts. One of these—the part most remote from matters of self-awareness—is all about that age-graded process by means of which young persons, situated as they are at different developmental way stations, differently think about the general process of belief entitlement; that is, in ways that are perhaps not unlike their building understanding of their own and others respiratory or digestive process, young children, who are not automatically born into the world with any such understanding, are necessarily obliged to slowly work out for themselves the general mechanics of how it is that anyone might end up knowing and believing anything at all. The second of these closely related matters concerns children's changing conceptions of the individualized, often highly personalized ways in which they and others go about gaining their own unique understanding of what, at least at one level, are often overlapping experiences. As it turns out, the contemporary theories-of-mind literature is all about the first of these matters but has remained strangely silent about the second. The critique that runs through this chapter is meant as a vehicle for approaching some better understanding of why the theories-of-mind literature has adopted this awkwardly narrow interpretive course, and what can be and is being done to better explore the questions of how and when young persons first come to see their own beliefs as interpretive achievements.

To best see how this broader understanding might be accomplished, it will prove useful to return to what it is that "commonsense," or "folk" psychology, ordinarily has to say about the route by which people are ordinarily thought to acquire knowledge about themselves in general, and about their distinctive pattern of beliefs in particular. Here, two different sorts of possibilities are commonly had in mind. One of these is that the mind's eye either naturally possesses, or can be made to acquire, the capacity to somehow turn back upon itself in a way that allows for some sort of "introspection"—some reflexive capacity for mental bootstrapping that allows the self to become the proper "object" of its own attempts at self-scrutiny. Although much admired, the practical details of how this sort of mental rubbernecking is thought to be accomplished have often proven difficult to specify, and so not everyone (many psychological theorists included) is entirely convinced that anything like such self-reflective acts are actually possible.

A second way in which it is commonly imagined that people often succeed in becoming aware of the unique character of their own thought processes involves the familiar idea that in the process of coming to know or believe this or that, the actual machinery of one's own epistemic efforts automatically leaves its telltale fingerprints or rifling marks on the shell casing of our mental products. Carefully scrutinizing such signature markings, it is widely imagined, provides a potentially public, and thus perhaps more trustworthy, means of working out what it is that separates one's own thoughts from those of one's fellows. While perhaps not the subject of regular discussion, few reach maturity without having learned through hard experience that one can betray more about the self than is intended by being too quick to tell others what that particular Rorschach inkblot looks like, or exactly what sort of animal one would like to be. Similarly, when Walt Whitman drew up his well-remembered list of things that he had loved, "peeled sticks, shiny and new; cracked blue willow bowls; the rough male kiss of blankets . . ." and so on, no one supposed that the object of such an exercise was to learn more about twigs or chinaware or bed linen. Instead, we take it, we are meant to be learning about what Whitman knew and chose to reveal about himself. Here, knowledge of one's self is no longer thought to be achieved through abusively trying to force the mind's eye to roll back on itself, but rather by taking seriously the Calvinistic proposal that "by their works ye shall know them." That is, like the traces left by subatomic particles shot through Wilson cloud chambers, or those silhouettes of holocaust victims etched on the walls of Hiroshima, people are ordinarily understood to reveal themselves by the effects that are left by their own passing as they move about the world transforming things as they come to know and use and love them. This is the well-understood reason that people are often shy to display their own artistic efforts, or why telling you what kind of tool they would like to be is appreciated to involve a sometimes unwanted act of self-disclosure. In short, because, as ordinary adults, we naturally understand that the job of meaning making necessarily involves leavening the "facts" with instances of self-construction, and because we appreciate that all of our intentional and creative acts automatically provide both a public and private window onto the our "Soul," anyone with things to hide automatically understands only too well that every truly frank or openhanded display of one's actual beliefs and desires automatically puts on public display otherwise private details about the self.

Quite apart, then, from whether one is or is not ready to commit to the extravagant possibility of introspection (where the subject is somehow imagined to self-reflexively become its own object), there would appear to be both room and precedent for assuming that a certain awareness of one's own person can be ordinarily achieved by simply summing across multiple occasions in which coming to believe "this" rather than "that" serves to backlight and thus make self-evident the particulars of one's own signature ways of knowing.

Even with only this much said about the natural relation between self-knowledge and the process of belief formation, students of the mechanisms of self-awareness would seem well within their rights to turn with high hopes to the new and burgeoning literature concerned with the development of so-called theories of mind. Here, it would seem natural to suppose, ought to be sedimented a whole cache of collective wisdom concerning how it is that young persons first become aware of their own and others' contributions to those beliefs that mark them out as singular persons. As it is, such hopes are likely to remain largely unfulfilled. Some part of the reason for this collective disappointment is to be found in the fact that the large bulk of the literature on children's developing theories of mind is all about what people in general, and individual others in particular, take having a belief—any belief—to mean, with specialized beliefs about beliefs regarding the self and other typically batch-processed and treated as being all of a singular piece. This is not, of course, an unreasonable place to begin, and if this narrowness of focus were all there were to say against the theories-of-mind literature as a possible avenue toward some better understanding of self-awareness, then it would presumably only require a biding of one's time until the subject of the self as subject gradually worked its way toward some more center-stage position.

Unfortunately, the problem runs deeper, and this is it. The process of belief formation, as understood within our usual "folk" or "commonsense" psychology, is ordinarily seen to involve what John Searle (1983) calls both a "mind-to-world" and a "world-to-mind" direction of fit. The first of these is all about what Baldwin and Piaget called "accommodation," and concerns the way in which the mind is made to "fit" or conform to the particulars of the world. There is obviously little in any such passive recording process that could inform one's search for self-awareness. The second—what Searle labels a "world-to-mind direction of fit"—is all about "assimilation," or how the mind acts upon or deforms the world in the process of interpreting or constructing it in relation to its own already-present structures. Self-awareness, to the degree that it can be achieved through attention to acts of knowing, necessarily arises through attention to those process of assimilation by means of which experiences are actively interpreted in the course of achieving some world-to-mind direction of fit between external events and their mental representation.

For reasons that this chapter aims to make plain, the contemporary theories-of-mind literature turns out to be almost exclusively about mind-to-world directions of fit, or the way that mental life is understood to be shaped in the process of passively accommodating to a world that is not of one's own mental making. As such, it is painfully short of details regarding matters of interpretation, and so is left largely empty of direct relevance to the study of self-awareness. Working out why this is so, and offering some examples of how this unfortunate state of affairs might be set right, will be the central task of this chapter. What follows, then, is a broad account of the theories-of-mind literature as it stands, plus a more or less detailed set of criticisms of this collective undertaking, all offered with the aim of finding some alternative conception that gives equal time to both mind-to-world and world-to-mind relations between people and their environments, and, in the process, opens up the possibility that self-awareness can be and, in fact, regularly is achieved through an exploration of those commonalities evident in one's own interpretive acts.

The agenda that we mean to follow here begins with an attempt to first get clear about what is ordinarily meant by talk of children's "developing theories-of-mind," and to report out upon certain of the banner headlines that have helped to shape public perceptions of this research program. We mean to do this, not by vying to become simply the next in a growing backlog of straight-ahead literature reviews, a species of publications already thick on the ground (e.g., Carpendale, 1995; Feldman, 1992; Lewis & Mitchell, 1994; Moses & Chandler, 1992; Russell, 1992), but by generally making the case that the large bulk of things so far written on the topic of children's theories of mind has been oddly and counterproductively marked by its awkwardly ahistorical and nondevelopmental character, its general disinterest in the causes and consequences of the age-graded changes that it seeks to document, and by its seemingly arbitrary focus upon that restricted subset of beliefs that just happen to be truth-conditional. All of this is, of course, a rather rude and heretical way of talking about what, for many good reasons, is regularly held up as the new jewel in the crown of contemporary cognitive-development research. The work currently being done inside the new theories-of-mind tradition is, after all, arguably more philosophically informed, more tastefully "non-Cartesian," more confidently anti-Piagetian, more consonant with the new nativism and the new "modularism," and altogether more "cognitive science-like" than is your ordinary, run-of-the-mill cognitive-developmental undertaking. Work in this tradition is generally closely reasoned, its findings are frequently catchy and replicable, and, more often than not, its results have provided new strength to the arms of all those committed to the increasingly popular neonativist idea that young persons are full of previously unrecognized, inborn competencies, too long hidden from view by an injudicious reliance on badly bungled and age-inappropriate assessment techniques. With all of this to be said in its favor, is not there some better place, you might well wonder, for us to do our fault finding? Our answer is the balance of this chapter, which attempts at least a partial census of those essential matters that contemporary theorist of mind have oddly and counterproductively set outside of the tent.

Before starting to frame the contents of any such "Salon des Refusés," however, there is a need to first begin with some necessarily brief account of those various matters that are presently and agreeably housed inside the broad tent of things already clearly marked with the theory-of-mind colors. The first section, to immediately follow, ventures such an overview by highlighting certain milestones that mark the path of the theory-of-mind enterprise's meteoric rise to prominence. Having risked these dangerously synoptic remarks, the chapter then proceeds across the subsequent sections, by taking up in turn our several suggestions that the theories-of-mind literature (1) is unnecessarily ahistoric; (2) is unacceptably nondevelopmental; (3) is inappropriately insulated from much of the rest of the developmental enterprise; (4) is unaccountably disinterested in the causes and consequences of supposed theory change; and (5) is arbitrarily restrictive in its designation of what should count as a "belief."

WHAT IS MEANT BY ALL OF THIS TALK ABOUT DEVELOPING THEORIES OF MIND, AND WHY SHOULD WE BE INTERESTED IN IT ANYWAY?

Although notable exceptions having to do with the relation between second- or higher-order beliefs and other things such as perceptions, pretense, desires, values, and intentions are there to be found, the theories-of-mind literature is, first and foremost, an account of preschool children's emerging beliefs about beliefs. More particularly, the great bulk of what has been written on this subject can be seen to reduce to a series of claims and counterclaims about *when* (and sometimes *why*) it is that such young persons come, when they do, to appreciate that closely held beliefs can actually be mistaken. In short, theory-of-mind research is primarily research about the onset of false-belief understanding.

Good procedural, as well as conceptual, reasons support the importance of (though not an exclusive preoccupation with) the topic of *false-belief* understanding. Procedurally, the problem is that it is extremely difficult to find any measurable light between actions based on true beliefs and other contingently formed counterpart behaviors that are carried out in the total absence of beliefs (e.g., it is hard to distinguish between my acting on the closely held belief that hot stoves are dangerous and the family cat's having unreflectively come to treat them in precisely the same way). False beliefs, by contrast, tend to shout out their presence from the rooftops by mistakenly promoting maladaptive actions that are hard to otherwise countenance (e.g., confusedly showing up a day early for my Tuesday appointment becomes newly understandable after it comes out that I am somehow caught in the grip of a wrongheaded belief that tomorrow is Wednesday).

Given the evident tactical advantages that come along with focusing research attention on false rather than true beliefs, it also turns out to be a special, added bonus that anything one might be led to say about true beliefs can usually be said equally, or even more convincingly, given evidence for the workings of false beliefs. This follows for the reason (evidently known to children of some but not other ages) that the very act of holding to a belief necessarily entails an understanding of the possibility that such a belief might be mistaken. As Davidson (1984) puts it, "Error is what gives belief its point" (p. 168). In light of the fact that truth and falsity are mutually constitutive in just this way, researchers are left free to focus their attention where they will, and so to choose to concentrate on false-belief understanding where the methodological lights are brightest.

For the reasons just outlined, contributors to the theory-of-mind literature have been quick to appreciate that unless children of a given age (or members of various other species) can be shown to appreciate the very possibility that beliefs could be false, then there is little justification for granting them a concept of belief at all. Armed with such assurances, and encouraged by the methodological niceties afforded by tests of false as opposed to true beliefs, investigators concerned with the study of children's developing theories of mind were quick to settle into a search pattern that has had as its primary aim trying to figure out who is and who is not capable of entertaining the possibility of false belief. If, some of our fellow travelers (e.g., ordinary adults and, perhaps, certain "higher" primates) are capable of honoring just such a distinction between true and false beliefs, while others (e.g., domestic pets, small children, certain psychologically troubled individuals) are not, then we need to quickly figure out ways of deciding who is and who is not with us in these matters. From the outset, then, the theories-of-mind literature has been centrally concerned with working out procedural ways of effectively deciding when it is that false-belief understanding first puts in an appearance on the ontogenetic scene.

Although precise talk of children's "theories of mind" is new, concern over the question of how and when persons first come to honor the distinction between the world as such and our beliefs about it is old and the subject of long-standing scientific concern. Within this century, the psychological study of this matter has played itself out on two largely independent stages. Among developmentalists, about whom a great deal more will shortly be said, touchstone figures such as James Mark Baldwin and Piaget and Vygotsky worried this problem in the context of their extensive writings on the subject of childhood egocentrism. Elsewhere, comparative psychologists, and evolutionary biologists from Darwin to Premack and beyond, have similarly struggled with the question of whether humans are alone in the animal kingdom with their thoughts about the invisible world of beliefs and other intentional states. Because it bears directly on our own present purposes, some brief part of this second story deserves being re-peated.

By common account (see, e.g., the edited volume by Whiten, 1991), the 19th-century forebears of modern comparative psychology reportedly felt far too few reservations about speaking not only about the beliefs and desires of other animals, but also about the second- and higher-order be-liefs that diverse species were imagined to entertain about one another's in-tentional states. Predators, for example, were regularly credited with be-liefs about the beliefs of their prey, who, in turn, were imagined to take ac-tive steps to cloud the minds of their pursuers. Although never without its critics, the whistle was eventually blown on such easy anthropomorphic talk by Lloyd Morgan in 1894, who, by way of his so-called "Canon of Parsimony," entreated his less-cautious contemporaries to never again in-terpret an action in terms of "a higher psychical faculty if it can be inter-preted as the outcome of the exercise of one that stands lower in the psy-chological scale" (cited in Baldwin, 1988, p. 245). With Morgan's can*n*on [sic] trained on its back, the comparative study of animal behavior entered into a long drought period of self-loathing and cautionary behaviorism, full of safe preoccupations with hardwired connections and blind contin-gencies. And that appears to be about how things tended to lay until some short time ago when, in the 1970s, a certain new breed of comparativists began to entertain anew the once-heretical prospect that at least some pri-mates and members of other "higher" species might, after all, actually sub-scribe to sufficiently organized thoughts about the thoughts of others to justify their being credited with something like an embryonic "theory of mind."

All of these undercurrents came to the surface of attention within the psychological world with the publication in 1978 of Premack and Woodruff's now-classic target article entitled "Does the Chimpanzee Have a Theory of Mind?" Among much else that made this keystone paper memorable is the fact that several of its commentators (Bennett, 1978; Dennett, 1978; Harman, 1978) seriously took up the task of trying to spell out what would count as a minimally complex experiment capable of demonstrating that anyone ("man" or beast) did in fact actually appreciate the possibility of false belief, and so qualify as holding to some fledgling "theory of mental life." Although Premack (1988), among others (e.g., Povinelli, 1996), has gone on subsequently to cast doubts on the possibility that chimpanzees actually do hold to some bona fide theory of mind, the effect of the original Premack and Woodruff article has proven to be enor-mous, effectively jump-starting the whole of the present theories-of-mind enterprise. Whatever may eventually prove to be the case with other pri-mates, human children older than a certain age do, however, inarguably end up subscribing to something sufficiently like a fully fledged theory of mind, so that a whole new generation of cognitive developmentalists have found it sensible to wonder about exactly how and when this happens.

Forget about the Animals—What about Children?

Even if only educated fleas did it, the comparative study of animals' possible beliefs about beliefs is sufficiently interesting in its own right to have potentially supported a thriving theories-of-mind industry. As it happened, however, the bottom-up pressure arising from the comparative study of other species has proven to be more of a leavening ingredient than a mainstay of what is being billed here as the theories-of-mind success story. An altogether more salient lure—at least to many cognitive developmentalists evidently weary of living in the shadow of giants and full of hope about the new "cognitive science"—was that the theories-of-mind enterprise held out the prospect of decisively proving Piaget wrong. After all, had Piaget not linked the decline of childhood egocentrism, which, by a certain way of reckoning, could be set as equivalent to having no theory of mind whatsoever, to the emergence of "concrete operational thought," and did newly emerging evidence concerning children's first beliefs about beliefs not clearly locate such accomplishments within the preschool years, a half a childhood lifetime away from the onset of operational thought? Here, at last, was a seemingly novel possibility, closely linked to what was most exciting about new thoughts in the philosophy of mind, continuous with the best of contemporary comparative psychology and all smartly turned out in a fresh cloak of methodological rigor. Best of all, there was the live prospect of taking over from where the now evidently moribund study of social role taking had confusedly left off, not only by linking the study of mind to building interests in "folk psychologies" more generally, but by otherwise getting with the broad program of the new "cognitive science." A program that seems largely dominated by the impulse to boot downstairs all those cognitive competencies once ungenerously thought to be the exclusive province of only school-age children. Given all of these incentives, it is perhaps only a small wonder that the study of children's developing theories of mind has taken off in quite the spectacularly incendiary fashion that it has.

Because the point of the present chapter is to find certain faults with, rather than to once again exhaustively summarize, the whole of this new theories-of-mind literature, only the most sketchy review of this work will be attempted here. If one must settle for caricature, then perhaps it is not too wide of the mark to suggest that the fundamental claim on which most of this literature has turned is that, by a much younger age than was once widely supposed, preschool children (but not their still-younger nursery school counterparts) already seem to appreciate that others are capable of holding to and acting upon beliefs that are actually false. In what is inarguably the landmark study of this new literature, Heinz Wimmer and Joseph Perner (1983) carefully followed Daniel Dennett's formula for a "minimally complex" demonstration of false-belief understanding. They did this, at least in the first instance, by running what were initially young, school-aged children through a procedure that tested their understanding of the fact that some target character, strategically deprived of certain key bits of essential information, would necessarily founder on that ignorance by coming to, and subsequently acting upon, beliefs that were false.

In the now-standard version of this procedure, a child and adult puppet (e.g., Maxi and his mother) are shown to participate in the storing of some desired item (e.g., a chocolate bar) in one of the two available containers. Later, after the child puppet has been removed to some other remote location and thus left in the dark about what might happen next, the adult puppet is given some pretext for shifting the target item from its original location A to new container B. The critical test question intended to measure children's understanding of the possibility of false belief concerns where it is that child subjects assume the returning puppet character will "look for" or "believe" the target item to be. Interestingly, children younger than the age of 6 (in the first studies), and later 5, and eventually 4, were found to confuse their own knowledge concerning the current location of the chocolate with the now-outdated belief appropriate to the returning puppet by predicting that this story character would somehow mysteriously know that the target object was now located in the new container B, despite having been kept ignorant of its "unexpected transfer." So-called "reality errors" of this sort have typically been counted as evidence of a failure on the part of young subjects to appreciate the possibility of false belief, whereas children who succeeded on such tasks are regularly credited not only with false belief understanding, but also with their first, and perhaps only, genuine "theory of mind."

The substantial body of research that has grown up in response to Wimmer and Perner's strong claims to have identified a critical "cognitive deficit" in the thinking of children younger than 4, and to have established the subsequent existence of a previously undiscovered watershed in the course of children's cognitive development has taken a variety of forms. Perhaps not surprisingly, a sizable portion of this new literature has been given over to what amounts to a long series of more or less redundant re-demonstrations of the same basic finding. Not only do young preschoolers have difficulty driving a proper wedge between belief and reality when things are surreptitiously moved about, but, for example, they show related confusions on so-called "unexpected contents tasks" where, unbeknownst to them, some novel items (e.g., pencils) are substituted for the familiar contents of various commercially packaged products such as

M&Ms. A still larger contingent of these many publications is made up of a group of primarily in-house studies that are all premised on the common assumption that false-belief understanding is an achievement of some age groups and not others, but that doubt the standard claim that only 4-year-olds, but not still younger children, deserve to be credited with something like such a novel theory of mind. Here, the common claims have been that standard "unexpected change" and "unexpected contents" tasks actually fall short of being "minimally complex" for the sundry reasons that these now "standard" measures are judged as being too verbally top-heavy, too computationally complex, too lacking in appropriate temporal markings, too dependent on coincidental memory or verbal abilities, too devoid of personal relevancy, or are otherwise too plagued with salience problems for their own good. Although these remain far from settled matters, the clear thrust of this work has been to show that, under what are variously described as either "most appropriate" or "optimizing" conditions, children as young as 2 or 3 can also succeed on tasks that require some apparent understanding of the possibility of false beliefs.

Beyond such efforts to "move the goalposts," still other studies have concerned themselves with possible relations between standard measures of false-belief understanding and other familiar markers of cognitive or social-cognitive competence, including the ability to correctly employ certain mental-state terms, to behave deceptively, to engage in social pretense, or to solve various appearance–reality problems. Finally, there has been a range of comparative studies that have inquired into the presence or absence of some "theory of mind" in children belonging to various clinical groups (e.g., the mentally challenged and children with autism), and among various primate species. These comparative efforts hold out the important promises of aiding in the difficult tasks of coming to understand the roots of certain psychopathologies and of better finding our own proper place in the phylogenetic order. All this obviously amounts to a great deal of work, most of which has deservedly landed in psychology's best developmental journals and publishing houses. What follows are some of our reasons for suggesting why you should, nevertheless, remain dissatisfied with these collective efforts.

DOING WITHOUT HISTORY

Notwithstanding the occasional bone toss to cohorts of still earlier philosophers of mind and the seemingly obligatory denunciation of Piaget—apparently for the agreed-upon reason that "almost all of [his] substantive claims about the child's conception of the mind have turned out to be wrong" (Gopnik, 1993, p. 14)—anyone fed on an exclusive diet of contemporary theories-of-mind research would be well within his or her rights to conclude that the relevant history of this field began in 1978, with the publication of Premack and Woodruff's seminal paper, or perhaps as late as 1983, with the appearance of Wimmer and Perner's classic article on "Beliefs about Beliefs."[1] What, more than anything else, seems especially odd about this collective dare that history should repeat itself is that, were you to try telling almost any otherwise-informed psychologist about the recent flurry of research into children's developing theories of mind, he or she would, we promise you, automatically assume that you were talking about some natural extension of the several decades of active research into the subjects of childhood egocentrism and the slow acquisition of role-taking competence. In fact, the question first on the lips of most of those living outside the largely impenetrable thicket of self-citations that encircles the theories-of-mind literature tends to be "How is this data different from all previous data, give or take a couple of years?" Inquiring minds aside, one can search in vain through large tracts of the theory-of-mind literature without finding references that could support answers to such questions. Henry Wellman's synoptic book *The Child's Theory of Mind* (1990), to take only a single example, does not even contain an entry on "role taking" in its index.

Clearly some curtain has been drawn between a present that is scarcely more than a dozen years old and a much longer past that is seen to no longer warrant any mention. Discounting the risks of somehow being turned into a pillar of salt, or otherwise losing one's place as new under the sun, why this seemingly arrogant reluctance to look back? Two answers (excuses?) recommend themselves. The first is that because the body of evidence collected under the banner of childhood egocentrism and social role taking is so often taken by theorists of mind to be simply wrong on its face, there simply may seem no point in dwelling on such a discredited past. The second is that because Piaget and all of those understood to be caught up in his orbit are so regularly characterized as having placed their bets on the failed horse of Cartesian introspectionism, ignominy may seem their natural and just desserts. Both of these dismissive possibilities are sufficiently bold and brassy to warrant closer inspection.

First, then, what about the seemingly straightforward empirical assertion that Piaget and those social role-taking theorists who came after him have simply condemned themselves out of their own mouths by wrongly supposing that egocentrism (an earlier generation's alleged proxy for the absence of false-belief understanding) goes on being a chronic and defining characteristic of the thinking of 6- and 7- and even 8-year-olds? Several things are wrong with this picture. These critics are right, of course, that Piaget (1923/1955) did in fact report, for example, that nearly half of the speech of a group of 6-year-olds was egocentric (in that it was poorly adapted to the needs of others), and similarly found that many 6- and 7-year-olds continued to respond egocentrically on tests of visual perspective taking (Piaget & Inhelder, 1948/1956). It is just as evident, however, although often overlooked, that the majority of the speech of Piaget's sample of 6-year-olds was not egocentric (Beilin, 1992; Piaget, 1923/1955), and that this is the same Piaget (e.g., 1932/1965, 1964/1967, 1965/1995) who also regularly insisted that egocentrism and counterpart problems in social role taking were features of *each and every* developmental period. Nor is there any particular shortage of more latter-day studies that have looked for and found evidence of one or another form of egocentrism in persons who are a good deal older than 6 or 7 (e.g., Chandler & Boyes, 1982; Chandler, 1988; Elkind, 1967; Selman, 1980).

Although it would seem, then, myopic at best to imagine that the discovery that 4-year-olds already possess an understanding of the fallibility of beliefs should somehow automatically erase from collective memory all that was previously written on the subject of childhood egocentrism, what is inarguably true about the immediately preceding decades of research into this topic is that it did, in fact, fall into incoherence. It did so, we argue, not because thousands of subjects in hundreds of earlier social-cognitive experiments had somehow misled us, but rather because of the collective inability of a whole research community to find any believable way of understanding the fact that so many children of essentially comparable levels of maturity nevertheless ended up responding to what were meant to be equivalent assessment strategies in incomprehensibly different ways. This happened, we hope to demonstrate, not because only subjects tested after 1983 have succeeded in revealing their true character, but rather because, without aiming to do so, an earlier cohort of social-cognitive theorists inadvertently managed to end up measuring a variety of different things and calling them all by the same name. In a subsequent section, "Doing without Subjects Younger or Older Than 4," we will attempt to make the case that an important part of this confusion is owed to the fact that false-belief understanding and the insight that knowledge is unavoidably interpretive ended up being hopelessly conflated. For the moment, however, our only point is to lay stress on the fact that no one's interests are well served by the self-congratulatory assumption that only past, but not present, research should be dismissed because of unresolved disagreements over the best choice of measurement strategies.

The second and more pernicious reason commonly offered up by contemporary theorists of mind for turning a blind eye toward earlier generations of evidence on the subject of children's changing beliefs about beliefs is that Piaget and the other like-minded investigators who produced this evidence are all seen to be equally guilty of having fallen under the sway of a deeply misguided form of Cartesian dualism—view according to which it is wrongly supposed that all knowledge of mental life is somehow owed to a once-mysterious, and now regularly discounted, capacity for introspection.

The suggestion that Piagetian theory, along with the whole history of prior research into the development of social role-taking competence, ought to be consigned to the dustbin because of its misguided allegiance to some mistaken "Cartesian assumption . . . that the mind is somehow transparent to itself via introspection" (Wimmer & Hartl, 1991, p. 126), can be traced, at least in part, to Paul Churchland (1988) and his "slash and burn" efforts to reduce all possible readings of the knowing process to what he has labeled the "traditional" and the not-so-traditional, or "theory view." On this reading, so-called "traditional views" are said to be Cartesian in character because of their supposed allegiance to the antique idea that knowledge of other minds is effectively owed to some inbuilt capacity for introspection. The permissible alternative said to be remaining, which Churchland labels the "theory view," is held out as championing the opposite possibility by maintaining that "the mind/brain . . . learns about itself . . . through a process of conceptual development . . . that parallels exactly the process by which it apprehends the world outside of it" (p. 80). High on the list of serious problems owed to this arbitrary two-way split is, of course, that Descartes (who, as Eagle [1982] suggested, may not have actually been a "Cartesian" either) has been dead for a very long time. In view of this evident fact, the best prospect to "counting coup" against some opponent who could be described as "still kicking" is seen to lie along a path aimed at finding some still-surviving practitioner of such ancient introspectionist arts. The preferred candidate for this scapegoat possibility has been Piaget, who, along with his "disciples," is characterized as having helped to put theory-theory on a solid "anti-Piagetian footing" (Perner & Astington, 1992, p. 146) by naively buying into the wrong or crippled half of Churchland's fictive dichotomy. In what is offered up as support of this strained attempt at guilt by association, Perner and Astington cite a heavily worked quotation from Piaget and Inhelder's *The Child's Conception of Space* (1948/1956), in which egocentrism is said "to encourage [the child] to accept [his own view] . . . as the only one possible . . . [and] turn it into a kind of 'false-absolute'" (p. 194)—a caricature that these authors sum up by stating that according to both Descartes and Piaget, "we know each mental state through our direct introspective experience" (pp. 150–151). In much the same fashion, Wimmer, Hogrefe, and Sodian (1988), and later Wimmer and Hartl (1991), also argue that "the well-known tradition originating from Piaget (1932/1965), which conceives social-cognitive development as a movement from egocentrism to perspective taking . . . shares the Cartesian assumption . . . that the mind is transparent to itself via introspection" (Wimmer & Hartl, 1991, pp.

125–126). In more generous moments, such as in their 1992 report to the Jean Piaget Society, Perner and Astington, at least, conceded that their account of "Piaget's notion of egocentrism" is not so much a true account of Piaget's own views on the subject as it is a critique of what they call "its popularized version" (p. 150). What such a qualifier still fails to make clear, however, is that the "popularity" of such a view is largely confined to that subset of contemporary theorists of mind eager to find some justification for their self-serving decision to rewrite history by equating any and all alternatives to their own "theory view" with a widely discredited brand of circular Cartesian dualism.

The difficulty here is that all such dismissive attempts to equate Piagetian ideas with those of Descartes are simply wrong on their face. Anyone not in the grip of eliminative materialism, or who had otherwise troubled to read Piaget's numerous attacks on introspectionist views (e.g., 1965/1972), would be quick to agree with Chapman (1988), who pointed out that Piaget explicitly rejected any and all attempts to ascribe direct knowledge of the self to young children. Instead, it seems clear enough, even from the most casual reading, that Piaget regularly argued in favor of the proposition that self-knowledge, like knowledge of every sort, is first and foremost the slow developmental consequence of the internalization of action.

None of this is to say, of course, that Piaget did not share in what Gopnik (1993) calls the "incontrovertibly true" assumption that people assert "first-person privileged knowledge as a matter of phenomenology" (p. 1), or that Piagetian theory does not make room for the possibility that, through processes of "reflective abstraction" (Piaget, 1974/1976), older individuals do sometimes add to their store of understanding by recombining lower-order knowledge already at their disposal. What is decidedly not the case, however, is that either Piaget, or most social-cognitive theorists that used his theory as a springboard, ever proceeded on the highly suspect assumption that the process of gaining knowledge about others is first or primarily the result of introspection. Quite to the contrary, the common assumption among all those working within a broad Piagetian tradition has been that young children are functionally blind to the fact that there is what Searle (1983) has called a mind to world direction of causality between external events and our knowledge of them, and that, as a result, they automatically fail to understand that mental events bear the personalized stamp of whoever it is that is doing the knowing. As such, if young persons do accept their own view as any kind of "false absolute," it is not, according to those whose theories are broadly constructivistic in character, because children are closet Cartesians who take themselves as the measure of all things, but, rather, because they can see no meaningful distinction between the way the world is and the way they automatically take it to be (Chapman, 1989).

Were this meant to be some close defense of Piaget against the motivated redescription of his views by contemporary theory-theorists, then a great deal more could and would need to be said in support of the proposition that all such attempts to equate constructivism with Cartesian brands of introspectionism are at best poorly conceived. For the moment, however, our less ambitious aim is only to point out that the ahistoric character of the current theories-of-mind enterprise cannot be fairly justified or excused on the faulty assumption that Piaget, and the whole of the social role-taking literature, can be safely discounted for the bogus reason that both are collectively lost in some discredited Cartesian past.

If this is enough said against the common practice among contemporary theorists of mind of dismissing out of hand everything that occurred before 1978, then what about the altogether more serious accusation that the theory-theory tradition is not only blind to history but is also decidedly antidevelopmental as well?

DOING WITHOUT DEVELOPMENT

In his 1962 lectures to the Menninger Foundation's Psychoanalytic Society, Piaget, perhaps ungraciously, singled out for special criticism Freud's account of the emergence of the superego by branding this particular piece of the larger theory as unacceptably "non-genetic." Quite apart from any clinical advantage or therapeutic leverage otherwise owed to Freud's assertions that the superego is incorporated as a piece and consequently goes on relentlessly being self-same despite subsequent experience, such a static portrayal counted, in Piaget's eyes, as indefeasible evidence against any possibility that this aspect of psychoanalytic theory deserves to be counted as legitimately developmental. Quite the opposite was true, Piaget maintained, for the reason that, notwithstanding its childhood origins, the superego was merely parachuted into place as part of an imported resolution to the Oedipal (or Electra) complex and, having arrived *de novo*, subsequently remained insulated from further structural change.

The moral of this old story is to help drive home the point that it is not enough, in order to qualify as legitimately developmental, for any aspect of one's psychic architecture to simply have some identifiable point of onset in the usual age-graded flow of child development. Rather, things that are meant to be understood developmentally—at least according to Piaget and subsequent generations of constructivist theorists—require some history and need to be understood as being at least subject to the *possibility* of further change.

Against this background of expectations, how are we best to understand the status of the different set of more contemporary claims currently being made in behalf of children's reportedly "developing" theories of mind? The essential assertion that we mean to back up in the paragraphs to follow is that, like Freud's notion of the superego, most of what is presently being marketed under the banner of studies of children's developing theories of mind is not, in the last analysis, developmental at all but more of a species of *one miracle* "child psychology," similar to those static accounts of maturational change promoted during most of the first half of this century.

It is, of course, true enough that the theories-of-mind literature is full to overflowing with claims and counterclaims about the particular age at which children first acquire this or that understanding about the nature of what are mostly beliefs about beliefs. It is equally true that a number of the principal players in this literature have generated two- and sometimes three- or four-stage accounts of children's changing conceptions of the relations between beliefs and other things such as perception or desire (e.g., Perner, 1991; Wellman, 1990). Nor is there a total shortage of research intended to work out what might stand as necessary prerequisites to the eventual achievement of some bona fide theory of mind (e.g., Baldwin & Moses, 1994; Wimmer & Hartl, 1991). All of these efforts notwithstanding, however, it nevertheless remains fundamentally the case that the theories-of-mind literature is, first and foremost, a search for some singular transitional moment before which children possess anything that could legitimately qualify as a true belief about a belief, and after which nothing lies in store but the gradual prospect of growing expertise in the exercise of such an essentially unitary "either/or" accomplishment.

Wellman and his colleagues (e.g., Wellman & Woolley, 1990), for example, have done a great deal of interesting work on the prospect that children younger than 4 are "desire" psychologists who understand their own and other's "wants" without yet understanding how these are related to their truth-conditional beliefs. Wellman (1990) also makes the case that such a desire psychology may also serve to prefigure what is to come later regarding beliefs about beliefs. Unfortunately, doing this is not the same thing as offering up an account of how the emergence of a genuine representational account of mental life is possible at all. Flavell (1988) and his colleagues have offered a similar account of how nursery school children first understand the "connections" between various mental events well in advance of their only later coming to grips with the true representational character of belief. Unfortunately, how such "connections" are different from and more than merely antecedent to fully fledged representations is left largely unspecified.

Perner (1991) has likewise developed a three-stage model in which children younger than 4 are credited with a theory of behavior, if not a theory of mind, and has even invented the prospect of their being something that he dubs "preliefs," (Perner, Baker, & Hutton, 1994) and that are alleged to predate, but not yet qualify, as real "beliefs" in any legitimate sense. Finally, Perner and Wimmer (1985) did briefly attempt to work out what it might mean to have third or still higher-order sorts of ideas involving beliefs about beliefs about beliefs, all on the recursive prospect that "even fleas have lesser fleas upon their backs to bite em, and on and on and on it goes and so *ad infinitum.*" This attempt to portray the future as a series of recursions or multiple embeddings was subsequently dropped in Perner's 1991 book, however, and, more recently, it has been demonstrated (Sullivan, Zaitchik, & Tager-Flusberg, 1994) that when tested using less complex story materials than those employed by Perner and Wimmer, even "preschoolers can attribute second-order beliefs" (p. 395) to others. In any case, the ability to deal with the complexity of multiple recursions is not the same thing as, or counts as evidence for, the existence of any qualitative change in children's understanding of the nature of the mind.

More recently, there has been some interest in developments in children's understanding of mind that might occur after the age of 4 (e.g., Flavell, Green, & Flavell, 1995; Wellman & Hickling, 1994). For example, Flavell, Green, and Flavell (1995) found that preschoolers still have much to learn about both *when* people are thinking and *what* they are thinking about. Although this represents an important line of research, we suggest that the new skills being documented in this work have more to do with matters of performance than any truly qualitative difference in children's understanding of the nature of the mind.

Unfortunately, what seems not to be the case, then, about each and every one of these otherwise interesting suggestions about what might come either before or after the singular achievement of a so-called "representational theory of mind" is anything that might legitimately qualify as an earlier- or later-arising "form" of what is understood to be fundamentally the same thing; that is, in order for an account to qualify as "developmental" in any really strong or nontrivial sense, it is not enough to simply point to another matter that purportedly stands either earlier or later in some age-graded queue. Instead, what we need to learn about in the process of comprehending the development of children's beliefs about beliefs is something about how it is that earlier- or later-arriving "forms" of *one and the same thing* succeed in lining themselves up and transiting one into the other. It is this need for some account of children's developing understanding of beliefs that is not satisfied by simply pointing to the fact that, at some still earlier date, children regularly ignore beliefs in favor of desires, or "connections," or find some recursive way of supercharging a fundamentally "one-off" account of belief to some still higher power. Rather, what needs to be looked for instead, we would suggest, is evidence to the effect that some particular age-specific rendering of the notion of beliefs *qua* beliefs are themselves predated by other, still less mature forms of something that still manages to qualify as decidedly belief-like, or is otherwise followed up by other subsequent and still more grown-up forms of belief.

While so-called "stage" theories of children's changing beliefs about beliefs are, then, already thick on the ground, none of these accounts would appear to qualify as being developmental in the "strong" sense currently under discussion. Nor is what is being called for here to be found elsewhere in that long list of studies that have sought to merely move the goalposts of false-belief understanding *X* number of months closer toward some zero point of conception. Rather, more often than not, the implicit or explicit purpose of most such early-onset studies has been to work toward the discovery of some "innate theory of mind module" that is simply said to get "turned on" at a particular point some months or years sooner than other investigators have so far supposed. Alan Leslie (1987), who has recently proposed two such innate modules (one supposedly maturing at 9 months and another at 18 months), has perhaps been the most explicit of such neonativist theorists. As Gopnik (1993) points out, however, "no one actively working in the field, not even Leslie, has suggested that the 3- to 4-year-old shift [toward supposedly genuine false belief understanding] is the result of the maturation of such a module" (p. 14).

It is not, of course, that no one has ever thought to wonder about what might precede 2- or 3- or 4-year-olds' ability to pass false-belief tasks or to suggested that there might actually be some "form" of understanding that goes beyond simply equating knowledge with unencumbered access to the facts. Josef Perner's (Perner et al., 1994) largely impenetrable notion of "preliefs" is perhaps a step in this needed direction. Although not billed as such, so too is Clements and Perner's (1994) recent observation that children who ordinarily fail any and all standard false-belief measures nevertheless begin by looking at the correct location when asked, "Where will [the local equivalent of] Maxi look for his chocolate when he returns?" Such evidence clearly seems to hint at something like a nonverbal form of false-belief understanding.

Arriving from a quite different quarter, there is another and differently inspired set of recent outlaw studies that have as their purpose the formulation of an account that would demote the achievement of false-belief understanding to the status of being only one in a potentially quite protracted list of alternative forms of changing beliefs about beliefs (e.g., Carpendale, 1995; Carpendale & Chandler, 1996; Chandler & Lalonde, 1996; Pillow, 1991; Pillow & Weed, 1995). A fuller account of this new and more self-consciously developmental work will be the focus of the section to immediately follow. For the moment, however, we mean only to lay final stress on the unhappy point that despite the existence of literally hundreds of studies concerned with the precise age of onset of false-belief understanding, the great bulk of this work has remained fundamentally nondevelopmental in character. As a consequence, the currently available "one miracle view" of false-belief understanding—the view that currently tends to be regarded as the be-all and end-all of children's maturing beliefs about belief—ends up looking more like Freud's notion of the superego: something that is merely parachuted in at some particular point along the ontogenetic course and is subsequently imagined to go on being itself without any alteration in its basic form.

What is especially perplexing, as well as disappointing, about this particular lost opportunity is that one of the virtues most commonly claimed for the abandonment of what Churchland (1988) has called "traditional theories," in favor of some "theory view" alternative, is that such nontraditional accounts, which are meant to profit from their reliance on scientific theories as a source model, are regularly advertised as paving a smooth path toward the study of the process of theory revision. As Gopnik (1993) points out, one of the common features of both scientific and commonsense theories is that they are "revisable," in that "they change, often quite radically, in response to new evidence" (p. 3). Although there are good reasons to doubt whether commonsense or "folk" theories are as corrigible or defeasible as Gopnik and others have suggested (e.g., Blackburn, 1991; McDonough, 1991), the steady diet of stories that Churchland and Stich have put out about phlogiston and the failed astronomies of ancient camel drivers contain enough promissory notes about theory change that consumers of this literature are well within their rights to expect a raft of analogous accounts about how children also keep on changing the form of their beliefs about beliefs (Russell, 1992). As it is, however, the contemporary theories-of-mind literature has fallen seriously behind its own rhetoric by reducing everything about children's beliefs about beliefs to a one-miracle model in which false-belief understanding is the only intellectual achievement and failures to appreciate the role of ignorance the only intellectual crime.

Why we have ended up with so little in the way of promised theory revision is not entirely clear. Some part of an answer may be contained in the especially narrow conception of belief that has characterized the study of children's theories of mind. This possibility is further taken up later in the section called "Doing without the Better Half of Belief." A further possibility, and the one on which this section on the nondevelopmental character of theory-theory needs to close, is potentially there to be found in the contrasting ways in which Piaget and contemporary theory-theorists have approached the relation between scientific theories and children's developing conceptions of mental life. Whereas theory-theorists from Churchland to Gopnik have proceeded on the promissory note that our knowledge of scientific theory change holds the potential of informing our understanding of transformations in commonsense, Piaget moved in just the opposite direction. His fundamental interest was in understanding scientific knowledge, and he turned, at least initially, to the study of children as a "temporary" expedient chosen on the conviction that only by first understanding the genetic course of children's developing cognitions could we later hope to arrive at a means of understanding scientific revolutions (Chapman, 1988; Garcia, 1987; Murray, 1979). Of course, Piaget may have been wrong in his presumption that more elementary forms of knowledge might be observable in child development but not in the development of science, but, at least, his own decision to fit science to children, rather than proceeding the other way around, did end up generating a rich and differentiated account of intellectual development that is still missing in the theory-theory's one-miracle account of false-belief understanding.

DOING WITHOUT SUBJECTS
YOUNGER OR OLDER THAN 4

Because most contemporary theorists of mind regard the ability to entertain beliefs about beliefs as a unitary capacity that, like Athena, is imagined to spring fully formed from the heads of 4-year-olds, there is little evident enthusiasm among them for any plan that would involve apparently squandering further energies on the presumably pointless study of still younger or older children, who are taken to either lack such talents entirely or to already possess them in full measure. A direct consequence of this either–or assumption is that, aside from the occasional study of pretense or the early use of mental state terms, only the smallest fraction of the large army of young people whose thoughts and actions make up the database of the theories-of-mind literature are younger than 40 or older than 60 months of age. Theory-theorists are not the first, of course, to limit the focus of their attention to a single age group, and their choosing to do so would be no special crime were it not for the extravagance of their broad claim that, by the end of the preschool years, most young persons already possess all of the fundamental capacities required for subscribing to the commonsense folk psychology of their elders.

Among the several things that are wrong with this picture, the most glaring, we mean to show, is the readiness of most contemporary theory-theorists to equate the passing of standard tests of false-belief understanding with also holding to a more fully fledged "interpretive" theory of mind. Although there are a number of conceptual and empirical reasons for challenging this easy assumption, forget about these for the moment and try imagining instead how the singular and small-caliber idea that all of life's misunderstandings can be reduced to matters of simple ignorance might actually play in Peoria, or have been received by generations perhaps less skeptical than our own.

This is us trying to sell such a limited bill of fare at the Old Globe:

Dear Bard,

A quick plot suggestion for your next light comedy. What about having some dark Moor stumble around in ignorance because a hand-kerchief that he left in one container was, unbeknownst to him, moved to some new location while he was off fighting the Turkish fleet.

Best,

M.C. & J.C.

Dear Drs. C. & C.,

Many thanks for the suggestion, but wouldn't it be altogether more interesting and true-to-life if circumstances could be devised so that much the same chain of events could be shown to read one way to any-one secure in their affections, and quite the opposite way if one's well of certainty had been somehow poisoned by the green-eyed snake of jeal-ousy?

Sincerely,

W. S.

Or, failing this, might it not be better to try moving back even further into some still darker age when the world was thought to be flat.

Dear Socrates,

Just to let you know that modern social science has gone on to demonstrate that the real reason that the world is understood differently by different people is that while some have all of the relevant facts at their disposal, others are still living in ignorance.

Best,

M. C. & J.C.

Dear Drs. C. & C.,

Your analysis has obviously fallen into error because, while intru-sive experience may be like a signet ring and people's minds not unlike waxen blocks, it is also the case, as I have undertaken to make clear to my young colleague *Theaetetus*, that "in some men, the wax in the soul is deep and abundant . . . and so the signs that are made in it are lasting, because they are clear and have sufficient depth . . . it is a different mat-ter [however] when a man's 'heart' is 'shaggy' or when it is dirty and of impure wax; or when it is very soft or hard . . . for then the impressions have no depth. All such people are liable to false judgments" (Plato, 1990, p. 329), not simply because they are more ignorant than others, but because they are differently struck by what amount to the same facts.

Sincerely,

Socrates

What is common to our own and to these and numerous other perhaps less than thoroughly modern accounts is the shared view that the theories of mind to which one could potentially subscribe can and do take different forms, some of which seem altogether more immature than others. Given the improbability of thinking otherwise, you might be inclined to think that we have somehow gotten things wrong by intimating that exponents of contemporary theory-theories seem all too ready to equate simple false-belief understanding with the more demanding idea that knowledge is an interpretive achievement. Were this not so.

As it is, many prominent theory-theorists do in fact explicitly dismiss the possibility of *any*, let alone an "interpretive," stage in children's understanding of mind beyond straightforward false-belief understanding (Ruffman, Olson, & Astington, 1991), and explicitly claim that by the age of 4 to 5 years, children have acquired "an interpretational or constructive understanding of representations" (Wellman, 1990, p. 244), or that even preschoolers see mental contents as being constructed "actively by the person, on the basis of inference and subject to biases, misrepresentations, and active interpretation" (Wellman & Hickling, 1994, p. 1578). As Meltzoff and Gopnik (1993) put it, "By five years old, children seem to . . . understand that a person's beliefs about the world are not just recordings of objects and events stamped upon the mind, but are active interpretations or construals of them from a given perspective" (p. 335). Similarly, Perner states that "around 4 years children begin to understand knowledge as representation, with all its essential characteristics. One such characteristic is *interpretation*" (1991, p. 275, emphasis in original). Clearly, then, the suggestion that contemporary theorists of mind do in fact subscribe to the idea that, by 4 years of age, young children already hold to a fully blown interpretive theory of mind is not some attempt at character assassination, at least not on our own part.

It needs to be acknowledged, of course, that there are some grounds for the argument that when children can pass false-belief tests, appearance–reality tests, and Level 2 perspective taking measures, they have achieved some understanding that the same event or object can be represented in multiple ways, and, consequently, that children as young as age 4 do already hold to a view of the mind that Perner and Davies (1991) characterize as "an active processor of information" that is, 4-year-olds do evidently already appreciate that minds privileged with more information (as in the case of standard measures of false-belief understanding) do, in fact, end up responding differently than those who have been kept in the dark about how the ground beneath them may have shifted in their absence. Equating such fundamentally different and "ignorance based" achievements with anything like an adult understanding of the interpretive nature of knowledge is, however, a possibility only available to those already in the grip of a picture that permits no more reasonable alternative (Carpendale & Chandler, 1996; Chandler, 1988). Realizing that others may go wrong by acting out of ignorance is simply not what is ordinarily had in mind by those who speak of the possibility that different persons can and do regularly interpret one and the same thing differently.

In order to better appreciate the sharp difference that divides our own, we hope, richer and more "true to life" version of what it could possibly mean to subscribe to a genuinely "interpretive" view of the knowing process, from the contrastive accounts of those who, like Perner and Davies (1991), equate the notion of "interpretation" with the capacity to see the mind as an "active processor of information," consider for a moment two different couples, each of which exits a movie theater with very different ideas about what went on there. In the first case (obviously meant to parallel the standard false belief test, in which Maxi's belief about the current location of the chocolate differs from that of his mother's because he was out of the room when it was moved), one member of the couple goes out for popcorn and ends up missing a key scene in the film. In the second, both watch the entire movie together, but like movie critics working for competing newspapers, nevertheless come away from the theater assigning very different meanings to what they both saw. Situations of this second sort, which we think come closest to what is ordinarily meant by the notion of interpretation, would obviously be impossible to understand by anyone whose current theory of mind involved no more than the simple appreciation of ignorance required by standard measures of false-belief understanding (Carpendale & Chandler, 1996).

Finally, some mention also needs to be made of what Flavell and his colleagues (Flavell, Everett, Croft, & Flavell, 1981) have called "Level 2 perspective-taking tasks" (i.e., tasks requiring an appreciation that when an object is placed between us, what I see front-to-back, you see back-to-front). Again, the standard finding is that 4- but not 3-year-olds, appreciate that a picture of, for example, a turtle that they see standing on its feet will be viewed as lying on its back by someone seated across the table from them. Here, again, the standard interpretation is that subjects who can pass such a task already understand what Flavell has called "representational diversity" and so deserve credit as already holding to an "interpretive theory of mind." Despite the fact that 4-year-olds are able to say one thing when asked "how" they see the turtle and manage to say something marginally different when asked "how" the same drawing might be seen by someone else with a different angle of regard, it would seem generous beyond reason to credit them with anything like a bona fide "interpretive theory of mind" for having done so; that is, Flavell's particular way of going about setting up such Level 2 perspective-taking problems would appear to conflate more legitimate questions about interpretation with different and altogether simpler ones about perceptual angles of regard. At least, one could not, we take it, usefully try to get to the bottom of how various people might be understood to differently "see" issues such as capital punishment or abortion by "turning the tables" in the same sort of literal way that Flavell and his colleagues seem to have in mind.

The finding that by approximately 4 years of age, children already understand that others will set aside what they take to be bad advice or misleading information has also been taken by Perner and Davies (1991) to demonstrate that such preschoolers already appreciate that individuals with different information at their disposal will evaluate incoming information differently, and so deserve to be credited with what they also call an "interpretive" theory of mind. While all of this is quite interesting, it still does not, in our view, tell us anything about the question of when children come to recognize that minds actually interpret in the special epistemic sense of differently assimilating. It seems unlikely that anyone would seriously doubt that children as young as 4 see their own or others' minds as "active" in the broad sense of evaluating new information in the light of previous beliefs. What is at real issue, but unfortunately left unilluminated by the Perner and Davies approach to the problem, is the question of when children first come to realize that minds not only somehow "crunch," in some procedural way, the evidence already at their disposal, but also go beyond this to actually influence how experience is interpreted or construed.

The upshot of all that has just been said is that, contrary to what you may have otherwise heard, standard appearance–reality measures, Level 2 perspective-taking tasks, and tests of false-belief understanding do not actually provide information that is useful in deciding when it is that children ordinarily first come to an appreciation of the interpretive character of the knowing process.

One study sequence that does speak more directly to this question is the work done by Pillow (1991; Pillow & Weed, 1995). Briefly, he has demonstrated that 5-, 6-, and 7-year-olds, but not younger children, seem to appreciate that people's likes and dislikes will dictate how they end up viewing a range of morally and factually ambiguous events. Such findings suggest that whatever else they might already know about beliefs and desires, young children's first inklings of the idiosyncrasies of other people's mental lives may arise during this 5- to 7-year period. While pointing in the right direction, Pillow's procedures, like most standard false-belief measures, still continue to depend upon the strategy of inculcating different story characters with different background information, and so, consequently, still fail to provide a direct test of the more complex insight that the same information is often legitimately interpreted in more than one way.

In our own prior research (Carpendale & Chandler, 1996; Chandler, 1988; Chandler & Lalonde, 1996), we have (1) argued that the minimally complex experiment necessary in order to properly evaluate who does and who does not subscribe to anything that might qualify as a legitimately "interpretive" theory of mind would be one in which passing marks could be earned only by those capable of demonstrating an appreciation of the fact that different persons can and do find different meaning in one and the same stimulus event; and (2) hypothesized that false-belief understanding is necessary but not sufficient for this still later arriving ability. In order to evaluate this hypothesis, what is needed, in addition to some more or less standard measure of false-belief understanding, is some second set of different procedures specifically designed to get at children's growing realization that different people are likely to construe ambiguous stimuli in different but still legitimate ways. To date, two variations of this general measurement strategy have been employed. In one of these, Chandler and Lalonde (1996) asked 5- to 7-year olds (all of whom had already passed a relatively complex measure of false-belief understanding) to consider the real and imagined responses of two target characters to a range of stimulus fragments that, while known by the subjects to be edge details of other more complete drawings, proved to be ambiguous in much the same way as do amorphous clouds, Rorschach ink blots, and puddles of spilled milk. Across a series of procedural variations on this same theme, the exceptionless finding was that, not until they were typically 7 did children, who were already experts in false-belief understanding, also demonstrate that it is possible to misinterpret the same ambiguous stimuli in a variety of equally defensible ways.

In a related study sequence (Carpendale & Chandler, 1996), we put to use a different class of ambiguous stimuli that is marked by the fact that its members tend to offer good evidence for two, but only two, different interpretations. These stimuli included ambiguous figures (e.g., Jastrow's [1900] famous "duck–rabbit"), or ambiguous messages (e.g., sentences that included homophones and other ambiguous referential communications). Five-year-old subjects who easily passed a standard false-belief test were convinced that only one of the interpretations of the ambiguous stimuli offered by puppet characters could possibly be legitimate, and alternative interpretations must, therefore, be mistaken. Although a smattering of 6-year-olds occasionally passed some of these procedures, more typically, only children of 7 or 8 years showed any real willingness to acknowledge the possibility that one and the same stimuli can afford more than one legitimate interpretation.

By focusing attention so pointedly on what would appear to be children's earliest transition to an interpretive theory of mind at about the age of 7 or 8, we do not mean to imply that this is the end of the story. Instead, there are good reasons to suppose that what young school-age children know about interpretation is only a first chapter in a continuing developmental course that has already been demonstrated to extend at least through adolescence and perhaps into early adulthood as well (e.g.,Chandler, 1987; Kuhn, Amsel, & O'Loughlin, 1988; Kuhn, Pennington, & Leadbeater, 1983; Perry, 1970). At least one such later-arriving transition point would appear to occur when adolescents first show signs of understanding that no automatic "stop rule" can be fitted to their earlier particularized or "retail" doubts, leaving each and every claim to knowledge open to a new and more abstract form of "wholesale" uncertainty in which nothing remains sacred or safe from the vertigo of relativism and paralyzing effects of skeptical doubt (Chandler, 1987). In a similar vein, while a beginning grasp of the interpretive character of the knowing process may also be a necessary first step toward the eventual insight that selves reveal themselves in the commonalities that cut across their various interpretive efforts, it seems likely that this fledgling insight will demonstrate itself to be only a necessary but by no means sufficient condition for the emergence of a working ability to enlist the patterns evident in one's own constructive efforts as an effective tool in the process of achieving self-awareness.

Whatever the final answer might prove to be concerning precisely when in the course of their development people first begin and finally end the process of coming to a mature conception of mental life, what now seems already clear enough is that there are no correct conceptual or empirical reasons for supposing that all or even most of what is really interesting about this process just happens to fall within a province exclusively owned by 40- to 60-month olds. As we have meant to show, the evident tendency to think otherwise is traceable, in important measure, to a mistaken tendency to overestimate the importance of false-belief understanding and is responsible, in large part, for the lamentably ahistorical and nondevelopmental character of the current theory of mind enterprise.

DOING WITHOUT CAUSES AND CONSEQUENCES

In addition to doing without history or development, the theory-of-mind literature has also proven itself to be remarkably disinterested in the causes and consequences of acquiring an understanding of what, for the most part, amounts to a grasp of the possibility of false belief. If acquiring an understanding of the fact that people can be led astray by their own ignorance actually deserves all of the good press that it has enjoyed, then what, you might well ask, are its causal antecedents, and how does the eventual accomplishment of this ability actually impact on the lives of those who are already so blessed? The unwelcome answers to both of these naturally arising questions is that we really do not know (which is perhaps forgivable), and that almost no one has bothered trying to find out (which is not). The contrast between this essentially know-nothing approach and the rather better track record built up during the ancestral period in which earlier generations studied childhood egocentrism and social role-taking competence is both well marked and deserving of some better understanding.

As is reflected in the usual pattern of their research interest, the now largely hidden cohort of investigators, who once filled out the ranks of an earlier generation of investigators similarly concerned with the study of children's developing conceptions of mental life, was, more often than not, made up of retooled child clinical psychologists who were rather quick to ask questions about the living arrangement conducive to and human consequences of acquiring this or that so-called social-cognitive ability. In clear contrast, most contemporary theory-theorists are pure cognitive types who cut their professional teeth trying to masticate the works of more recent "cognitive scientists" such as Fodor or Churchland, and for whom the study of familial antecedents and practical consequences of false-belief understanding seems other than a first priority. As a partial consequence of these differing lines of descent, we currently know next to nothing about the antecedents and consequences of coming to a first theory of mind. There are, of course, occasional exceptions to this general rule, and, because they exist in such short supply, they deserve our special attention. In taking up this embarrassingly small literature, it will prove useful to consider in order those few programs of research aimed at uncovering first the "mechanisms" or causal antecedents, and then the few known practical consequences of first coming to a fledgling theory of mind.

Origins of Theories of Mind

The empirical study of possible causes or origins of theories of mind is only recently beginning to attract some of the attention it deserves (see, e.g., the recent edited book by Lewis & Mitchell, 1994). There has been some theoretical interest, of course, in such possible causal chains, most of which, it turns out, has centered on the drawing of certain loose analogies between the "theories" being produced by young children and those generated by scientists themselves. In this literature, children are usually said to be like little scientists in that they too collect data and somehow form theories. In both cases, the actual process of theory change is said to be motivated by the accumulation of counterevidence. How, more precisely, this is meant to occur would appear to be as much of a mystery for cognitive developmentalists as it is for philosophers of science. About all we seem free to conclude in either instance is that more "data" are better than less.

Empirical contributions to our understanding of the causal course of theory construction or change are in even shorter supply and are largely confined to a handful of studies meant to trace out the possible contribution of simply having more rather than less social experience. Some of these (e.g., Dunn, 1994; Dunn, Brown, Slomkowski, Tesla, & Youngblade, 1991) have been largely observational and have shown, for example, that young children who were noted to have more rather than less conversations with their mothers about feelings and beliefs were the same subjects who, 7 months later, demonstrated some better understanding of false beliefs. Other of these investigations have involved efforts to backtrack on subjects who performed differently on standard measures of false-belief understanding in an effort to search out demographic characteristics that might distinguish those who succeeded either early or late on these laboratory procedures. One such recent study (Perner, Ruffman, & Leekam, 1994) turned, for example, upon the rather crude measure of simply counting the number of siblings "owned" by subjects who happened to have done either well or badly on standard tests of false belief. Again, what we learn from this experiment is largely that more rather than less social experience tends to be a good thing. Subsequently, Jenkins and Astington (1996) report having replicated Perner et al.'s original finding, but also go on to state that this effect for family size was greater for preschoolers who were less linguistically competent. A further step removed from the initial level of simplicity is the more recent effort of Lewis, Freeman, Kyriakidou, Maridaki-Kassotaki, and Berridge (1996), who have attempted to work out the possible consequences of interacting with older (and so, presumably, more socially experienced) rather than younger social interactional partners. In this context, it was found that interactions with older siblings and adults were more consistently related to children's false-belief understanding than interactions involving individuals still younger than one's self.

Although necessary first steps in the direction of eventually finding a way to the causes or mechanisms responsible for accomplishing some fledgling theory of mind, what is perhaps most notable about this small handful of studies is just how small a collection they end up making. Hardly what one might reasonably expect, you say with some professional embarrassment, of a research enterprise already large enough to have eclipsed much else on the current cognitive-developmental scene. Of course, there is not much point in overly agonizing about still hidden causes or mechanisms if the real truth of the matter being hoped for is that there must exist, lurking behind the scene, some factory-built theory-of-mind module that, in all but the most catastrophic of situations, manages to turn itself on in its own sweet time. Neonativist ambitions of this caliber appear to exist in sufficient number to help account for the fact that there is no evident stampede to quickly close the gap in our understanding about causes and mechanisms.

Doing without Consequences

A key reason for the popularity of the theory-of-mind enterprise is the shared expectation that the acquisition of a working understanding of mental life is not only an essential aspect of normal social development but also that any developmental failures in this quarter will prove diagnostic of various forms of psychopathology. If the theory-of-mind enterprise proves itself to be on the right track, then there is every reason to expect that evidence regarding these matters will be forthcoming. Although these are perhaps still "early days," both of these hopes have, at least to date, gone largely unrequited. With particular reference to the expectation that the acquisition of an increasingly sophisticated understanding of the nature of mind should somehow underpin an increasingly sophisticated understanding of social life, there is little to report beyond the already-cited studies of Dunn and her colleagues suggesting a correlation between children's development of a more mature understanding of mind and more successful social functioning. Work coming out of our own laboratory (Lalonde & Chandler, 1995) offers what may be the only remaining published evidence concerning the ordinary consequences of false-belief understanding for children's social life. In this short study, 3-year-olds who were quick to achieve false-belief understanding, relative to their age-mates, were also rated by their preschool teachers as more socially competent in areas thought to require an understanding of others' mental states.

The broad contrapuntal intuition that any delays or failures in the process of coming to a mature understanding of mental life will naturally eventuate in serious problems in social adaptation is similarly underresearched. The single exception to this generalization is to be found within a thriving, if controversial, program of research (e.g., Baron-Cohen, Leslie, & Frith, 1985; Baron-Cohen, Tager-Flusberg, & Cohen, 1993) that has demonstrated that autistic (but not mentally retarded) children may entirely lack the capacity to grasp the possibility of false belief. As Hobson (1993) has pointed out, however, these failures on the part of children with autism may well prove to be more the consequence than the cause of these children's profound social deficits. Whatever proves to be the fact of the matter there is no room for doubt that a brace of correlational studies demonstrating special social competence in those especially quick off the theories-of-mind mark, or a narrow gauge sequence of studies linking autism to failures in false-belief understanding constitute a rather disappointing showing that falls importantly short of a competent demonstration that all of the energy and expense so far invested in the theory-of-mind enterprise is compensated for by the evident practical relevance of such efforts.

DOING WITHOUT THE BETTER HALF OF BELIEF

Theory-theory's singular preoccupation with false-belief understanding has served not only to seal off the field from its own history, to cost it its status as a legitimate developmental enterprise, to blunt its interest in the study of expected antecedents and consequences, and to impose unnatural limitations on the age groups deemed worthy of study, but has also, we now mean to show, artificially narrowed the focus of inquiry to that restricted subset of beliefs that just happen to be truth conditional. By a certain widely practiced way of reasoning, this can be seen to amount to essentially no limitation at all, for the reason that, by such lights, beliefs are just seen to be the sorts of things that are automatically, and as a matter of absolute necessity, either true or false. This, without a doubt, is "true" of at least some of the things we call beliefs. The moon is or is not made of green cheese, standing in a draft will or will not cause you to catch your death of cold, and so on, proposition after proposition, world without end. A whole philosophical tradition has, in fact, grown up to defend the idea that there are only facts that are true or false and values that are neither, and that the mental attitudes we maintain about these neatly divide into beliefs in the first case and matters of taste or opinion or value in the second. The earlier quotation from Davidson (the one according to which "error" was said to be "what gives belief its point" [1984, p. 168]) gives voice to a part of this broadly shared framework of understanding, as does, for example, the whole tradition of carving up academia into the "natural" and "human" sciences. The alternative possibility to be pursued here, however, is that there are more kinds of belief under the sun than are considered within any such neatly bifurcated philosophy.

As a way of potentially getting in the mood to hear more about such an inherently untidy possibility, think for a moment about what you and others *know* you actually *believe*. Some believe *in* "the Holy Christian church, the communion of saints, the forgiveness of sins and life everlasting." Others believe *that* capital punishment or abortion or taxation without representation is wrong. Still others believe that red wine has more character, that Picasso is overrated, and that sometimes a cigar is just a cigar. And, yes, we do also sometimes believe in our own fair share of simple, truth-conditional propositions. Consideration of such wide-ranging possibilities, all of which would seem to deserve some legitimate place within the ordinary language conception of belief, has prompted a small number of theorists of mind to side with common sense in hypothesizing the existence of a broad continuum bracketed on one end by straightforward factual beliefs and on the other by undefended opinions about individualized matters of personal preference.

Although such views bear some repeating, it is important, before taking up the space required to do so, to try first to get really clear about why it might actually matter one way or another whether all beliefs do or do not actually fit neatly into piles labeled "true" or "false," with everything else left strewn about the floor of arbitrary opinion. One such available reason already alluded to is that arbitrarily dichotomizing the world of mental attitudes into true or false beliefs, on the one hand, and matters of arbitrary taste, on the other, does serious violence to the way that ordinary people ordinarily think about the propositional contents of their minds. Beyond whatever prompting might be contained in this bit of phenomenology, there is, in addition, the embarrassing fact that the theories-of-mind literature continues to narrowly focus almost exclusively on easily settled questions of the sort having to do with whether Maxi's chocolate is in container *A* or *B*, or whether the Smarties tube actually contains candies or pencils, leaving effectively untouched all those more heartfelt matters involving beliefs about, for example, what is right and what is wrong. Consequently, the overlooked opportunity presents itself that, if we could somehow see our way clear also to begin investigating children's beliefs about beliefs of this currently excluded caliber, then the theories-of-mind literature might find itself newly opened up to all those axiological issues currently walled off as being about moral reasoning or value-laden beliefs about ourselves or other nonepistemic considerations traditionally assumed to be without any decidable fact of the matter. On the strength of such expansive possibilities, it seems well worth the effort to pause to consider some of the claims of those who set themselves in opposition to the more standardized dichotomy between fact-bearing beliefs and unwarrantable opinion.

Actually, the woods are rather full of recent philosophers of record who find no pleasure at all in attempts to divide things up into beliefs that are truth bearing and those remaining mental attitudes that are presumably not. Putnam (1987), for example, among others (e.g., Hanson, 1958; Kuhn, 1970; Rorty, 1979) has argued convincingly against this classical fact–value dichotomy, suggesting, among other things, that, with the demise of logical empiricism, any sharp division between certain or foundational "facts," and subjective and controversial "values" needs to be rejected in favor of a view that seeks to arrange various forms of knowledge along a rough continuum from facts to values.

John Searle (1969) has added importantly to this more contemporary view by stressing the distinction between what he refers to as "institutional" as opposed to merely "brute" facts. As he puts it:

> There are many kinds of facts which obviously are objective facts and not matters of opinion or sentiment or emotion at all, which are hard, if not impossible to assimilate to [the classical fact–value dichotomy]. . . . Any newspaper records facts of the following sort: Mr. Smith married Miss Jones; the Dodgers beat the Giants three to two in eleven innings; Green was convicted of larceny; and Congress passed the Appropriations Bill. There is certainly no easy way that the classical picture can account for facts such as these. That is, there is no simple set of statements about physical or psychological properties of states of affairs to which the statements of facts such as these are reducible. (pp. 50–51)

Rather, he insists, the existence of such social or institutional facts depends on the regular relations that exist between the people that constitute such institutions. Armed with such an account of social and institutional facts, it becomes potentially more understandable why some of the things once written off as mere values or opinions may actually have enough of a factual nature to begin bracketing them with beliefs about brute fact rather than throwing them into the bone pile of mere opinion.

For the moment, there exists only the smallest handful of three or four empirical studies (one of which is our own) that have self-consciously taken up the task of trying to sample different kinds of beliefs occurring at more than a single point along this proposed fact–value continuum. In one of these, Flavell, Flavell, Green, and Moses (1990) undertook to explore the prospect that preschoolers understand the possibility of "representational diversity" in the case of so-called "value beliefs" (e.g., cat food tastes "yummy" to cats, but "yucky" to them) a number of months before they were willing to countenance the possibility of such divergent views about more standard "fact beliefs." If there is a problem to be found with this promising first effort, it is that the particular sorts of "value beliefs" that Flavell and his colleagues undertook to examine would appear to lie so near the traditional "value" end of their proposed "fact to value dimension" that there was little room for supporting such propositions with any of the "good reasons" that Searle claims back those more social of "institutional" facts that legitimately vie for status as things about which there is some arguable fact of the matter. As it is, any personalized matter of individual taste (e.g., "I *like* chocolate better than vanilla") can be given a semantical twist to make it appear more "belief-like" (e.g., "I *believe that* chocolate is better than vanilla"), but doing so scarcely comes to grips with the serious problem of value-impregnated belief.

In a second and related series of studies coming out of our own laboratory (Carpendale, 1995; Carpendale & Chandler, 1996), we also took up this same problem, eventually demonstrating that some years before young school-age children first begin entertaining the possibility that people are entitled to a diversity of beliefs about single matters of fact, their younger preschool counterparts are already practicing the more democratic view that different persons are inclined to like or value different things, all for the reasons that are seen to have more to do with their own makeup than anything "objectively" true about states of the world. While these research efforts may make some contribution to the task of eventually working out where pure facts leave off and unadulterated values take over, they, like the findings of Flavell and his colleagues, cannot be easily counted as having seriously taken up the challenge of broadly sampling from the little understood continuum that runs between brute facts and personalized values.

Finally, Bradford Pillow (1991; Pillow & Weed, 1995) has also undertaken work that is meant to explore the ways in which people's likes and dislikes influence how they end up viewing a range of morally and factually ambiguous events. These studies, while useful in demonstrating that 5-, 6-, and 7-year-olds (but not still younger children) understand something of how values shape our beliefs, are, nevertheless, like our own and Flavell's related efforts, insufficient in that they still leave largely untouched the broad question of how young children's developing beliefs about beliefs might unfold across the broad reaches of the continuum linking beliefs and values.

What seems missing from the studies just cited, and what would appear to be required in order to more fully take up the task we are stumping for, is some better way of thinking about where frank matters of value leave off and where some more legitimate range of belief-like propositions might reasonably begin. Without access to some such conceptual tools, future research, like that already reported, will end up merely circling but never really coming to terms with that class of beliefs that is neither truth conditional, nor a matter of undefended personal preference.

The closest thing we have come upon that promises to offer any real assistance in solving this interpretive problem is provided by Jürgen Habermas (1981/1984), who insists that the broad category of beliefs is not, after all, defined by the fact that its members are either true or false, but rather whether they can be legitimately held out as *valid* as opposed to *invalid*. In his theory of communicative action, Habermas makes the point that every time someone makes a statement that is not clearly marked as a simple pronouncement of personal preference, there is an implicit expectation that, if challenged or questioned, he or she could potentially back or warrant such a claim by pointing to appropriate evidence or some chain of good reasons for assuming its validity. In this context, Habermas distinguishes what he calls "cognitive instrumental rationality" from so-called "communicative rationality." In the case of instrumental rationality, the relation of first importance is that which holds between the subject and some object in the material environment, and the validity claims that are asked for and offered tend to involve pointing to various lines of *evidence* that support the truthfulness of the claim in question. As Habermas points out, however, "We do not use the term "rational" solely in connection with expressions that can be true or false" (p. 10). Rather, within the sphere of actions that feature claims about the shared social world, the "knowledge embodied in [such] normatively regulated actions . . . does not refer to the existence of states of affairs but to the validity of norms" (p. 16). In such socially regulated contexts, narrow matters of truth or falsity are often not at issue at all, and, instead, "the agent makes the claim that his behavior is *right* in relation to a normative context recognized as legitimate" (p. 15).

The whole point of rehearsing this much of Habermas's broad claim that what makes a thing a belief is not limited to whether it is true or false, but whether it is, in his wider sense, "valid," is that the broader horizon of this construction effectively works also to make room for that large class of strongly held beliefs that have more to do with what is right or wrong, and not merely whether it is truth conditional. If people tells you that they believe that capital punishment or abortion or taxation without representation is *wrong*, it is rarely the case that they mean to be understood as commenting on what they believe to be an arbitrary matter of personal taste. Rather, such normative beliefs are generally understood by those who subscribe to them as valid in the sense that they judged such issues to be open to being rationally backed by some chain of good reasons that are meant to warrant these convictions as *right*. Any candidate account of children's developing beliefs about beliefs that makes no room for beliefs of this sort needs to be seen as only a partial and incomplete account.

One obvious point of contact between these necessarily abstract matters of truth and validity, and the more concrete world of research into children's social cognitive development is suggested by the evident similarity between Habermas's distinction between warrantable validity claims about rightness as opposed to undefended claims about personal likes and dislikes, on the one hand, and the homologous distinction drawn by Turiel and others interested in moral reasoning (e.g., Nucci & Turiel, 1978; Turiel, 1983) between supposedly universal moral norms and nonuniversal cultural values on the other. If by trading on Habermas's interpretive strategy for broadening the notion of belief, it proves possible to begin a process that would allow the theories-of-mind literature to expand into the still-separate arena of moral development, then an important step will have been taken toward tearing down some of the walls that have left theory—theory strangely isolated from the rest of the developmental world. It also seems likely that one's beliefs about matters other than factual states of the world may prove to be especially rich resources in our efforts to approach the problem of self-understanding through an examination of those patterns of interpretive commonality that cut across our assembled beliefs.

SUMMARY AND CONCLUSION

The same general point, now made in a variety of different ways throughout this chapter, is that the theories-of-mind enterprise suffers under a series of self-imposed limitations that have effectively cut it off from its own history, rendered it more a champion of the permanent than an advocate of development, restricted its focus to a narrow bandwidth of age-graded change, cost it any measured concern with the likely causes and natural consequences of cognitive growth, and so truncated the meaning of belief that much of what is most promising about children's changing conceptions of mental life ends up being left lying on the cutting-room floor. Having read this much, you are perhaps wondering all over again why an undertaking filled with so much raw promise has managed to deliver so little. Some part of an answer to this question is provided, we suggest, by the fact that, while bent upon exposing changes in the various epistemic stances of their own child subjects, the epistemological orientation of most theory-theorists themselves has continued to be both opaque and often poorly differentiated from the claims being made on behalf of young persons. Perner (1991), for example, is very explicit in his description of the representational theory of mind that he means to attribute to 4-year-olds—a view that openly relies on a brand of naive or metaphysical realism that is completely unapologetic about its frankly objectivist assumptions. Perhaps because such views are so regularly promoted as "mature" or adult-like, it never becomes quite clear, however, whether Perner believes that, as psychologists, we too should join young children in such views. Bickhard (1992a), at least, argues that the problematic view of knowledge as "encoding" that Perner attributes to 4-year-old children is fundamentally the same as Perner's own, leaving both child and theorist so much the worse for wear. Although perhaps more explicit, Perner would seem not to be alone in his allegiance to what amounts to an untempered correspondence theory of truth that equates knowledge with having mental representations that "match" reality *as it really is* (Chapman, 1991/in press). At least, easy talk about beliefs being "false" because they fail to correspond with "ob-

184 DEVELOPMENT OF SELF-AWARENESS ACROSS THE LIFESPAN

jective" matters of fact, or "appearances" that stray from "reality," lends itself to being understood as expressive of such a realist view.

Although this is not the time or place to imagine arbitrating a contest about the merits or demerits of such broadly philosophical matters, it is perhaps fair to say that there is a building consensus to the effect that such objectivist theories end up presupposing precisely what they are meant to explain (e.g., Bickhard, 1992b; Lakoff, 1987; Putnam, 1988; von Glasersfeld, 1979), and, as a result, have "virtually disappeared" (Overton, 1994) from more contemporary philosophical accounts.

Whatever other good or bad things one might otherwise find to say about such objectivist views, what is, for present purposes, perhaps their most unfortunate feature is that they offer very little in the way of toeholds where development might gain any real purchase. Within such accounts, interpretation, for example, reduces to misinterpretation, and reality, once seen for what it really is, can hardly be improved upon by access to more and more powerful "forms" of knowing. Little wonder, then, that most theorists of mind have tended to simply discount as uninteresting the failed efforts of anyone younger than age 4 to get things exactly right and have found it hard to imagine how anyone who is already able to see reality with clear eyes could be expected to improve upon a picture-perfect match between the world and its unencumbered representation. For anyone in the grip of such a picture, all those fundamental abilities necessary for accurate self-awareness are seen to be well in place before the end of the preschool years. By contrast, anyone persuaded by the contents of this chapter will have come to see children's growing conceptions of mental life as having much longer legs, capable of carrying them well into adolescence and early adulthood.

If, as we argue, false-belief understanding is only an early way station en route toward an increasingly interpretive view of mental life, and if an essential part of this developmental course ordinarily involves a growing appreciation of the constructive ways in which minds routinely and characteristically shape the subjective nature of their own experience, then this process of epistemic development clearly prepares the way for a whole new class of insight concerning one's self as a knower.

ACKNOWLEDGMENTS

The writing of this chapter was supported by a Natural Sciences and Engineering Research Council of Canada grant to Michael J. Chandler and by a Social Sciences and Humanities Research Council of Canada Post Doctoral Fellowship to Jeremy I. M. Carpendale. We would like to thank Chris Lalonde, Charlie Lewis, Brian Sokol, and the editors of this volume for their helpful comments on earlier drafts of this chapter.

NOTE

1. For an almost singular exception to this broad generalization, see Flavell's (1992) chapter written as part of a Symposium of the Jean Piaget Society, in which he argues strongly in favor of the importance of seeing continuities between the egocentrism, social role taking, metacognition, and theories-of-mind literatures.

REFERENCES

Baldwin, D. A., & Moses, L. J. (1994). Early understanding of referential intent and attentional focus: Evidence from language and emotion. In C. Lewis & P. Mitchell (Eds.), *Children's early understanding of mind: Origins and development* (pp. 133–156). Hove, England: Erlbaum.

Baldwin, J. D. (1988). Learning to deceive. *Behavioral and Brain Sciences, 11,* 245–246.

Baron-Cohen, S., Leslie, A. M., & Frith, U. (1985). Does the autistic child have a "theory of mind"? *Cognition, 21,* 37–46.

Baron-Cohen, S., Tager-Flusberg, H., & Cohen, D. J. (Eds.). (1993). *Understanding other minds: Perspectives from autism.* New York: Oxford University Press.

Beilin, H. (1992). Piaget's enduring contributions to developmental psychology. *Developmental Psychology, 28,* 191–204.

Bennett, J. (1978). Some remarks about concepts. *Behavioral and Brain Sciences, 4,* 557–560.

Bickhard, M. H. (1992a). Commentary. *Human Development, 35,* 182–192.

Bickhard, M. H. (1992b). How does the environment affect the person? In L. T. Winegar & J. Valsiner (Eds.), *Children's development within social contexts: Vol. 1. Metatheory and theory* (pp. 63–92). Hillsdale, NJ: Erlbaum.

Blackburn, S. (1991). Losing your mind: Physics, identity, and folk burglar prevention. In J. D Greenwood (Ed.), *The future of folk psychology* (pp. 196–225). Cambridge, England: Cambridge University Press.

Carpendale, J. I. M. (1995). *On the distinction between false belief understanding and the acquisition of an interpretive theory of mind.* Unpublished doctoral dissertation, University of British Columbia, Vancouver, Canada.

Carpendale, J. I. M., & Chandler, M. J. (1996). On the distinction between false belief understanding and subscribing to an interpretive theory of mind. *Child Development, 67,* 1686–1706.

Chandler, M. J. (1987). The Othello effect: Essay on the emergence and eclipse of skeptical doubt. *Human Development, 30,* 137–159.

Chandler, M. J. (1988). Doubt and developing theories of mind. In J. W. Astington, P. L. Harris, & D. R. Olson (Eds.), *Developing theories of mind* (pp. 387–413). New York: Cambridge University Press.

Chandler, M. J., & Boyes, M. (1982). Social cognitive development. In B. B. Wolman (Ed.), *Handbook of developmental psychology* (pp. 387–402). Englewood Cliffs, NJ: Prentice Hall.

Chandler, M. J., & Lalonde, C. (1996). Shifting to an interpretive theory of mind:

5- to 7-year-olds' changing conceptions of mental life. In A. Sameroff & M. Haith (Eds.), *Reason and responsibility: The passage through childhood* (pp. 111–139). Chicago: University of Chicago Press.

Chapman, M. (1988). *Constructive evolution: Origins and development of Piaget's thought*. New York: Cambridge University Press.

Chapman, M. (1989). Understanding a primitive mentality. *Linguistics, 27,* 1152–1156.

Chapman, M. (in press). Constructivism and the problem of reality. In I. Sigel (Ed.), *The development of representation thought: Theoretical perspectives.* Mahwah, NJ: Erlbaum. (Originally presented as a talk in 1991)

Churchland, P. M. (1988). *Matter and consciousness* (Rev. ed.). Cambridge, MA: MIT Press. (Original edition published 1984)

Clements, W. A., & Perner, J. (1994). Implicit understanding of belief. *Cognitive Development, 9,* 377–395.

Davidson, D. (1984). Thought and talk. In D. Davidson (Ed.), *Inquiries into truth and interpretation* (pp. 155–170). Oxford, England: Oxford University Press.

Dennett, D. C. (1978). Beliefs about beliefs. *Behavioral and Brain Sciences, 4,* 568–570.

Dunn, J. (1994). Changing minds and changing relationships. In C. Lewis & P. Mitchell (Eds.), *Children's early understanding of mind: Origins and development* (pp. 297–310). Hove, England: Erlbaum.

Dunn, J., Brown, J., Slomkowski, C., Tesla, C., & Youngblade, L. (1991). Young children's understanding of other people's feelings and beliefs: Individual differences and their antecedents. *Child Development, 62,* 1352–1366.

Eagle, M. (1982). Privileged access and the status of self-knowledge in Cartesian and Freudian conceptions of the mental. *Philosophy of the Social Sciences, 12,* 349–373.

Elkind, D. (1967). Egocentrism in adolescence. *Child Development, 38,* 1025–1034.

Feldman, C. F. (1992). The new theory of theory of mind. *Human Development, 35,* 107–117.

Flavell, J. H. (1988). The development of children's knowledge about the mind: From cognitive connections to mental representations. In J. W. Astington, P. L. Harris, & D. R. Olson (Eds.), *Developing theories of mind* (pp. 244–267). New York: Cambridge University Press.

Flavell, J. H. (1992). Perspectives on perspective taking. In H. Beilin & P. B. Pufall (Eds.), *Piaget's theory: Prospects and possibilities* (pp. 107–139). Hillsdale, NJ: Erlbaum.

Flavell, J. H., Everett, B. A., Croft, K., & Flavell, E. R. (1981). Young children's knowledge about visual perception: Further evidence for the Level 1–Level 2 distinction. *Developmental Psychology, 17,* 99–103.

Flavell, J. H., Flavell, E. R., Green, F. L., & Moses, L. J. (1990). Young children's understanding of fact beliefs versus value beliefs. *Child Development, 61,* 915–928.

Flavell, J. H., Green, F. L., & Flavell, E. R. (1995). Young children's knowledge about thinking. *Monographs of the Society for Research in Child Development, 60*(1, Serial No. 243).

Garcia, R. (1987). Sociology of science and sociogenesis of knowledge. In B. Inhelder, D. de Caprona, & A. Cornu-Wells (Eds.), *Piaget today* (pp. 127–140). Hove, England: Erlbaum.

Glasersfeld, E. von (1979). Radical constructivism and Piaget's concept of knowledge. In F. B. Murray (Ed.), *The impact of Piagetian theory: On education, philosophy, psychiatry, and psychology* (pp. 109–122). Baltimore: University Park Press.

Gopnik, A. (1993). How we know our minds: The illusion of first-person knowledge of intentionality. *Behavioral and Brain Sciences, 16,* 1–14.

Habermas, J. (1984). *The theory of communicative action: Vol. 1. Reason and the rationalization of society.* Boston: Beacon Press. (Original work published 1981)

Hanson, N. R. (1958). *Patterns of discovery.* London and New York: Cambridge University Press.

Harman, G. (1978). Studying the chimpanzee's theory of mind. *Behavioral and Brain Sciences, 4,* 576–577.

Hobson, P. (1993). Understanding persons: The role of affect. In S. Baron-Cohen, H. Tager-Flusberg, & D. J. Cohen (Eds.), *Understanding other minds: Perspectives from autism* (pp. 204–227). New York: Oxford University Press.

Jastrow, J. (1900). *Fact and fable in psychology.* Boston: Houghton Mifflin.

Jenkins, J. M., & Astington, J. W. (1996). Cognitive factors and family structure associated with theory of mind development in young children. *Developmental Psychology, 32,* 70–78.

Kuhn, D., Amsel, E., & O'Loughlin, M. (1988). *The development of scientific thinking skills.* San Diego: Academic Press.

Kuhn, D., Pennington, N., & Leadbeater, B. (1983). Adult thinking in developmental perspective. In P. Baltes & O. Brim Jr. (Eds.), *Life-span development and behavior* (Vol. 5, pp. 157–195). New York: Academic Press.

Kuhn, T. (1970). *The structure of scientific revolutions* (2nd ed.). Chicago: University of Chicago Press.

Lakoff, G. (1987). *Women, fire, and dangerous things: What categories reveal about the mind.* Chicago: University of Chicago Press.

Lalonde, C. E., & Chandler, M. J. (1995). False belief understanding goes to school: On the social–emotional consequences of coming early or late to a first theory of mind. *Cognition and Emotion, 9,* 167–185.

Leslie, A. M. (1987). Pretense and representation: The origins of "theory of mind." *Psychological Review, 94,* 412–426.

Lewis, C., Freeman, N. H., Kyriakidou, C., Maridaki-Kassotaki, K., & Berridge, D. (1996). Social influences on false belief access: Specific sibling influence or general apprenticeship? *Child Development, 67,* 2930–2947.

Lewis, C., & Mitchell, P. (Eds.). (1994). *Children's early understanding of mind: Origins and development.* Hove, England: Erlbaum.

McDonough, R. (1991). A culturalist account of folk psychology. In J. D Greenwood (Ed.), *The future of folk psychology* (pp. 263–288). Cambridge, England: Cambridge University Press.

Meltzoff, A., & Gopnik, A. (1993). The role of imitation in understanding persons and developing a theory of mind. In S. Baron-Cohen, H. Tager-Flusberg, & D. J. Cohen (Eds.), *Understanding other minds: Perspectives from autism* (pp. 335–366). Oxford, England: Oxford University Press.

Moses, L. J., & Chandler, M. J. (1992). Traveler's guide to children's theories of mind. *Psychological Inquiry, 3*, 286–301.

Murray, F. B. (1979). Preface. In F. B. Murray (Ed.), *The impact of Piagetian theory: On education, philosophy, psychiatry, and psychology* (pp. ix–xiii). Baltimore: University Park Press.

Nucci, L., & Turiel, E. (1978). Social interactions and the development of social concepts in preschool children. *Child Development, 49*, 400–407.

Overton, W. F. (1994). The arrow of time and cycles of time: Concepts of change, cognition, and embodiment. *Psychological Inquiry, 5*, 215–237.

Perner, J. (1991). *Understanding the representational mind*. Cambridge, MA: MIT Press.

Perner, J., & Astington, J. W. (1992). The child's understanding of mental representation. In H. Beilin & P. B. Pufall (Eds.), *Piaget's theory: Prospects and possibilities* (pp. 141–160). Hillsdale, NJ: Erlbaum.

Perner, J., Baker, S., & Hutton, D. (1994). Prelief: The conceptual origins of belief and pretense. In C. Lewis & P. Mitchell (Eds.), *Children's early understanding of mind: Origins and development* (pp. 261–286). Hove, England: Erlbaum.

Perner, J., & Davies, G. (1991). Understanding the mind as an active information processor: Do young children have a "copy theory of mind"? *Cognition, 39*, 51–69.

Perner, J., Ruffman, T., & Leekam, S. R., (1994). Theory of mind is contagious: You catch it from your sibs. *Child Development, 65*, 1228–1238.

Perner, J., & Wimmer, H. (1985). "John thinks that Mary thinks that . . .": Attribution of second-order beliefs by 5- to 10-year-old children. *Journal of Experimental Child Psychology, 39*, 437–471.

Perry, W. G. (1970). *Forms of intellectual and ethical development in the college years*. New York: Holt, Rinehart & Winston.

Piaget, J. (1955). *The language and thought of the child*. New York: Meridian Books. (Original work published 1923)

Piaget, J., & Inhelder, B. (1956). *The child's conception of space*. London: Routledge & Kegan Paul. (Original work published 1948)

Piaget, J. (1962). The relation of affectivity to intelligence in the mental development of the child. *Bulletin of the Menninger Clinic, 26*, 129–137.

Piaget, J. (1965). *The moral judgment of the child*. New York: Free Press. (Original work published 1932)

Piaget, J. (1967). *Six psychological studies*. New York: Vintage. (Original work published 1964)

Piaget, J. (1972). *Insights and illusions of philosophy*. London: Routledge & Kegan Paul. (Original work published 1965)

Piaget, J. (1976). *The grasp of consciousness*. Cambridge, MA: Harvard University Press. (Original work published 1974)

Piaget, J. (1995). *Sociological studies*. New York: Routledge. (Original work published 1965)

Pillow, B. H. (1991). Children's understanding of biased social cognition. *Developmental Psychology, 27,* 539–551.

Pillow, B. H., & Weed, S. T. (1995). Children's understanding of biased interpretation: Generality and limitations. *British Journal of Developmental Psychology, 13,* 347–366.

Plato. (1990). *Theaetetus* (Trans. M. J. Levett). Indianapolis, IN: Hackett.

Povinelli, D. (1996). Chimpanzee theory of mind? The long road to strong inference. In P. Carruthers & P. K. Smith (Eds.), *Theories of theories of mind* (pp. 293–329). Cambridge, England: Cambridge University Press.

Premack, D. (1988). "Does the chimpanzee have a theory of mind?" revisited. In R. W. Byrne & A. Whiten (Eds.), *Machiavellian intelligence: Social expertise and the evolution of intellect* (pp. 160–179). Oxford, England: Oxford University Press.

Premack, D., & Woodruff, G. (1978). Does the chimpanzee have a theory of mind? *Behavioral and Brain Sciences, 4,* 515–526.

Putnam, H. (1987). *The many faces of realism.* LaSalle, IL: Open Court.

Putnam, H. (1988). *Representation and reality.* Cambridge, MA: MIT Press.

Rorty, R. (1979). *Philosophy and the mirror of nature.* Princeton, NJ: Princeton University Press.

Ruffman, T., Olson, D. R., & Astington, J. W. (1991). Children's understanding of visual ambiguity. *British Journal of Developmental Psychology, 9,* 89–102.

Russell, J. (1992). The theory theory: So good they named it twice? *Cognitive Development, 7,* 485–519.

Searle, J. R. (1969). *Speech acts: An essay in the philosophy of language.* London: Cambridge University Press.

Searle, J. R. (1983). *Intentionality: An essay in the philosophy of mind.* Cambridge, England: Cambridge University Press.

Selman, R. L. (1980). *The growth of interpersonal understanding: Developmental and clinical analyses.* New York: Academic Press.

Sullivan, K., Zaitchik, D., & Tager-Flusberg, H. (1994). Preschoolers can attribute second-order beliefs. *Developmental Psychology, 30,* 395–402.

Turiel, E. (1983). *The development of social knowledge: Morality and convention.* New York: Cambridge University Press.

Wellman, H. M. (1990). *The child's theory of mind.* Cambridge, MA: MIT Press.

Wellman, H. M., & Hickling, A. K. (1994). The mind's "I": Children's conceptions of the mind as an active agent. *Child Development, 65,* 1564–1580.

Wellman, H. M., & Woolley, J. D. (1990). From simple desires to ordinary beliefs: The early development of everyday psychology. *Cognition, 35,* 245–275.

Whiten, A. (Ed.). (1991). *Natural theories of mind: Evolution, development and simulation of everyday mindreading.* Oxford, England: Blackwell.

Wimmer, H., & Hartl, M. (1991). Against the Cartesian view on mind: Young children's difficulty with own false beliefs. *British Journal of Developmental Psychology, 9,* 125–138.

Wimmer, H., Hogrefe, J., & Sodian, B. (1988). A second stage in children's conceptions of mental life: Understanding informational accesses as origins of knowledge and belief. In J. W. Astington, P. L. Harris, & D. R. Olson (Eds.),

Developing theories of mind (pp. 173–192). New York: Cambridge University Press.

Wimmer, H., & Perner, J. (1983). Beliefs about beliefs: Representation and constraining function of wrong beliefs in young children's understanding of deception. *Cognition, 13*, 103–128.

CHAPTER SEVEN

Self-Awareness and Social Intelligence

WEB PAGES, SEARCH ENGINES, AND NAVIGATIONAL CONTROL

MARTIN E. FORD
MICHELLE A. MAHER

Social intelligence is influenced by a broad range of psychological, contextual, and developmental processes (M. Ford, 1986). The purpose of this chapter is to explore how a subset of these processes associated with self-awareness can facilitate or constrain the effectiveness of social thought and action.

To accomplish this objective, we outline a conceptual framework that maps out the component processes contributing to social intelligence and the organization of these processes in unitary patterns. This framework—the Living Systems Framework, or LSF (D. Ford, 1987, 1994)—suggests that self-awareness is a manifestation of a more general *information–consciousness–attention (ICA) arousal* function that plays a critical role in maintaining coherent and coordinated social activity. We use the metaphor of navigating the World Wide Web to explain how this arousal function works and to illustrate the ubiquity and importance of self-awareness as a factor in socially intelligent and unintelligent functioning. We also note how the LSF is compatible with and able to extend other theories of self-awareness that have been built on a control systems model. We conclude by further explicating the concept of "navigational control" and by calling for research directly addressing the hypothesized link between this capability and social intelligence.

BASIC CONCEPTS AND PRINCIPLES OF THE LIVING SYSTEMS FRAMEWORK

The LSF (D. Ford, 1987, 1994; M. Ford & D. Ford, 1987) is a comprehensive theory of human functioning and development designed to represent, at the person-in-context level of analysis, all aspects of being human. The LSF also provides a framework within which more specific theories about subsets of phenomena can be developed, as illustrated by M. Ford's (1992) Motivational Systems Theory (MST) and D. Ford and Lerner's (1992) Developmental Systems Theory (DST).

The central organizing concept of the LSF is that of a self-organizing, self-constructing, adaptive control system, or, in simpler terms, a *living system*. The control system metaphor—a familiar one in the literature on self-awareness (e.g., Carver & Scheier, 1981, 1990; Duval & Wicklund, 1972)—refers to the capacity of humans to set goals and to construct and execute plans for accomplishing those goals, with all phases of the process regulated by information about current conditions (feedback information) and possible future circumstances (feedforward information). The term "self-organizing" refers to continuously operating processes that serve to interrelate the various "parts" of a person to create unified patterns of functioning. The term "self-constructing" refers to the capacity of humans to transcend the limitations of mechanistic control systems by constructing, elaborating, and repairing their own "hardware" or biological structure (e.g., through biological growth and maturation, and repair of damaged tissue), and by constructing, elaborating, and revising their own "software" or behavioral repertoire (e.g., through learning and skill development).

Functional Components

Each person-in-context living system is composed of a biological structure and a variety of interdependent component processes organized in complex patterns to support the effective functioning of the system as a whole (see Figure 7.1). These processes include the following:

1. *Biological functions*: (a) life-sustaining biological processes—including growth, maintenance, operation, and repair of the biological structure; energy production; and maintenance of biological steady states; and (b) providing a biological foundation for psychological and behavioral functioning.

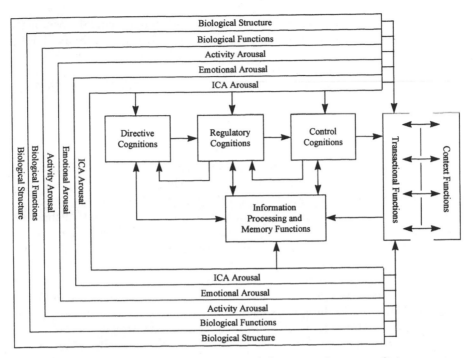

FIGURE 7.1. The functional components of the person-in-context living system (for more information, see D. Ford, 1987, 1994; Ford & Urban, in press; M. Ford, 1992; M. Ford & D. Ford, 1987).

2. *Transactional functions*: (a) exchange of materials essential for biological functioning (i.e., ingestive and eliminative actions); (b) information collection (i.e., sensory–perceptual actions); (c) body movement (i.e., motoric actions); and (d) information transmission (i.e., communicative actions).

3. *Governing (cognitive) functions*: (a) system organization and coordination—including the directive function (i.e., personal goals); regulatory function (i.e., evaluative thoughts); and control function (i.e., planning and problem-solving thoughts); and (b) information-processing and memory functions.

4. *Arousal functions*: (a) management of energy production, distribution, and use—including ICA arousal (i.e., energizing of perceptual and cognitive functions); activity arousal (i.e., energizing of transactional functions); and emotional arousal (i.e., energizing and regulating attention and activity arousal patterns in contexts involving prototypical adaptive problems and opportunities).

5. *Context functions*: (a) compatibility with the individual's personal goals and biological, transactional, and cognitive capabilities; (b) availability of material and informational resources necessary to sustain and support psychological, behavioral, and biological functioning; and (c) presence of a social and emotional climate supporting and facilitating effective functioning.

Behavior Episodes

A basic premise of the LSF is that, as long as a person is alive, all of these functions are operating, both semiautonomously and as an organized unit (the Principle of Unitary Functioning). In the immediate context of ongoing activity, these patterns of unitary functioning are called "behavior episodes." A behavior episode is a context-specific, goal-directed pattern of behavior that unfolds over time until one of three conditions is met: (1) the goal organizing the episode is accomplished, or accomplished well enough (sometimes called "satisficing"); (2) the person's attention is preempted by some internal or external event and another goal takes precedence (at least temporarily); or (3) the goal is evaluated as unattainable, at least for the time being. Daily life is essentially a continuous stream of behavior episodes, although different episodes can vary dramatically in terms of their significance, duration, and impact.

Behavior episodes can also vary in terms of the degree to which different kinds of transactional functions are involved in the pattern of ongoing activity. In an *instrumental episode*, a person is actively engaged in some motor or communicative activity designed to influence the environment in some way, and is actively seeking feedback information from the environment about the results of that activity (e.g., playing a basketball game). In an *observational episode*, a person is seeking relevant informational "input" from the environment through sensory–perceptual processes such as looking, listening, or touching, but there is no "output" to speak of, because the person is not trying to influence the environment (e.g., watching a basketball game). In a *thinking episode*, both output- and input-related transactional processes are inhibited, and there is only "throughput"; that is, instead of trying to influence or obtain information from the environment, the goal is to try to improve the mental organization of some existing informational pattern, either for the intrinsic pleasure of doing so or because one anticipates using that information in future behavior episodes (e.g., thinking about an upcoming basketball game).

An important premise of the LSF is that the most enriching learning experiences are generally those involving a combination of instrumental, observational, and thinking episodes—especially when efforts are made to establish conceptual connections among those episodes. Many mentoring and apprenticeship arrangements exemplify this way of organizing learning episodes (e.g., Gardner, 1991; Rogoff, 1990). Research on how to facilitate learning productivity in academic contexts also emphasizes the importance of integrating different kinds of behavior episode experiences (Walberg & Haertel, 1997).

Behavior Episode Schemata

Behavior episode experiences provide the raw materials for constructing a complex repertoire of enduring behavior patterns through the self-construction of *behavior episode schemata*, or BES. A BES is a generalized, integrated, internal representation of a particular kind of behavior episode, or more commonly, a set of similar behavior episodes (including episodes that have only been imagined or observed). Behavior episodes are most likely to be perceived as similar when they involve similar goals and contexts (D. Ford, 1994).

A BES provides guidance about what one should pay attention to and how one should think, feel, and act in an ongoing or anticipated behavior episode by representing both content (e.g., specific kinds of thoughts and actions) and propositional information (e.g., procedural knowledge and cause–effect information). Of course, the quality of this guidance depends on the range and relevance of behavior episode experiences contributing to the BES, as well as the extent to which key information from these behavior episodes was learned and remembered. The influence of a BES also depends on the extent to which its generalized guiding properties are appropriate and useful for current and future circumstances.

The BES concept is similar to concepts such as motor schema (e.g., Schmidt, 1975), perceptual schema (e.g., Arbib, 1989), cognitive schema (e.g., Neisser, 1976), and self-schema (e.g., Cantor & Harlow, 1994; Markus, Cross, & Wurf, 1990). However, it is broader in that a BES represents the functioning of the whole person-in-context (i.e., it integrates cognitive, perceptual, motor, self, emotional, and biological schemas). This also makes the BES concept a useful tool for conceptualizing personality functioning (M. Ford, 1994). Cantor and her colleagues (Cantor & Harlow, 1994; Cantor & Kihlstrom, 1987) have also used a broad version of the schema concept to help explicate the nature of personality and social intelligence.

A potentially useful metaphor for a BES is that of a World Wide Web page. Both a BES and a Web page represent an organized pattern of information on a particular "topic" given meaning by the purposes of the user and the content of the informational components. The utility of the BES or Web page as a guide for thought and action depends on the amount and quality of learning or programming effort that has been invested in the construction of this informational pattern. Just as a Web page may be poorly constructed or filled with inaccurate or useless information, a BES may provide only vague or otherwise flawed notions of how one might go about dealing with a particular situation. For example, a teenager learning to drive a car may display poor judgment or react slowly to a situation that a more experienced driver could handle easily and automatically. The experienced driver's repertoire of car-driving BES provides a much more powerful and reliable source of guidance because of the repetition and variety of behavior episode experiences contributing to those BES.

Except when actively being processed, Web pages and BES exist only as potential informational patterns, not as continuously available "structures." As a result, they cannot be directly altered or improved until they are activated and put "on screen" or "in mind" (D. Ford, 1994). When a BES or Web page is activated as a result of some triggering intention or event, each is activated as an entire unit. However, only a small portion of the overall pattern of information is available for "on-screen inspection" or conscious processing at any one time. It is possible in each case to shift one's attentional focus from one element to another in the overall pattern, or to another BES (or Web page) linked to that pattern of information. Nevertheless, the amount of information that can be attended to and processed at any one time is severely limited (D. Ford, 1994).

Analogous to the use of Hypertext to link Web pages into broader informational networks, a BES can also be elaborated or combined with other BES and BES components to increase its scope and effectiveness. Over time, this can yield a very powerful BES encompassing a diverse repertoire of optional behavior patterns organized around a related set of goals and contexts. By combining a number of such BESs together, a qualitatively superior kind of expertise called "generative flexibility" can emerge. This ability to quickly generate effective options for dealing with particular kinds of problems and opportunities is a hallmark of social and practical intelligence (M. Ford, 1986; Sternberg & Wagner, 1986), as well as a major goal of Web page designers.

LIVING SYSTEMS FRAMEWORK CONCEPTUALIZATION OF SOCIAL INTELLIGENCE

The LSF is concerned with the impact of various features of the person and context—both individually and collectively—on the effectiveness of human functioning. The primary criterion for defining and assessing effective functioning is *goal attainment within some setting or domain of activity.* This means that only after the relevant goals and contexts have been identified for a particular research or intervention question can one proceed to investigate whether particular processes or patterns contribute to effective functioning within these parameters.

In the LSF, the concept of "achievement" is used to describe effective functioning at the level of a particular behavior episode. In other words, achievement represents the attainment of a personally or socially valued goal in a particular context. At the BES or personality level of analysis, the concept of "competence" is used to describe generalized patterns of effective functioning. A competent person has both a demonstrated pattern of achievement across a range of relevant behavior episodes and strong potential for achievement in similar future episodes as a consequence of having developed a BES repertoire that is well adapted to the goals and contexts anchoring the episodes of interest.

The concept of "intelligence" is also used to refer to broad patterns of effective functioning. However, a variety of different meanings for the concept of intelligence have emerged in the literature as a result of differences in the particular qualities of an effective BES repertoire emphasized by different theoretical approaches (see Table 7.1). Nevertheless, in each case, intelligence is closely linked to the central theme of attaining important goals in some task or social–cultural setting (Cantor & Harlow, 1994; Sternberg, 1985). "Social intelligence" refers to the effective pursuit of goals focusing primarily on interactions or relationships with other people (M. Ford, 1986, 1995).

Psychological Processes Contributing to Social Intelligence

Since the person-in-context system always functions as a unit, all of the component processes depicted in Figure 7.1 can be expected to contribute to the attainment of valued social goals at some level. For example, poor health or social isolation may make it difficult for people to engage in any meaningful social activity. Most research on social intelligence, however, has focused on the cognitive and emotional processes influencing social activity in contexts assumed to be responsive to motivated, skillful individuals.

Pioneers in the field of social intelligence were primarily interested in exploring people's ability to construct accurate social representations using perceptual cues or contextual descriptions containing incomplete or ambiguous social information (e.g., Keating, 1978; O'Sullivan & Guilford, 1975; Thorndike & Stein, 1937; Walker & Foley, 1973). However, this emphasis on informational input processes has turned out to be surprisingly unproductive, as there is little evidence that skills such as social perceptiveness, social understanding, social insight, or social inferencing ability account for significant amounts of variance in broad assessments of social competence or social intelligence in normal populations (M. Ford, 1984; Nisbett & Ross, 1980; Shantz, 1983; Sternberg & Smith, 1985; Wong,

TABLE 7.1. Seven Prototypical Meanings of Intelligence: LSF Interpretation with Example (Organizational Consultant)

---

1. Breadth of knowledge in a general domain of expertise

   LSF interpretation: Quantity of accurate, useful information represented in the BES (and associated concepts and propositions) relevant to some general set of goals and contexts.

   Example: The consultant has extensive knowledge about different kinds of consulting techniques and their utility for different kinds of organizations and problems.

2. Depth of knowledge in an area of specialization

   LSF interpretation: BES (and associated concepts and propositions) relevant to a relatively circumscribed set of goals and contexts.

   Example: The consultant has the expertise and experience needed to understand a particular kind of organization or problem in great detail.

3. Performance accomplishments in a general domain or area of specialization

   LSF interpretation: Degree to which BES enactments (i.e., actual performances) meet objective standards representing success, mastery, excellence, or high levels of achievement.

   Example: The consultant successfully facilitates effective or improved organizational performance.

4. Automaticity or ease of functioning in a general domain or area of specialization

   LSF interpretation: Degree to which BES enactments meet criteria representing smooth, polished functioning such as effortlessness, grace, or elegance.

   Example: The consultant is able to recognize and deal with organizational problems in a polished, efficient, and professional manner.

5. Skilled performance under highly challenging conditions

   LSF interpretation: Probability of successfully enacting relevant BES under highly evaluative, arousing, difficult, or distracting conditions.

   Example: The consultant is able to work effectively with people in the organization despite a chaotic organizational environment and/or difficult economic conditions.

6. "Generative flexibility"—ability to alter behavior patterns in response to varying circumstances

   LSF interpretation: Degree to which relevant BES are rich and varied with regard to potential combinations or optional components and their demonstrated use.

   Example: The consultant can work with many different kinds of clients and organizations, and is able to generate alternative strategies or novel approaches to challenging problems.

TABLE 7.1. *Continued*

7. Speed of learning and behavior change

LSF interpretation: Degree to which existing BES can be incrementally improved in rapid fashion, or readily replaced in favor of more adaptive patterns.

Example: The consultant can quickly adjust strategies or implement new approaches in response to feedback or to match changing organizational needs.

Day, Maxwell, & Meara, 1995). Deficits in these skills typically must be rather severe before they manifest themselves as major obstacles to effective social behavior (M. Ford, 1984).

On the other hand, substantial evidence has emerged during the past two decades suggesting that a variety of social cognitive processes involved in the direction, regulation, and control of social behavior are excellent predictors of socially intelligent functioning (Cantor & Kihlstrom, 1987; Cantor & Harlow, 1994; M. Ford, 1984, 1986, 1995). These social cognitions include motivational processes such as those involved in social goal setting, social self-regulation, and social self-evaluation (Bandura, 1997; M. Ford, 1982, 1987, 1996; Ford, Wentzel, Siesfeld, Wood, & Feldman, 1986; Marlowe, 1986; Seligman, 1991; Sternberg & Kolligian, 1990; Wentzel, 1994; Wheeler & Ladd, 1982), as well as skill-related processes such as those involved in social planning and problem solving (M. Ford, 1982, 1987; Foster, Krumboltz, & Ford, in press; Goldstein, 1988; Goldstein & Glick, 1987; Rathjen & Foreyt, 1980; Spivack, Platt, & Shure, 1976; Shure & Spivack, 1980).

During this same time frame, several productive lines of research have emerged illustrating strong links between certain emotional arousal processes and social intelligence. Perhaps most notably, research on empathic arousal has provided strong evidence for the importance of emotions as motivators of prosocial accomplishments (Batson, 1990; Eisenberg & Miller, 1987; M. Ford, 1996; Ford, Wentzel, Wood, Stevens, & Siesfeld, 1989; Hoffman, 1982). Numerous other self- and other-focused emotions, including guilt, pride, anger, affection, anxiety, and depression have also been implicated as critical factors in facilitating or inhibiting the attainment of social goals (Cantor & Harlow, 1994; D. Ford, 1994; M. Ford, 1992; Ford et al., 1989; Goldstein, 1988; Goleman, 1995; Hoffman, 1982; Marlowe, 1986). The concept of "emotional intelligence" has also emerged as an integrative conceptualization emphasizing the use of emotion management skills and information about one's own and others' feelings to guide social thought and action in effective ways (Goleman, 1995; Mayer & Salovey, 1993; Salovey & Mayer, 1990).

What about ICA Arousal?

Empirical work on the psychological processes contributing to effective social behavior now encompasses virtually the entire range of human functioning summarized in Figure 7.1. However, one major gap remains in the
social intelligence literature. The missing link is research on the role of ICA
arousal in socially intelligent and unintelligent functioning. This is a rather
surprising omission given the widespread interest in this aspect of human
functioning among psychological scientists (e.g., Baumeister, 1990, 1991;
Csikszentmihalyi, 1990; Eisenberg & Fabes, 1992; D. Ford, 1994; Rogoff,
1990; Smirnov, 1994), and the emphasis placed on arousal processes in
psychotherapy theories (Ford & Urban, in press). It seems likely that this
empirical gap is attributable, at least in part, to the lack of an integrative
conceptual framework that clarifies the role of the ICA arousal function in
the overall person-in-context system, and that distinguishes this function
from other information-based processes. The LSF addresses both of these
concerns and thus serves as a useful vehicle for encouraging empirical
work on the contribution of ICA arousal to social intelligence.

LIVING SYSTEMS FRAMEWORK
CONCEPTUALIZATION OF
INFORMATION–CONSCIOUSNESS–ATTENTION
AROUSAL

Effective coordination of a person's cognitive and transactional functions
(goal setting, planning, evaluation, decision making, motor and communicative activity, perception, and learning) requires an organized, carefully
timed pattern of neural excitation. ICA arousal processes are responsible
for this complex coordinating task. These processes organize the flow of
behavior episodes by selectively energizing informational components in
the system as needed to maintain effective, goal-directed activity. More
specifically, this arousal function "variably energizes the nervous system to
support the selective collection, transformation, construction, and use of
information in directing, controlling, and regulating behavior" (D. Ford,
1994, p. 236).

From the perspective of the LSF, the terms "consciousness" and "attention" refer to different aspects of the same basic function. The concept
of "consciousness" refers to a general state of readiness for awareness,
whereas "attention" refers to a more specific state of preparedness to carry
out information-based functions. Specifically, attention refers to the *targeted* process of energizing informational content through selective information monitoring and collection (perception), cognitive reconstruction (re-

membering, thinking), or motoric or communicative action. When the focus of attention is particularly sharp or the magnitude of energy expended is unusually high, terms such as "vigilance," "alertness," and "concentration" are often used.

When ICA arousal is directed toward informational content pertaining to the self, the term "self-awareness" is typically used. In other words, when a person becomes self-aware, a perception, thought, or feeling (or an integrated combination of informational components, as in a BES) is being selectively energized to facilitate the potential use of that information to help guide goal-directed activity. For example, when people expend attentional energy thinking about their goals and the possible consequences of pursuing those goals, they are likely to improve their chances of behaving in an intelligent manner (M. Ford, 1992; Locke & Latham, 1990), as well as their ability to negotiate major life obstacles and opportunities (Ford & Nichols, 1991; Sternberg & Spear-Swerling, Chapter 8, this volume).

The Multiple Meanings of Self-Awareness

Much of the literature on ICA arousal is unnecessarily confusing because it fails to clearly distinguish between (1) the process of selectively energizing thoughts and perceptions relevant to a person's current goals and contexts (ICA arousal), (2) the content of the information being energized (the targeted thoughts and perceptions), and (3) the process of using or manipulating the energized information (cognitive processing of targeted thoughts and perceptions) (D. Ford, 1994). For example, the ill-defined concept of "metacognition" is typically defined as thinking about thinking. But in many cases, discussions about metacognition are not about cognition at all. Rather, metacognition is simply being used as a synonym for ICA arousal, that is, the process of *attending to one's thoughts* in an effort to make them more available for guiding goal-directed activity.

The concept of self-awareness further illustrates this problem, as it is commonly used (e.g., in this volume) to refer not only to the process of energizing self-relevant information, but also to the information being energized and subsequent cognitive processing of that content. The distinction is similar to the difference between a flashlight (the energizing arousal process) and the objects illuminated by the flashlight (e.g., self-perceptions and self-referent thoughts). Turning on a flashlight in a dark room is analogous to activating a BES and indicates a need to monitor or locate potentially relevant information to guide goal-directed activity. The illuminating (energizing) function itself is analogous to awareness, with some aspects of the activated BES in focal attention, some aspects in peripheral awareness, and others outside of consciousness but still potentially able to have an impact on the person's activity. The information about the self or environ-

202 DEVELOPMENT OF SELF-AWARENESS ACROSS THE LIFESPAN

ment that is selectively energized by the "flashlight" of ICA arousal is perceived and evaluated in terms of its relevance and utility for the person's current concerns. If the information highlighted in awareness is unrelated to those concerns, attentional energy can be redirected to other potentially useful BES and BES components. If the information in awareness is relevant but not in an optimal or usable form, ICA arousal processes can energize further cognitive processing and informational self-construction activities (e.g., learning, retrieving, categorizing, analyzing, integrating, reasoning, concept formation, theory development).

Search Engines

Moving from the "low-tech" metaphor of a flashlight back to the "high-tech" metaphor introduced earlier of BES as mental web pages, it is useful to think of the ICA arousal function as being analogous to a Web browser. Awareness shifts in dynamic fashion from one BES (Web page) to another, and to different informational components within a particular BES, in an effort to get the most useful information possible "on screen" or "in mind" at any given point in time.

It is also useful to distinguish between three different manifestations of this "browsing" capability as it operates in the three different kinds of behavior episodes noted earlier: instrumental, observational, and thinking episodes. That is because it is only possible to browse one type of episode at a time in an efficient manner (D. Ford, 1994). It is as if people have three different kinds of "search engines" that they use selectively to energize information in these different kinds of behavior episodes.

The function of the *observational search engine* is to coordinate the operation of the person's information-monitoring and information-collection functions. In this case, energy is being directed primarily outward toward the context. The function of the *thinking search engine* is to coordinate the operation of the person's information-processing and memory functions and the cognitions responsible for directing, regulating, and controlling behavior patterns. In these kinds of episodes, energy is primarily directed inward toward one's mental processes. Finally, the function of the *instrumental search engine* is to coordinate the operation of the person's cognitive and transactional functions with a context that is continually changing as a result of the person's instrumental activity. This is the most demanding kind of arousal function, because it requires flexible shifting of awareness between two dynamic sources of information (person and context). In contrast, observational and thinking search engines can concentrate on either contextual or psychological patterns of information.

The degree of difficulty one is likely to experience in conducting these searches and in shifting from one type of search to another depends on the level of familiarity and expertise one has developed with respect to current behavior episodes. When an activated BES provides detailed, accurate information and highly automated guides for behavior (as is often the case when driving a car or engaging in casual conversation), only a small amount of "browsing" time and energy will be needed to stay on track and behave effectively. In such situations, there may be ample attentional capacity left over for daydreaming, problem solving, or exploratory scanning of the environment.

In contrast, both the need to conduct extensive searches of different kinds and the difficulty in doing so will be greater in situations that involve unfamiliar elements, emotionally arousing qualities, or challenging standards of performance. For example, a quarterback who is trying to decide whether to change to a new play (an "audible") after looking at the positioning of the defense will need to quickly and flexibly shift from an observational search engine ("What defense are they playing?") to a thinking search engine ("What does our playbook say will work in this situation?") to an instrumental search engine ("This play isn't working. Now what do I do?!").

RELATIONSHIP TO OTHER THEORIES OF SELF-AWARENESS

The information-based arousal processes characteristic of an instrumental search engine are similar to those described by Duval and Wicklund (1972) in their theory of self-awareness. This theory highlights the shifting focus of attention between internal and external sources of information, as well as the role of goals and contexts (i.e., personal standards and situational factors) in shaping attentional strategies. Of particular interest is the idea that self-awareness may highlight an attribute of the self that is discrepant with some preferred conception of the self. Building on a classic discrepancy reduction model of motivation, the theory posits that people will try to resolve the discomfort resulting from awareness of personal imperfections or transgressions by either refocusing attention outward and actively avoiding a state of self-awareness, or by maintaining an internal focus and taking the necessary steps to reduce or eliminate the discrepancy (Wicklund, 1980).

Although there is some empirical support for the notion that self-awareness can, under certain circumstances, facilitate the encoding of self-relevant information (Hull & Levy, 1979), more accurate self-evaluations (e.g., Pryor, Gibbons, Wicklund, Fazio, & Hood, 1977), the resolution of goal conflicts (e.g., Diener & Srull, 1979; Gibbons & Wright, 1983), and stronger links between attitudes and behavior (e.g., Greenberg & Musham,

1981; Hutton & Baumeister, 1992), inconsistencies in the literature suggest that the role of ICA arousal processes in shaping thought and action is much more complex than a simple discrepancy reduction model would predict (as illustrated by concepts such as "repression" and "unconscious defense mechanisms" in psychotherapy) (Ford & Urban, in press). In particular, it appears that the state of self-awareness, by itself, is not reliably associated with socially intelligent outcomes (Gibbons, 1990). Attentional focus on the self can either facilitate or inhibit effective social behavior depending on the information being highlighted and the criteria for defining effectiveness.

The self-regulation framework proposed by Carver and Scheier (1981, 1990) expands on a simple control system model in a number of important ways. Perhaps most notably, the concept of relevant standards, defined by Baumeister (1991) as the "essence of self-awareness" (p. 89), is pragmatically redefined in this framework to include a diversity of goals at different levels of abstraction. Carver and Scheier also introduce the important idea that awareness of a discrepancy between actual and preferred conduct will not lead to efforts to reduce that discrepancy if the individual does not believe that the higher standard is attainable. In other words, goals and feedback are not sufficient to motivate action in the absence of positive capability and context beliefs (M. Ford, 1992).

Consistent with the LSF's use of the BES concept to describe how goal-directed behavior patterns are organized, Carver and Scheier (1981, 1990) rely heavily on the concept of "schema" to explain how information about goals, standards, and other elements of self-regulated functioning are encoded in memory. In particular, they suggest that awareness focused internally may increase the accessibility of an individual's general self-schema, or self-concept, which in turn can facilitate the collection and processing of self-relevant information. This is similar to the LSF concept of selectively energizing BES to promote effective coordination of a person's cognitive and transactional functions, except that the LSF places more emphasis on goal- and context-specific thought patterns than on broad self-concepts as functional elements that might be energized by ICA arousal processes.

Although self-awareness is not a motivational process per se (i.e., it energizes motivational processes rather than serving a motivational function itself), several contemporary theories of motivation have highlighted the role of ICA arousal in effective personal and social functioning (Bandura, 1997; Baumeister, 1991; Csikszentmihalyi, 1990; M. Ford, 1992; McCombs, 1991; Ridley, 1991). Csikszentmihalyi (1990) and Baumeister (1990, 1991) have been particularly eloquent in emphasizing how the ability to control the informational content in consciousness can dramatically alter mental health and behavioral effectiveness. Consistent with this view,

ICA arousal is a key component of all psychotherapy procedural models (Ford & Urban, in press).

SELF-AWARENESS AND SOCIAL INTELLIGENCE: THE CONCEPT OF NAVIGATIONAL CONTROL

There appears to be consensus among theorists interested in ICA arousal that awareness is constantly shifting in dynamic fashion between different components of the person-in-context system, with shifts governed primarily by current concerns and intentions (i.e., people will pay attention to things that are important to them) and by contextual factors (i.e., people will pay attention to information that is perceptually salient or discrepant from expectations) (Bargh, 1982; D. Ford, 1994). There is also general agreement that people may be able to control at least some aspects of this shifting process (Baumeister, 1990, 1991; Csikszentmihalyi, 1990; McCombs, 1991; Ridley, 1991). The ability to exercise intentional control over the selective targeting of ICA arousal—the primary process responsible for maintaining organized, coherent patterns of perception, thought, and action—can be metaphorically described by the term "navigational control."

Just as the World Wide Web is cluttered with a seemingly endless number and variety of home pages, a person's BES repertoire includes the informational residue of thousands of behavior episodes ranging across a diversity of goals and contexts. Given the tremendous potential for cognitive and behavioral disorganization resulting from this mental "clutter," it stands to reason that the ability to navigate through this information in a selective, targeted fashion is likely to play a very important role in socially intelligent and unintelligent functioning. This process is facilitated by a variety of mental heuristics (e.g., Fiske & Taylor, 1984; Nisbett & Ross, 1980) that work something like the "bookmarking" function that makes it possible to quickly locate and access selected Web pages known to be of general utility.

Of course, the ability to selectively "pull up" different BES (or Web pages) and to efficiently maneuver within those informational units will obviously be of little help if the activated information is impoverished or poorly constructed. Nevertheless, even the most elegant creations will be of little utility if they are not available when they are needed. In many social situations, timing is everything. Regretfully ruminating about what "I could have done. . ." or what "I should have said. . ." may enable one to minimize the effects of some mistakes and transgressions, but social challenges typically must be navigated as they are unfolding, not after the ship has run aground (see also Sternberg & Spear-Swerling, Chapter 8, this volume).

Navigational control is less likely to be a critical issue in highly familiar or informationally redundant behavior episodes than in novel or unpredictable circumstances. In well-learned social situations that remain true to expectations, one can behave intelligently with minimal investment of attentional energy by relying on automated "scripts" (Abelson, 1981) or habits (i.e., well-developed BES). Social scripts are particularly useful in contexts that require close conformity to a set of rules or conventions (e.g., behaving properly in school or church) and in repetitive situations, where efficiency is highly valued (e.g., greeting neighbors and coworkers, putting the kids to bed).

However, it is precisely because the mundane episodes of daily life place few demands on people's cognitive and attentional resources that assessments of social intelligence rarely focus on these kinds of situations. As is the case with assessments of academic or general intelligence, the primary contexts of interest are typically those involving some degree of difficulty or challenge, at least for the ordinary person. For example, M. Ford's (1982) Social Competence Nomination Form is designed to measure a person's ability to handle challenging social situations involving the attainment of self-assertive and integrative social-relationship goals. In such situations, navigational control is likely to play a critical role in facilitating intelligent behavior—perhaps even more so than in situations involving nonsocial challenges, as social contexts tend to be more dynamic and unpredictable than most task contexts.

Classroom teaching provides a particularly good example of a context in which navigational control capabilities are likely to have a strong impact on socially intelligent functioning. Such contexts involve multiple, competing goals, pressing time demands, and complex learning episodes involving unexpected events, challenges, and opportunities. Teachers must strategically allocate and reallocate attentional resources in rapid, flexible fashion based on feedback pertaining to their instructional objectives and the ongoing dynamics of the classroom. It is no wonder that educational researchers and practitioners have begun to emphasize qualities related to navigational control in their descriptions of effective teaching (e.g., Clark & Peterson, 1986; Osterman & Kottkamp, 1993).

Navigational control may be a particularly important skill in behavior episodes in which an individual is experiencing one or more strong, negative emotions. Such emotions indicate the presence of a significant challenge and intense concern about one's efforts to meet this challenge (Frijda, 1988). These emotions are functional in that they motivate people to try to overcome perceived threats, obstacles, and problems (D. Ford, 1994; M. Ford, 1992); however, they must be effectively managed to avoid potentially dysfunctional consequences (Ford & Urban, in press; Goleman, 1995). ICA arousal processes appear to play a very important role in helping people manage intense emotions (Baumeister, 1990, 1991; Csikszentmihalyi, 1990; Eisenberg & Fabes, 1992). For example, Eisenberg and her colleagues (Eisenberg et al., 1997) recently conducted a study of young children's self-regulatory capabilities in which they concluded that

> children who could regulate their attention appeared to be resilient to stress and, perhaps as a consequence, were better liked by peers and viewed by adults as being more socially appropriate. . . . Level of attentional control was particularly important for predicting social functioning for children prone to negative emotion. (p. 307)

The importance of navigational control in complex, challenging situations is largely due to the inherent inability of humans to attend closely to more than a small amount of information at any one time (perhaps due to short-term-memory limitations) or to do so for lengthy periods of time (Bargh, 1982; D. Ford, 1994; Kahneman, 1973; Miller, 1956). Moreover, the kind of information that can be brought into awareness from one moment to the next is limited by the kind of "search engine" in operation. In other words, people can actively monitor events in the environment using an observational search engine, *or* they can actively monitor internal thoughts and feelings using a thinking search engine, *or* they can actively monitor the results of their actions using an instrumental search engine. While it may be possible with strong navigational control skills to shift quickly and efficiently from one search engine to another, it is not possible to conduct all three kinds of searches simultaneously except at the level of diffuse or peripheral awareness. As D. Ford (1994) explains, "Attention may be focused primarily upon organizing either input (information-collection tasks), throughput (remembering and thinking tasks), or output (implementation of action plans), but not more than one at a time" (p. 274). Poor navigational control capabilities can thus leave one susceptible to phenomena such as ruminating about negative affective states, being fixated on problems or possible negative consequences rather than seeking solutions to those problems, and perseverating with ineffective problem-solving strategies (Bandura, 1997; Baumeister, 1984, 1990; Ford & Urban, in press; Kuhl & Beckmann, 1993; Tobias, 1979, 1985).

Stimulus Control versus Intentional Control

A basic premise of the LSF is that behavior episodes are organized primarily by goals and contexts, that is, by what people are trying to do, and where and when they are trying to do it. When personal goals are the primary factors organizing a behavior episode, the resulting behavior can be described as *intentionally controlled behavior*. When elements of the context are the primary factors organizing a behavior episode, the resulting behavior can be described as *stimulus-controlled behavior* (D. Ford, 1994). The concept of navigational control implies a high level of intentionally controlled behavior, with thoughts about desired futures (goals/intentions) instructing the person's navigational control system (ICA arousal function) to selectively energize the mental "Web pages" (BES) most likely to facilitate progress. This initiates a process of scanning and focusing (D. Ford, 1994) that energizes the informational components of greatest significance and utility for the current episode. This activity may include observational and thinking episodes designed to help people prepare for future instrumental episodes.

Individual differences in navigational control capabilities suggest that these skills may play an important role in socially intelligent and unintelligent behavior. At one extreme are at-risk and diagnostically labeled individuals with symptom patterns highlighted by generalized attentional and social deficits and difficulty in maintaining intentionally controlled behavior (e.g., attention-deficit/hyperactivity disorder, schizophrenia, narcissism, various conduct and character disorders) (Frick & Lahey, 1991; Garmezy, 1977; Landau & Moore, 1991; Rodriguez, Mischel, & Shoda, 1989; Winsler, in press). Other cases of socially ineffective behavior may be more specific to particular contexts or domains of functioning. For example, Dodge (1986) has demonstrated that highly aggressive children—who often use aggressive behavior very deliberately to accomplish self-assertive and social-management goals—develop habits of selective attention that cause them to miss or ignore information inconsistent with their assumption that others have hostile intentions toward them. Similarly, many couples involved in high-conflict divorce and custody battles appear to be inordinately focused on their own concerns, feelings, and beliefs, with little inclination or ability to shift awareness either to others' needs or to practical strategies for maximizing the attainment of their own goals (Johnston, 1994). Along these same lines, Foster et al. (in press) found that training shy adult males to focus their attention on their social partners' perspectives and concerns was an effective technique for enhancing their ability to attain social-relationship goals.

At the other extreme are descriptions of exemplary individuals that link sophisticated navigational control capabilities with optimal psychological and behavioral functioning (e.g., Csikszentmihalyi, 1990; Csikszentmihalyi & Csikszentmihalyi, 1988; Kobasa, 1979). Indeed, Csikszentmihalyi's use of the term "flow" to describe optimal experience fits nicely with the concept of navigational control (i.e., skillful navigation leads to a sense of flow). As Csikszentmihalyi (1990) explains,

> A person can make himself happy, or miserable, regardless of what is happening "outside," just by changing the contents of consciousness. We

all know individuals who can transform hopeless situations into challenges to be overcome. . . . This ability to persevere despite obstacles and setbacks is the quality people most admire in others, and justly so; it is probably the most important trait not only for succeeding in life, but for enjoying it as well. To develop this trait, one must find ways to order consciousness so as to be in control of feelings and thoughts. (p. 24)

Controlling the Contents of Consciousness: Horizontal and Vertical Forms of Navigational Control

Precisely what feelings and thoughts are being referred to in the preceding quotation? What is being controlled when people exercise their navigational control capabilities? The answer can be summarized by referring back to Figure 7.1, which portrays all of the functions of the person-in-context living system. Specifically, in addition to people, objects, and events in the environment, ICA arousal processes can produce self-awareness of aspects of an individual's

Biological structure (e.g., "Oh no, I have a zit!")
Biological functioning (e.g., "I'm not feeling too well")
Material-exchange transactions (e.g., "This tastes funny")
Information-collection transactions (e.g., "Hey, check it out!")
Body-movement transactions (e.g., "I can't reach that far")
Information-transmission transactions (e.g., "No, that's not what I meant")
Directive/goal thoughts (e.g., "I want you here with me")
Regulatory/evaluative thoughts (e.g., "Things don't seem to be going too well")
Control/planning thoughts (e.g., "How am I going to do that?")
Information-processing and memory functions (e.g., "That sure seems familiar")
ICA arousal (e.g., "What was I thinking?")
Activity arousal (e.g., "Sorry, I'm too tired")
Emotional arousal (e.g., "You make me so mad!")

This list makes it clear that navigational control of ICA arousal processes is much more complex than a simple choice of focusing awareness on either the self or others. To maintain effective functioning in challenging social situations, an individual will typically need to monitor many different aspects of the self and environment in a manner analogous to clicking on different Hypertext links within one or more Web pages. Navigational control thus entails not only the ability to maneuver between different sources of information, but also the ability to focus selectively on precisely those sources of information most likely to facilitate goal progress. For example, someone who chronically ruminates about social conflicts may be quite skilled at shifting attention from one system component to another but fail to attend to the information that could actually help resolve those conflicts (e.g., awareness of misunderstandings, skill deficits, misaligned goals, or helpful friends). Understanding habitual patterns of ICA arousal that prevent people from thinking constructively about social problems and opportunities is a major focus of several intervention approaches in clinical and counseling psychology (Bandura, 1997; Ford & Urban, in press).

In addition to this "horizontal" form of navigational control, several theorists have proposed that a "vertical" form of navigational control may also play an important role in maintaining and promoting effective functioning. For example, Powers (1973, 1989) posits 10 hierarchically organized levels of perception (i.e., levels of informational feedback related to some goal) that people may try to control in the course of goal-directed activity, with all but the lowest levels potentially under one's conscious control. In this hierarchy, the goal of one episode is simultaneously the means for attaining some broader goal at the next-higher level of perception. The ability to maneuver skillfully and strategically from a focus on lower-level subgoals (i.e., specific outcomes or consequences in the here and now) to awareness of the higher-order purposes served by those achievements appears to be an effective way to facilitate motivation, performance, and a sense of meaningfulness (Bandura & Schunk, 1981; Baumeister, 1990, 1991; Harackiewicz & Sansone, 1991; Locke & Latham, 1984, 1990; Powers, 1989; Schunk, 1991).

Baumeister's (1990) escape theory provides another compelling argument for further investigation of vertical forms of navigational control. In this formulation, some people who experience highly negative life events (usually involving consequential social goals) blame themselves for real and imagined failures, resulting in a deep sense of inadequacy, worthlessness, and self-condemnation. These evaluative thoughts in turn lead to powerful negative emotions such as depression, loneliness, or guilt, and an aversive state in which "the person is acutely aware of self as inadequate, incompetent, unattractive, or guilty" (Baumeister, 1990, p. 91). This aversive self-awareness motivates the person to try to escape this unhappy state, primarily by "navigating downward" to a narrower and less demanding level of activity—a process Baumeister calls "cognitive deconstruction." This process refers to the filling of consciousness with concrete, meaningless content (e.g., trivial, moment-to-moment task accomplishments) and to an intentional—but ultimately misguided—failure to attend to the broader goals and contexts that normally provide meaning to a person's thoughts, feelings, and actions. The resulting sense of meaninglessness can result in a variety of social and emotional disorders and, when combined with a sense of hopelessness, is potentially fatal. "Suicide thus emerges as an escalation of the person's wish to escape from meaningful awareness and their implications about the self" (p. 91).

People Can Only Be Conscious of Information in Perceptible Form

A basic premise of the LSF is that most cognitive processing occurs outside awareness (i.e., "unconscious" processing is normal rather than abnormal as Freud proposed). Moreover, only some thoughts are in a form that is accessible to awareness (D. Ford, 1994). Specifically, humans cannot attend to something or be conscious of it unless they can perceive it; that is, it must be something they can see, hear, smell, taste, touch, or feel (cf. Searle, 1992). Thus, navigational control processes, both horizontal and vertical, can only search for and access information that is in some perceptible form.

Mental images commonly serve this purpose (i.e., we can see things in our "mind's eye"). However, a more flexible tool for bringing thoughts into consciousness is language. Because words (and other symbols) can be seen or heard or touched and yet do not have any intrinsic meaning, they can be attached to any idea that the user desires, thereby making that concept accessible to consciousness (i.e., because the idea is now conveyed by something in perceptible form). Moreover, if a group of people can agree about what words to attach to particular ideas, they can communicate about abstract concepts from a shared frame of reference. That is why language is such an important human capability, and why having a shared language is such an important factor in facilitating learning and cooperative group functioning. Words are no substitute for ideas, but they can provide the vehicle for energizing concepts, propositions, goals, plans, beliefs, and other thought patterns, thus making it possible to expand them, change them, learn from them, and use them more effectively to guide social thought and action (e.g., Ford & Nichols, 1991; Vygotsky, 1986).

CONCLUSION

The LSF provides a way of conceptualizing the nature of self-awareness and its role in promoting effective functioning that expands on previous control system models by providing powerful new tools for understanding the role of ICA arousal processes in the overall person-in-context system. Life is an ongoing stream of behavior episode experiences (D. Ford, 1994). We remember these episodes by constructing and integrating mental "Web pages" (BES) that enable us to anticipate how events might unfold and to identify possible obstacles, opportunities, and strategies for pursuing our goals. To the extent that our informational "search engines" (ICA arousal processes) can quickly locate relevant BES and BES components for organizing our functioning in observational, thinking, and instrumental episodes, we increase our chances of behaving in an intelligent manner.

In this chapter, we have introduced the concept of "navigational control" to describe the process of exercising intentional control over the selective targeting of ICA arousal. We also cited a broad range of theoretical and empirical work relevant to the hypothesis that navigational control plays an important role in facilitating socially effective behavior and suggested that research focused directly on this hypothesis would fill a significant gap in the literature on social intelligence. In doing so, we pointed out that not all situations require high levels of navigational control. Just as effective navigation of a boat can be accomplished with ease in calm, open waters ("smooth sailing"), socially intelligent functioning may be virtually effortless in friendly, familiar surroundings. It is when the going gets rough that the ability to target useful informational content in a timely, selective manner becomes crucial to the successful pursuit of one's social goals.

There appear to be long-term consequences of deficits in navigational control capabilities that make this a particularly important area of research for developmental and educational scholars and practitioners. For example, the ability to delay gratification in the preschool years—an ability that represents the triumph of intentionally controlled behavior over stimulus-controlled behavior—is closely related to attention deployment (navigational control) skills and a broad range of social and cognitive competencies in later childhood and adolescence (Funder, Block, & Block, 1983; Mischel, Shoda, & Peake, 1988; Rodriguez et al., 1989). Research on children at risk for attentional and emotional difficulties also points to the importance of navigational control capabilities in the development of social and academic competence (Eisenberg & Fabes, 1992; Eisenberg et al., 1997; Frick & Lahey, 1991; Landau & Moore, 1991; Winsler, in press).

Psychological practitioners may also find it useful to consider the possible benefits of focusing directly on their clients' navigational control capabilities. In most cases, there is a gap between an individual's actual and optimal (potential) performance. To the extent that people are not fully utilizing their capabilities by systematically activating their best available BES for a particular set of circumstances, there are opportunities to use self-awareness as a lever for activating the search engines of our mind and producing new behavioral options. Awareness of the limitations associated with current BES can also motivate efforts to develop more intelligent ways of interacting with significant people in our lives.

REFERENCES

Abelson, R. P. (1981). Psychological status of the script concept. *American Psychologist, 36,* 715–727.

Arbib, M. A. (1989). *The metaphorical brain 2: Neural networks and beyond.* New York: Wiley.

Bandura, A. (1997). *Self-efficacy: The exercise of control.* New York: Freeman.

Bandura, A., & Schunk, D. H. (1981). Cultivating competence, self-efficacy, and intrinsic interest through proximal self-motivation. *Journal of Personality and Social Psychology, 41,* 586–598.

Bargh, J. A. (1982). Attention and automaticity in the processing of self-relevant information. *Journal of Personality and Social Psychology, 43,* 425–436.

Batson, C. D. (1990). How social an animal? The human capacity for caring. *American Psychologist, 45,* 336–346.

Baumeister, R. F. (1984). Choking under pressure: Self-consciousness and paradoxical effects of incentives on skillful performance. *Journal of Personality and Social Psychology, 46,* 610–620.

Baumeister, R. F. (1990). Suicide as escape from the self. *Psychological Review, 97,* 90–113.

Baumeister, R. F. (1991). *Meanings of life.* New York: Guilford Press.

Cantor, N., & Harlow, R. E. (1994). Social intelligence and personality: Flexible life task pursuit. In R. J. Sternberg & P. Ruzgis (Eds.), *Personality and intelligence* (pp. 137–168). New York: Cambridge University Press.

Cantor, N., & Kihlstrom, J. F. (1987). *Personality and social intelligence.* Englewood Cliffs, NJ: Prentice-Hall.

Carver, C. S., & Scheier, M. F. (1981). *Attention and self-regulation: A control theory approach to human behavior.* New York: Springer-Verlag.

Carver, C. S., & Scheier, M. F. (1990). Origins and functions of positive and negative affect: A control-process view. *Psychological Review, 97,* 19–35.

Clark, C. M., & Peterson, P. L. (1986). Teachers' thought processes. In M. Wittrock (Ed.), *Handbook of research on teaching* (3rd ed., pp. 255–296). New York: Macmillan.

Csikszentmihalyi, M. (1990). *Flow: The psychology of optimal experience.* New York: Harper & Row.

Csikszentmihalyi, M., & Csikszentmihalyi, I. (Eds.). (1988). *Optimal experience: Psychological studies of flow in consciousness.* New York: Cambridge University Press.

Diener, E., & Srull, T. K. (1979). Self-awareness, psychological perspective, and self-reinforcement in relation to personal and social standards. *Journal of Personality and Social Psychology, 37,* 413–423.

Dodge, K. A. (1986). A social information processing model of social competence in children. In M. Perlmutter (Ed.), *The Minnesota Symposium on Child Psychology* (Vol. 18, pp. 77–125). Hillsdale, NJ: Erlbaum.

Duval, S., & Wicklund, R. A. (1972). *A theory of objective self-awareness.* New York: Academic Press.

Eisenberg, N., Guthrie, I. K., Fabes, R. A., Reiser, M., Murphy, B. C., Holgren, R., Maszk, P., & Losoya, S. (1997). The relations of regulation and emotionality to resiliency and competent social functioning in elementary school children. *Child Development, 68,* 295–311.

Eisenberg, N., & Fabes, R. A. (1992). Emotion, regulation, and the development of social competence. In M. S. Clark (Ed.), *Review of personality and social psychology: Vol. 14. Emotion and social behavior* (pp. 119–150). Newbury Park, CA: Sage.

Eisenberg, N., & Miller, P. (1987). Empathy and prosocial behavior. *Psychological Bulletin, 101,* 91–119.

Fiske, S. T., & Taylor, S. E. (1984). *Social cognition.* Reading, MA: Addison-Wesley.

Ford, D. H. (1987). *Humans as self-constructing living systems: A developmental perspective on behavior and personality.* Hillsdale, NJ: Erlbaum.

Ford, D. H. (1994). *Humans as self-constructing living systems: A developmental perspective on behavior and personality* (2nd ed.). State College, PA: Ideals.

Ford, D. H., & Lerner, R. M. (1992). *Developmental systems theory: A synthesis of developmental contextualism and the living systems framework.* Newbury Park, CA: Sage.

Ford, D. H., & Urban, H. B. (in press). *Systems of psychotherapy.* New York: Wiley.

Ford, M. E. (1982). Social cognition and social competence in adolescence. *Developmental Psychology, 18,* 323–340.

Ford, M. E. (1984). Linking social-cognitive processes with effective social behavior: A living systems approach. In P. C. Kendall (Ed.), *Advances in cognitive-behavioral research and therapy* (Vol. 3, pp. 167–211). New York: Academic Press.

Ford, M. E. (1986). A living systems conceptualization of social intelligence: Outcomes, processes, and developmental change. In R. J. Sternberg (Ed.), *Advances in the psychology of human intelligence* (Vol. 3, pp. 119–171). Hillsdale, NJ: Erlbaum.

Ford, M. E. (1987). Processes contributing to adolescent social competence. In M. E. Ford & D. H. Ford (Eds.), *Humans as self-constructing living systems: Putting the framework to work* (pp. 199–233). Hillsdale, NJ: Erlbaum.

Ford, M. E. (1992). *Motivating humans: Goals, emotions, and personal agency beliefs.* Newbury Park, CA: Sage.

Ford, M. E. (1994). A living systems approach to the integration of personality and intelligence. In R. J. Sternberg & P. Ruzgis (Eds.), *Personality and intelligence* (pp. 188–217). Cambridge, England: Cambridge University Press.

Ford, M. E. (1995). Intelligence and personality in social behavior. In M. Zeidner & H. Saklofske (Eds.), *International handbook of personality and intelligence* (pp. 125–142). New York: Plenum.

Ford, M. E. (1996). Motivational opportunities and obstacles associated with social responsibility and caring behavior in school contexts. In J. Juvonen & K. Wentzel (Eds.), *Social motivation: Understanding children's school adjustment* (pp. 125–153). New York: Cambridge University Press.

Ford, M. E., & Ford, D. H. (Eds.). (1987). *Humans as self-constructing living systems: Putting the framework to work.* Hillsdale, NJ: Erlbaum.

Ford, M. E., & Nichols, C. W. (1991). Using goal assessments to identify motivational patterns and facilitate behavioral regulation. In M. Maehr & P. Pintrich (Eds.), *Advances in motivation and achievement: Vol. 7. Goals and self-regulatory processes* (pp. 57–84). Greenwich, CT: JAI Press.

Ford, M. E., Wentzel, K. R., Siesfeld, G. A., Wood, D., & Feldman, L. (1986, April). *Adolescent decision making in real-life situations involving socially responsible and irresponsible choices.* Paper presented at the annual meeting of the American Educational Research Association, San Francisco.

Ford, M. E., Wentzel, K. R., Wood, D., Stevens, E., & Siesfeld, G. A. (1989). Processes associated with integrative social competence: Emotional and contextual influences on adolescent social responsibility. *Journal of Adolescent Research, 4,* 405–425.

Foster, S., Krumboltz, J., & Ford, M. E. (in press). Teaching social skills to shy adult males. *The Family Journal.*

Frick, P. J., & Lahey, B. B. (1991). The nature and characteristics of attention-deficit hyperactivity disorder. *School Psychology Review, 20,* 163–173.

Frijda, N. H. (1988). The laws of emotion. *American Psychologist, 43,* 349–358.

Funder, D. C., Block, J. H., & Block, J. (1983). Delay of gratification: Some longitudinal personality correlates. *Journal of Personality and Social Psychology, 44,* 1198–1213.

Gardner, H. (1991). *The unschooled mind: How children think and how schools should teach.* New York: Basic Books.

Garmezy, N. (1977). The psychology and psychopathology of attention. *Schizophrenia Bulletin, 3,* 360–369.

Gibbons, F. X., & Wright, R. A. (1983). Self-focused attention and reactions to conflicting standards. *Journal of Research in Personality, 17,* 263–273.

Gibbons, F. X. (1990). Self-attention and behavior: A review and theoretical update. In M. Zanna (Ed.), *Advances in experimental social psychology* (Vol. 23, pp. 249–303). San Diego, CA: Academic Press.

Goldstein, A. P. (1988). *The Prepare Curriculum: Teaching prosocial behavior.* Champaign, IL: Research Press.

Goldstein, A. P., & Glick, B. (1987). *Aggression replacement training.* Champaign, IL: Research Press.

Goleman, D. (1995). *Emotional intelligence: Why it can matter more than IQ.* New York: Bantam Books.

Greenberg, J., & Musham, C. (1981). Avoiding and seeking self-focused attention. *Journal of Research in Psychology, 15,* 191–200.

Harackiewicz, J. M., & Sansone, C. (1991). Goals and intrinsic motivation: You *can* get there from here. In M. L. Maehr & P. R. Pintrich (Eds.), *Advances in motivation and achievement: Vol. 7. Goals and self-regulatory processes* (pp. 21–49). Greenwich, CT: JAI Press.

Hoffman, M. (1982). Development of prosocial motivation: Empathy and guilt. In N. Eisenberg (Ed.), *The development of prosocial behavior* (pp. 281–314). New York: Holt, Rinehart & Wilson.

Hull, J. G., & Levy, A. S. (1979). The organizational functions of the self: An alternative to the Duval and Wicklund model of self-awareness. *Journal of Personality and Social Psychology, 37,* 756–768.

Hutton, D. G., & Baumeister, R. F. (1992). Self-awareness and attitude change:

Seeing oneself on the central route to persuasion. *Personality and Social Psychology Bulletin, 18,* 68–75.

Johnston, J. R. (1994). High-conflict divorce. *The Future of Children, 4,* 165–182.

Kahneman, D. (1973). *Attention and effort.* Englewood Cliffs, NJ: Prentice-Hall.

Keating, D. P. (1978). A search for social intelligence. *Journal of Educational Psychology, 70,* 218–233.

Kobasa, S. C. (1979). Stressful life events, personality, and health: An inquiry into hardiness. *Journal of Personality and Social Psychology, 37,* 1–11.

Kuhl, J., & Beckmann, J. (1993). *Volition and personality: Action and state orientation.* Toronto: Hogrefe.

Landau, S., & Moore, L. A. (1991). Social skill deficits in children with attention-deficit hyperactivity disorder. *School Psychology Review, 20,* 235–251.

Locke, E. A., & Latham, G. P. (1984). *Goal setting: A motivational technique that works.* Englewood Cliffs, NJ: Prentice-Hall.

Locke, E. A., & Latham, G. P. (1990). *A theory of goal-setting and task performance.* Englewood Cliffs, NJ: Prentice-Hall.

Markus, H., Cross, S., & Wurf, E. (1990). The role of the self-system in competence. In R. J. Sternberg & J. Kolligian, Jr. (Eds.), *Competence considered* (pp. 205–225). New Haven, CT: Yale University Press.

Marlowe, H. A. (1986). Social intelligence: Evidence for multidimensionality and construct independence. *Journal of Educational Psychology, 78,* 52–58.

Mayer, J. D., & Salovey, P. (1993). The intelligence of emotional intelligence. *Intelligence, 17,* 433–442.

McCombs, B. L. (Ed.). (1991). Unraveling motivation: New perspectives from research and practice. *Journal of Experimental Education, 60,* 3–88.

Miller, G. A. (1956). The magical number seven, plus or minus two: Some limits on our capacity for processing information. *Psychological Review, 63,* 81–97.

Mischel, W., Shoda, Y., & Peake, P. K. (1988). The nature of adolescent competencies predicted by pre-school delay of gratification. *Journal of Personality and Social Psychology, 54,* 687–696.

Neisser, U. (1976). *Cognition and reality.* San Francisco: Freeman.

Nisbett, R. E., & Ross, L. D. (1980). *Human inference: Strategies and shortcomings of social judgement.* Englewood Cliffs, NJ: Prentice-Hall.

Osterman, K. F., & Kottkamp, R. B. (1993). *Reflective practice of educators: Improving schooling through professional development.* Newbury Park, CA: Corwin Press.

O'Sullivan, M., & Guilford, J. P. (1975). Six factors of behavioral cognition: Understanding other people. *Journal of Educational Measurement, 12,* 255–271.

Powers, W. T. (1973). *Behavior: The control of perception.* Chicago: Aldine.

Powers, W. T. (1989). *Living control systems.* Gravel Switch, KY: Control Systems Groups.

Pryor, J. B., Gibbons, F. X., Wicklund, R. A., Fazio, R., & Hood, R. (1977). Self-focused attention and self-report validity. *Journal of Personality, 5,* 513–527.

Rathjen, D. P., & Foreyt, J. P. (Eds.). (1980). *Social competence.* Elmsford, NY: Pergamon.

Ridley, D. S. (1991). Reflective self-awareness: A basic motivational process. *Journal of Experimental Education, 60,* 31–48.

Rodriguez, M. L., Mischel, W., & Shoda, Y. (1989). Cognitive person variables in the delay of gratification of older children at risk. *Journal of Personality and Social Psychology, 57,* 358–367.

Rogoff, B. (1990). *Apprenticeship in thinking: Cognitive development in social context.* New York: Oxford University Press.

Salovey, P., & Mayer, J. D. (1990). Emotional intelligence. *Imagination, Cognition, and Personality, 9,* 185–211.

Schmidt, R. A. (1975). A schema theory of discrete motor skill learning. *Psychological Review, 82,* 225–260.

Schunk, D. H. (1991). Goal setting and self-evaluation: A social cognitive perspective on self-regulation. In M. L. Maehr & P. R. Pintrich (Eds.), *Advances in motivation and achievement: Vol. 7. Goals and self-regulatory processes* (pp. 85–113). Greenwich, CT: JAI Press.

Searle, J. R. (1992). *The rediscovery of the mind.* Cambridge, MA: MIT Press.

Seligman, M. E. P. (1991). *Learned optimism.* New York: Knopf.

Shantz, C. U. (1983). Social cognition. In J. H. Flavell & E. M. Markman (Eds.), *Cognitive development,* in P. H. Mussen (Ed.), *Carmichael's manual of child psychology* (4th ed., pp. 495–555). New York: Wiley.

Shure, M. B., & Spivack, G. (1980). Interpersonal problem solving as a mediator of behavioral adjustment in preschool and kindergarten children. *Journal of Applied Developmental Psychology, 2,* 211–226.

Smirnov, S. D. (1994). Intelligence and personality in the psychological theory of activity. In R. J. Sternberg & P. Ruzgis (Eds.), *Personality and intelligence* (pp. 221–247). Cambridge, England: Cambridge University Press.

Spivack, G., Platt, J. J., & Shure, M. B. (1976). *The problem-solving approach to adjustment.* San Francisco: Jossey-Bass.

Sternberg, R. J. (1985). *Beyond IQ: A triarchic theory of human intelligence.* New York: Cambridge University Press.

Sternberg, R. J., & Kolligian, J., Jr. (Eds.). (1990). *Competence considered.* New Haven, CT: Yale University Press.

Sternberg, R. J., & Smith, C. (1985). Social intelligence and decoding skills in nonverbal communication. *Social Cognition, 3,* 168–192.

Sternberg, R. J., & Wagner, R. K. (Eds.). (1986). *Practical intelligence: Nature and origins of competence in the everyday world.* New York: Cambridge University Press.

Thorndike, R. L., & Stein, S. (1937). An evaluation of the attempts to measure social intelligence. *Psychological Bulletin, 23,* 275–285.

Tobias, S. (1979). Anxiety research in educational psychology. *Journal of Educational Psychology, 71,* 573–582.

Tobias, S. (1985). Test anxiety: Interference, defective skills, and cognitive capacity. *Educational Psychologist, 20,* 135–142.

Vygotsky, L. S. (1986). *Thought and language* (2nd ed.). Cambridge, MA: MIT Press.

Walberg, H. J., & Haertel, G. D. (Eds.). (1997). *Psychology and educational practice.* Berkeley, CA: McCutchan.

Walker, R. E., & Foley, J. M. (1973). Social intelligence: Its history and measurement. *Psychological Reports, 33,* 839–864.

Wentzel, K. R. (1994). Relations of social goal pursuit to social acceptance, classroom behavior, and perceived social support. *Journal of Educational Psychology, 86,* 173–182.

Wheeler, V. A., & Ladd, G. W. (1982). Assessment of children's self-efficacy for social interactions with peers. *Developmental Psychology, 18,* 795–805.

Wicklund, R. A. (1980). Group contact and self-focused attention. In P. B. Paulus (Ed.), *Psychology of group influence* (pp. 189–208). Hillsdale, NJ: Erlbaum.

Winsler, A. (in press). *Parent–child interaction and private speech in boys with ADHD.*

Wong, C. T., Day, J. D., Maxwell, S. E., & Meara, N. M. (1995). A multitrait–multimethod study of academic and social intelligence in college students. *Journal of Educational Psychology, 87,* 117–133.

CHAPTER EIGHT

Personal Navigation

✧

ROBERT J. STERNBERG
LOUISE SPEAR-SWERLING

Perhaps one reason for the enduring popularity of fortune-tellers and tarot cards is the difficulty of predicting how well any individual's life will turn out and the desire to foresee and even influence that future life. Of course, sometimes one's predictions are right on the mark. Frequently, however, people surprise us. Individuals who seemed extremely talented in childhood may fail to realize their early promise; others, who at first appeared doomed to failure, may turn out amazingly well as adults.

Consider, for instance, William James Sidis, who is often given as an example of intellectual giftedness gone bad. According to Winner (1996), Sidis's father, Boris, had the highest ambitions for his son, born in 1898, even to the point of naming the son after the famous Harvard philosopher and psychologist, William James. Boris Sidis taught William to read and spell when his son was just 2 years of age. By 3, William was typing in French as well as English. At 10, he was speaking six languages. Boris sought media attention for his son, and for obvious reasons, got it. By the age of 15, William was graduated *summa cum laude* from Harvard.

From there, it was downhill all the way. After 1 year of graduate school, Sidis had completely lost interest in math, saying that the sight of mathematical formulas made him physically ill. He took a series of clerical jobs and devoted his attention to increasingly trivial pursuits. For example, at the age of 28, William wrote a book on the classification of the transfer slips that streetcar conductors give to passengers. Sidis became extremely bitter and described himself in an article in *The New Yorker* as having had great and ironic enjoyment leading a life of wandering irresponsibility after a childhood of scrupulous regimentation. He died of a brain hemorrhage at the age of 46.

Sidis is not a unique case. Adragon De Mello, another highly gifted child cited by Winner (1996), was pushed ruthlessly by his father. By age 9, he was doing calculus and writing screenplays. He was graduated from college with a math major at age 11. Eventually, his mother left his father and gained custody. De Mello's mother, unlike his father, had no interest in pushing her son to the limit. Nevertheless, De Mello quickly lost interest in academic pursuits.

Bad outcomes are not merely the result of parental pushiness, because there have been talented children who have been equally pushed and have turned out quite differently. John Stuart Mill is often given as an example. His father also pushed him mercilessly, allowing him no holidays (Winner, 1996). Norbert Wiener grew up in the same way and by 18 had earned a doctorate from Harvard. Both Mill and Wiener experienced some severe depressions, yet Mill became one of the great philosophers of all time, and Wiener invented the field of cybernetics.

Other youngsters experience serious difficulties rather than spectacular achievement in early childhood but nevertheless go on to find success in adult life. Charles Pellegrino, a scientist at Brookhaven National Laboratory, has worked in fields as diverse as paleontology, crustaceology, and marine archeology. His work on ancient DNA was the basis for the Michael Crichton novel and Steven Spielberg film *Jurassic Park*. Unlike the cases we have mentioned thus far, however, Pellegrino struggled with serious reading difficulties in childhood. He alludes to this fact in the dedication to his book *Unearthing Atlantis* (1991), an archeological account of the massive volcanic eruption on the Greek island of Thera: "To five who believed in a 'retarded' boy who could not read, but thought he would . . . Thank you for not believing in test scores, for not browbeating, and for believing instead that you could bring the boy out by encouraging his love of science" (p. vii).

Gregory Howard Williams surmounted another kind of adversity to achieve adult success. Williams was a child of mixed race who did not learn of his African American ancestry until later childhood, when his biracial father, who had long "passed" as white, took Gregory and a younger brother to live with African American relatives. In his autobiography, Williams (1995) describes a childhood marked by extreme poverty, racial discrimination, his father's alcoholism, and his mother's abandonment. Always an excellent student, Williams ultimately became Dean of the Ohio State University College of Law. Things did not turn out so well for Williams's younger brother, who did poorly in school, was chronically in trouble, and eventually was blinded in a shooting.

Given this extreme variation in outcomes, what, exactly, is the lesson to be learned from these children and their lives? Of course, these stories could be taken to illustrate the difficulty of generalizing from single-subject case studies (Stanovich, 1996), or conversely, to illustrate that even single-subject case studies do not provide sufficient information on the basis of which firm inferences about children's futures can be made. We believe, however, that an important principle of development can be inferred from these case studies. Moreover, the view we present is consistent with all of the case studies, suggesting that there is a moderator variable that distinguishes those who find success (however they and society define it) from those who do not. The moderator variable is the construct that serves as the basis of this chapter, namely, personal navigation.

We should say something at the outset regarding our use of the word "success." Obviously, there are varying degrees of success in life. One may go down in history as a famous philosopher, like Mill, or one may achieve success on a more modest scale; one may be highly successful in one's professional life but much less so in one's private life, or vice versa. In this chapter, we use the term "success" broadly to encompass both superlative and more ordinary degrees of success, and we include success in the private as well as a wide variety of professional domains.

We begin this chapter by explaining what we mean by "personal navigation" and why we became interested in navigation as a metaphor for understanding why some people find success in adult life and others do not. Next, we contrast the concept of personal navigation with existing psychological constructs, such as intelligence, personality, and motivation. We argue that although personal navigation draws upon many of these existing constructs, it is not identical to any of them. We also suggest that personal navigation provides a better way of understanding success in life than do any of these psychological constructs by themselves. Third, we discuss the main emphases of the navigational metaphor as compared with other psychological constructs. And finally, we propose some elements that may contribute to skill in personal navigation.

WHAT IS PERSONAL NAVIGATION?

Broadly speaking, *personal navigation* (PN) refers to a person's control of his or her voyage through life. The formal definition of the verb "to navigate" is "to plan, record, and control the course and position of (a ship or aircraft)" (*American Heritage Dictionary*, 1985, p. 833). Thus, personal navigation involves certain kinds of goals, plans, and beliefs (e.g., knowing where one wants to go and believing that one is capable of getting there), as well as dealing effectively with events in one's life (e.g., riding out "storms" such as a variety of personal crises). More specifically, among other things, PN involves finding a direction in life; maintaining this direction when appropriate and changing it when appropriate; moving in the direction at a velocity that is appropriate for the circumstances; using navigational aids in order to maintain the desired direction; and overcoming the obstacles that inevitably present themselves in any voyage. One's PN is enhanced to the extent that one is self-aware with respect to needs, desires, and goals. We believe that navigation is an apt metaphor for what people actually need to do in order to gain, maintain, and, at times, reestablish control over their lives. It is the means by which self-awareness is translated into a plan of action for one's life, in the context of the cultural milieu in which one lives as mediated by one's knowledge of this context. One may navigate through some domains of life (say, the work domain) more effectively than through other domains of life (say, the personal domain), much as sailors may be better able to navigate through some kinds of seas than through others.

Our interest in the construct of personal navigation has several sources. Perhaps the most immediate source for a construct such as PN is the wealth of personal life stories such as those that are briefly summarized at the beginning of this chapter. Just what is it that distinguishes those who ultimately become successful in adult life—as defined both by society and by the individuals themselves—from those who do not? We argue in this chapter that it is, in large part, expert personal navigation. Clearly, excellent school achievement is an asset, perhaps all the more so for children who, like Williams, grow up in particularly adverse life circumstances (see Werner & Smith, 1992). However, as the stories of Sidis and De Mello illustrate, even extremely precocious academic achievement provides no guarantee of success in adulthood. Furthermore, youngsters who do not excel in school early on, like Pellegrino, may end up as successful and accomplished adults. Neither does uniformly high IQ characterize all of the successful adults in our opening case studies; Pellegrino was thought to be "retarded," and Williams (1995) describes being crushed by disappointment when he received only a modest score on a school-administered IQ test.

One obvious feature that does distinguish the successful individuals in our case studies from the unsuccessful ones is that the former found "destinations" in their professional life, and set their sights on those destinations, whereas the latter failed to do so. Eventually, finding themselves drifting and at sea, they lost interest in the voyage. Gruber (1974/1981) has spoken of the evolving systems in the lives of creative people. But all people seek to have evolving systems; it just seems that some evolve toward some desired destination, whereas others evolve toward no destination in particular.

Part of finding a destination in one's professional life involves discovering work that brings intrinsic satisfaction and pleasure. When children are young, they may work hard at something primarily to please their parents, but success in adulthood requires, or at least is greatly facilitated by, pursuits that are intrinsically engaging and rewarding. Mihaly Csikszentmihalyi (1990) calls the experience of pleasure and total concentration during periods of peak performance "flow" and finds that it tends to characterize individuals who are successful in a wide variety of domains. For instance, Csikszentmihalyi found that among a group of 200 art students, the ones who were most likely to become serious painters were the ones who loved to paint, rather than those who were motivated primarily by the hope of fame or financial gain. Similarly, the successful adults in our case studies all appeared to find work that was intrinsically rewarding and a good "fit" to their interests, abilities, and personalities.

Literature is a second source of our interest in the navigational metaphor. The notion of life as a voyage to be navigated is an old and time-honored one in literature. Indeed, in one of the oldest surviving works of literature, *The Odyssey*, the story of the voyage of Ulysses can be seen as a metaphor for life's voyage, as shown in the words of Joachim du Bellay in *Les Regrets*, "Happy he who like Ulysses has made a glorious voyage."

Perhaps the greatest author of English literature, Shakespeare, used the metaphor of a voyage many times. For example, in *Julius Caesar*, Shakespeare spoke of "all the voyage of their life, Is bound in shallows and in miseries." And in *Timon of Athens*, Shakespeare spoke of "life's uncertain voyage."

One of the most lyrical uses of the voyage metaphor was that of Walt Whitman in *Aboard at a Ship's Helm*, in which Whitman wrote, "But O the ship, the immortal ship! O ship aboard the ship! Ship of the body, ship of the soul, voyaging, voyaging, voyaging." And perhaps Whitman's most well-remembered words in *O Captain! My Captain!* also spoke of a voyage: "Oh Captain! my Captain! our fearful trip is done, The ship has weathered every rack, the prize we sought is won, The port is near, the bells I hear, the people all exulting."

Other authors have also spoken of life as a journey, although not necessarily a maritime one. Dante Alighieri, in *Inferno*, suggested, "In the middle of the journey of our life I came to myself within a dark wood where the straight way was lost." Robert Frost also contemplated woods in *Stopping by Woods on a Snowy Evening*, but needed to resume his journey: "The woods are lovely, dark, and deep, but I have promises to keep, and miles to go before I sleep." Christina Georgina Rossetti asked, in *Up-Hill*, "Does the road wind uphill all the way? Yes, to the very end. Will the day's journey take the whole long day? From morn to night, my friend."

Hundreds if not thousands of references to life's journey—along roads, seas, paths, or whatever—adduce to the importance of the navigational metaphor in literature. Here, we argue that more important than the journey itself is how one navigates through it. These references make clear the importance of life's journey to literary authors. Why, then, have psychologists not done more with this metaphor? In particular, might this metaphor tell us something about a major source of individual differences in what leads some people to attain success in adult life, and others to fail?

EMPHASES OF THE METAPHOR
OF PERSONAL NAVIGATION

The metaphor of personal navigation, like any metaphor, serves to emphasize certain things and deemphasize others. What are some of its main emphases?

The Importance of Direction

This emphasis places the construct of PN in contrast with many other psychological constructs, which seem more to emphasize either abstract mental power (e.g., as measured by tests of maximum performance) or attributes underlying mundane, everyday functioning (e.g., as measured by tests of typical performance). The quality of a person's PN is something that unfolds over the course of the person's life. It is probably not something that is easily measured by conventional kinds of testing, such as testing of abilities, achievements, or personality. Yet without this kind of broad direction in life, an individual may function at a very high level on conventional tests, and may indeed be extremely talented but—like Sidis or De Mello—never fully realize those talents.

The Importance of Flexibility and Change in Course

Often, in traveling, one sets subgoals in order to reach one's main goal, and one may even change one's ultimate destination as one discovers new things about the various places to which one might go. Very few people set a life course in their early childhood and then stick exactly to it; rather, they need to keep modifying their course in order better to take into account what they learn about themselves, what they learn about others, and what they learn about the environment. The need for flexibility (see also Cantor & Harlow, 1994) becomes especially clear when one considers that all three of these elements are moving targets: One changes; other people change; and the environment changes as well. The unhappy veterinarian, whom we mention later, might have found a different and (for her) more suitable job if she had recognized sooner that she could not deal with euthanizing animals, and if she had used this knowledge to modify the course of her career.

Velocity Is Always in the Service of Direction

Jean Piaget sometimes referred, only half-jokingly, to the American problem, namely, the desire to rush people into later and later stages of cognitive development at earlier and earlier ages. We are a society preoccupied with speed, and our emphasis in assessments of abilities on speed of functioning is one way we show this preoccupation. Indeed, mental quickness is seen as essential to intelligence (see Sternberg, Conway, Ketron, & Bernstein, 1981).

Many people view as a suitable educational program for very bright youngsters one that substantially accelerates their educational progress (see, e.g., Stanley, 1974, 1976). But the result of such an emphasis on speed can be people who discover, sooner or later, that they are on the fast track to nowhere, or to nowhere they want to go. William James Sidis, discussed earlier, was an extreme example of such a discovery. In our hurry to get places, we often fail fully to consider whether they are places we truly want to reach.

The Importance of Overcoming Obstacles

Much of the difficulty of any kind of navigation is in overcoming the obstacles one encounters along the way to one's destination. Few long trips are smooth sailing all the way. Moreover, people will differ in the kinds and numbers of obstacles they encounter. Some people, especially the poor and the infirm, encounter many more obstacles in life than do other people.

The navigational metaphor suggests several types of obstacles. Barrier obstacles, analogous to mountains in land journeys or land-mass barriers in sea journeys, block one from proceeding any further. In life, barriers take many forms. Some of them, such as rejections due to lack of credentials, may represent legitimate attempts by society to ensure that opportunities go to the most qualified; others of them, such as racial or ethnic prejudice, may represent illegitimate attempts by interest groups to block those who are not members of societally preferred groups. Whatever the kind of obstacle, people have to decide whether to allow themselves to be blocked (e.g., give up on trying to get a certain kind of job), to attempt to get around the barriers (e.g., to try to get hired despite the lack of the usual educational qualifications), or break through the barriers (e.g., expose the racial prejudice and hope that one can thereby remove it). The decision one makes as to what to do about a barrier obstacle will powerfully affect whether one is able to reach one's desired destination, or perhaps, how one changes one's desired destination.

Resistance obstacles, analogous to heavy snow or rain in a journey, do not block one's path but, rather, make it harder to follow the path. For example, a man may be told that a job can be his, but be told that the job will be available only after he has done so many other things to get the job that the attractiveness of the job starts to decrease.

Visibility obstacles, analogous to darkness or fog, occur when one finds oneself unable to see where one is going. The route is there, but one cannot see it. For example, one may wish to advance in one's firm. One looks at the people who have advanced in order to figure out what they have done in order to get where they wanted to go. Clearly, each found a path. What is not clear is what the paths were.

Sabotage obstacles are attempts by others to render more difficult one's progress along the path of one's choosing. At times, a number of people are competing for the same or similar resources, or people may perceive themselves as in a zero-sum game, whether they actually are or are not. In such cases, attempts by some people to sabotage the progress of other people are fairly common. In politics, "dirty tricks" are basically attempts to sabotage opponents, often through the spread of false information. In business, false information may be spread about a competitive company. In academia, people reviewing articles or grants may downgrade their evaluations of others in an attempt to get for themselves the resources they are trying to have withheld from others.

Finally, resource obstacles refer to the lack of resources needed to get where one wants to go. They are analogous to lacking proper navigational equipment, or the vehicle one would need in order to make progress in the terrain one will encounter. For example, one might not have the capital to start the business one would like to start, or the funds to start the research one wishes to undertake.

People tend often to view obstacles as, in some sense, unnatural, as things that block their way that should not be there. In fact, though, as would be true of almost any trip, the presence of obstacles tends to be the natural course of things. The question is not whether they will be there, but what one will do when one encounters them (see also Sternberg & Lubart, 1995).

The Importance of a Long-Term Perspective

Most people's life voyages span many years and many stages of development. Like Williams or Pellegrino, one may get off to an inauspicious start but nevertheless be highly successful later in the voyage; conversely, like Sidis or De Mello, one may begin spectacularly but end up snatching defeat from the jaws of victory. Furthermore, the kinds of difficulties one faces in life vary greatly depending on one's stage in life. For example, the abilities necessary for success in formal schooling, such as reading ability, are tremendously important in childhood but are relatively less important in adult life. Thus, a person who appears to be in trouble at one stage in life may function much more successfully at another stage.

ELEMENTS OF PERSONAL NAVIGATION

Thus far, we have focused on differentiating PN from other psychological constructs. For the most part, we have alluded only briefly to certain abilities and personality characteristics that may contribute to success in personal navigation. But what, exactly, are these abilities and personality traits? Although our discussion must remain somewhat speculative, here we attempt to elaborate in more detail.

Abilities

Earlier, we suggested that PN involves "direction" in the most macroscopic sense, applied to one's life, and we discussed examples of a lack of this kind of direction, as in the case of Sidis or De Mello. In everyday parlance, we sometimes refer informally to someone's having a "sense of direction." What would be the analogue in a person's life?

We believe that the most essential elements of sense of direction involve self-understanding, in intellectual terms (what Gardner, 1983, refers to as "intrapersonal intelligence"), in emotional terms (as described in Salovey & Mayer, 1990), and in the integration of the two. A person has to know him- or herself, which Aristotle viewed as central to life's journey. Self-understanding seems to be a key to PN. Of course, one may be extremely gifted intellectually in many ways, like Sidis or De Mello, yet lack self-understanding.

In terms of the triarchic theory of human intelligence (Sternberg, 1985, 1996), one needs to (1) know one's strengths, (2) know one's weaknesses, (3) find a way to capitalize on the strengths, and (4) find a way to compensate for or remediate weaknesses. Consider people who have navigated their way to the top of their fields—highly successful poets or lawyers or doctors or scholars or politicians or whatever. Would they have been as successful had they chosen some other career? Perhaps, but on the face of it, more likely not. Why? Because those who navigate their way to the top have presumably found a career niche that represents a close to ideal fit between what they have to offer and what their field values. So, for example, the politician who is a master at gaining people's confidence and votes often does not, in fact, turn out to be the best president or governor, as we have seen many times in this country and others. Boris Yeltsin ran an extraordinary campaign for the Russian presidential election, but there are no indications that his leadership and management abilities in any way have matched his craftiness in gaining votes.

Furthermore, in order to find the right direction in life, one must not only understand one's intellectual strengths and weaknesses, but also evidence self-understanding in other ways. For instance, as we discussed earlier, someone might have all of the intellectual abilities required for an academic career, but be miserable because of other things that tend to go along with the profession, such as frequent travel and job changes. Someone might be a talented musician, but be unable to tolerate working nights and weekends, the lack of a stable income, and the like.

Also, people are not always interested in what they are good at. For example, someone might have the intellectual talents to excel in medicine, but not especially enjoy the study or practice of that field. It is our impression that successful people are almost always highly engaged in their work, usually to the point that one cannot imagine them doing anything else. Successful writers sometimes say that they "just have to write," but we think analogous statements could be made about successful scientists, elementary school teachers, musicians, and so on. In other words, finding direction in life involves applying one's self-understanding broadly, to one's personality, emotions, and interests, as well as to one's intellectual talents.

This point of view has strong implications for education. A major function of education should be to help students understand their strengths and weaknesses, make the most of their strengths, and also find ways to compensate for or correct weaknesses (Cronbach & Snow, 1977). In addition, educators should attempt to find students' unique interests, and, where possible, tap those interests in designing instruction, for example, through the project method. Educators should also help students develop new interests. Such an approach may even be a way to help children with serious academic difficulties to succeed in school, and eventually, in life— like Pellegrino. Unfortunately, however, it is not clear to us that our educational system fulfills these functions very well. Many of the students we teach, even at the graduate level, have only the foggiest idea of what their intellectual strengths and weaknesses are, much less how to manage them; have no special passion for their chosen field of study; or have given little thought to whether their chosen field suits their personalities. The result is not only a loss in their personal lives—if they enter fields that are not good fits to them—but also losses of productivity to society of having people mismatched to jobs.

We believe that one factor that contributes to this mismatching is an overemphasis among some psychologists on the importance of so-called general ability (*g*) in predicting job performance (e.g., Schmidt & Hunter, 1981; Ree & Earles, 1993) at the expense of other kinds of measures, such as, for example, other aspects of intelligence (Bandura, 1997; McClelland, 1973; Sternberg & Wagner, 1993; Sternberg, Wagner, Williams, & Horvath, 1995). Similarly, formal schooling usually emphasizes academic intelligence over other kinds of intelligence, such as creative or practical intelligence.

In terms of PN, however, creative intelligence will be important in creating a direction in which to navigate, and practical intelligence will be important in figuring out how to get there (Sternberg, 1985, 1996). Indeed, in our own work, we have found measures of practical intelligence to predict on-the-job success at least as well as, or better than, do measures of conventional academic analytical abilities.

The measures of practical intelligence we have used employ simulations of the kinds of problems people face in real-world, on-the-job problems. For example, a business executive might be asked to rate how useful each of several pieces of information would be in deciding upon a contractor to complete a job. Or a salesperson might be asked to rate the value of each of a number of different tips for success in racking up sales. Or an academic might be asked how he or she would allocate time among various tasks, given that there is not time to get them all done. We have used paper-and-pencil tests in all these domains. In the case of sales, we have also used a performance test, where an individual has to simulate trying to sell advertising space to a potential customer over the phone. We have found that both the paper-and-pencil test and the performance test predict sales performance, but that they are themselves relatively independent of each other. They measure different aspects of potential performance on the job.

Although there are generalized directions in which one can go in life, every person has to forge his or her own path. This generalization is especially true in countries, like Russia, that are undergoing periods of rapid change, and in which conventional paths to success may now be dead ends. In these countries, the combination of creative and practical intelligence that is important everywhere may be especially important.

All of these abilities—intrapersonal intelligence, emotional intelligence, creative intelligence, and practical intelligence—involve a kind of mental power or potential. But, as we have repeatedly emphasized, PN involves actually applying one's intellectual powers in life. Thus, like the most beautifully designed ship, even the most varied and exceptional intellectual talents will be for naught if they are not used effectively in navigating one's life course. Toward this end, we believe that certain personality traits may be especially important.

Personality

The personality traits central to PN might be thought of as the development and coordination of the "4Rs": resilience, relentlessness, restlessness, and risk taking. Let's consider each of these characteristics in turn.

Resilience

There is a considerable research literature on children who do well in spite of growing up under a variety of adverse conditions, including children who experience economic hardship (Elder, 1974; Werner & Smith, 1992), psychotic parents (Anthony, 1987), and being raised in foster institutions (Rutter, 1987, 1989). As we mentioned earlier, these kinds of children often are characterized as "resilient" and tend to share a number of characteristics, including an easy temperament. But, of course, everyone faces a certain amount of adversity in life, even individuals who grow up in relatively ideal circumstances. In order to succeed in life, everyone must be able to recover from failure or loss.

For instance, in academia, people must be able to bounce back from having papers and proposals rejected. If one is inordinately discouraged by this kind of short-term failure, to the point where one is deterred from trying again, then success will be impossible. Resilience is also important in one's personal life, of course. For example, at one time or another, virtually everyone experiences a failed romance or rejection from a prospective romantic partner; success in the romantic realm is unlikely if one is unable to recover from these kinds of setbacks.

Relentlessness

We use "relentlessness" to mean dogged determination or persistence, and researchers interested in resilient children and youth have also found these youngsters to be highly persistent in the pursuit of their goals. Americans sometimes seem to emphasize the importance of ability relative to effort (Stevenson & Stigler, 1992), but sheer persistence may often outweigh ability in success. Williams, whom we profiled in our opening, had considerable academic aptitude, but in his autobiography, he describes many obstacles that required relentless determination to overcome. For instance, in sixth grade, one memorable disappointment involved losing out on a coveted academic award, one that he clearly deserved, because of racial prejudice. Someone with academic aptitude equal to Williams's, but with less relentlessness, might well have given up in the face of obstacles like these and ended up a failure.

Like resilience, relentlessness is important even if one does not grow

up under particularly adverse circumstances. To return to our previous
ample, one characteristic that may separate successful from unsuccessful
academics is the willingness of the former to persist in paper submissions,
grant submissions, and the like, even in the face of repeated failure.

Restlessness

In order to navigate well through life, one cannot be too content to stay in
one place, but rather, must have the desire to keep moving on. In other
words, successful people are not too easily satisfied with their achievements; they continue to set meaningful goals for themselves. This characteristic may have been a problem for our examples of failed intellectual
giftedness, Sidis and De Mello, who were unable to maintain their prodigious early achievements into adult life. On the other hand, the successful
individuals in our case studies all kept seeking (and meeting) new challenges. Pellegrino, for instance, has made contributions in a diverse array
of scientific domains.

Successful individuals also appear to set their goals at the right level of
difficulty. For example, people who rank high in the need for achievement,
such as successful entrepreneurs, seek out moderately challenging tasks,
persist at them, and are especially likely to pursue success in their occupa-
tions. Why would these people seek out tasks that are only moderately
challenging? These are the tasks in which they are likely both to succeed
and to extend themselves. They do not waste their time on tasks so challenging that they have little probability of accomplishing them, nor do they
waste time on tasks so easy that they pose no challenge at all.

Risk Taking

When we think of risks, we often think of dramatic ones, as in the case of
the entrepreneur who stakes a tremendous amount of money on a single
business decision, or the stuntman who chances serious injury or even
death. However, success in any domain, whether personal or professional,
always entails risk, because the very attempt to succeed involves the potential for failure or rejection. The only alternative is to opt out of the attempt
altogether. For example, successful writers and artists must accept the risk
of rejection of their work, and romantic relationships entail making oneself
vulnerable to rejection by another person. Furthermore, the higher one
strives to achieve, usually, the higher the risk of failure. For instance, a particularly talented physician may seek out patients with more challenging
problems to treat, or a talented teacher may seek out more challenging
children to teach, but in each case, greater challenge tends to involve
greater risk of failure.

Of course, by "risk taking," we do not mean foolish or pointless risks. Even the best navigator would not deliberately sail into a hurricane if he or she could possibly avoid it. Rather, successful people take the kinds of meaningful risks that are necessary to advance in their work or their personal lives.

All of these personality traits—resilience, relentlessness, restlessness, and risk taking—must be applied in the right direction, or they are worse than useless. Most of us have known people, for example, who are relentless in their pursuit of the wrong romantic partner or the wrong career goals. But success is not just the automatic result of sheer intellectual brilliance or lofty goals, even the right kinds of goals. Rather, we believe that personality traits such as the "4Rs" are essential for even the most impressive intellectual abilities to be realized, and for successful navigation through life.

COMPARING PERSONAL NAVIGATION WITH OTHER PSYCHOLOGICAL CONSTRUCTS

Much of our interest in the concept of personal navigation stems not only from individual case studies and literature, but also from existing psychological theory and research. When it comes to predicting outcomes in adult life, psychologists may have a somewhat better track record than do fortune-tellers, but this track record is far from perfect. For instance, it has proven difficult to predict real-world performance from various kinds of tests; usually the correlations between scores on these tests and real-world achievement are no better than .3 or so (e.g., Hunt, 1980; Deary & Stough, 1996). Other investigators (e.g., Werner & Smith, 1992) have weighed multiple risk factors (and multiple protective factors) in attempting to predict adult outcomes in groups of children followed longitudinally. However, even this approach, as Weissbourd (1996) points out, is far from perfect in its predictive power.

Of course, it may be the measures that psychological researchers have used, and not the psychological constructs themselves, that are lacking. Indeed, there are several broad psychological constructs that would appear to relate strongly to PN. Is PN just a different name for one of these other constructs? Next, we attempt to address this question for the following constructs: intelligence, planning, wisdom, motivation and self-efficacy, and personality.

Intelligence

Our case studies illustrate that PN is not identical to intelligence conceived in any narrow way as simply IQ test performance or purely academic intelligence. However, many contemporary investigators have conceptualized intelligence in broader or more multidimensional terms, and we are certainly sympathetic to these views of intelligence. Could PN be another type of intelligence, or intelligence very broadly conceived?

For instance, practical intelligence refers to a person's ability to apply the intellectual skills they have in order to adapt to, shape, and select real-world environments (Sternberg, 1985). At first glance, this construct surely sounds somewhat like PN. But practical intelligence, like other kinds of intelligence, deals with a sort of power—the power, in this case, to apply intelligence to everyday life. Having intelligence, on the one hand, and actually using one's intelligence effectively in one's life, on the other hand, are two different things. PN is more about the latter than about the former.

In our own measures of practical intelligence, we assess people's "tacit knowledge" for how to succeed in management, sales, academia, or other professions. But what we are assessing is whether people have and know how to use tacit knowledge, not whether they actually choose to apply that knowledge effectively in their lives. For instance, some individuals with a fledgling academic career may have excellent tacit knowledge about what they need to do to be a successful academic, such as establishing a program of research, obtaining grants, getting papers published, and the like. They may even be intellectually capable of doing stellar research, writing wonderful papers, and so on. But, for a variety of reasons, they may not actually get any of these things done! Perhaps they are the kind of people who are always getting bogged down in personal crises that keep them from getting work accomplished. Perhaps they are too easily discouraged by failure, such as by getting papers rejected. Perhaps they are miserable because of certain requirements that tend to go along with an academic career, such as frequent travel and job changes. Or perhaps they are unable to organize and reconcile competing goals in their lives (see, e.g., Sheldon & Emmons, 1995). The point is that, if these kinds of individuals fail in their academic careers, it is not because they lack the intellectual abilities required for success. Rather, we suggest that PN—choosing the right direction, being able to overcome obstacles, and so on—may be a more useful way of understanding these individuals' difficulties.

Perhaps closer to PN than practical intelligence is social intelligence, which Cantor and Kihlstrom (1987) define as the knowledge repertoire underlying the interpretation of events and the making of plans in everyday life situations. We believe PN is somewhat different again, involving a person's total configuration of the self with respect to the self, and also all the people and events in one's life.

Another type of intelligence that has received considerable attention recently is emotional intelligence (Goleman, 1995; Salovey & Mayer, 1990). Emotional intelligence involves knowledge, regulation, and control of emotion. The ability to understand and to control one's own emotions is definitely useful in life and, as we will discuss later, likely plays a role in PN. Nevertheless—just as in the case of practical intelligence—having a set of abilities and actually using these abilities effectively to guide one's life are two different things. Thus, one might be emotionally intelligent, yet still, for a variety of reasons, be lacking in personal navigation.

Alfred Binet, whose test of intelligence formed the prototype for almost all intelligence tests currently used, was arguably one of the most influential researchers of all time. Binet's intelligence test was so influential a contribution to psychology that it overshadowed what might and perhaps should have been an even more important contribution: Binet's theory of intelligence. Binet's theory contains within it the seed of the concept of PN.

Binet and Simon (1905/1916) argued that intelligent thought comprises three distinct elements: adaptation, criticism, and direction. Adaptation refers to people's ability to adjust to the environments in which they find themselves. Criticism refers to people's ability to critique their own thoughts and actions. Direction involves knowing what has to be done and how to do it.

Adaptation has always played a major role in definitions and theories of intelligence. Indeed, in an early symposium on the definition of intelligence (Buckingham, 1921), adaptation was mentioned by many respondents (who were leading figures in the field of intelligence of the day) as important to intelligence. And adaptation forms the core not only of Binet's theory and test, but of Wechsler's (1939) as well. Adaptation continued to play an important role in a later survey of experts on definitions of intelligence (Sternberg & Detterman, 1986), although the frequency of its mention decreased to 13%.

Criticism, or the ability to monitor and critique one's own thinking and problem solving, did not play much of a role in the 1921 definitions of intelligence, but did play a more important role in the 1986 definitions, where "metacognition," which involves such criticism, was mentioned by 17% of respondents. Metacognition has also come to play a major role in cognitive theory, in general (see, e.g., Nelson, 1996).

At a microscopic level, direction has also played an important role in theories and measures of intelligence. To succeed on intelligence tests, test-takers need to know what strategies to use, and how to use them. But at a macroscopic level, intelligence theorists and measurement experts have been less vocal. The PN construct brings into the spotlight Binet's concept of direction.

PN involves direction at a highly macroscopic level, that is, the direction one takes in one's life, either toward the successful realization of one's abilities or toward their lack of realization. As we have already noted, personal navigation is more about using one's intelligence effectively than it is about intelligence itself. It may also be heavily involved in coordinating the kinds of intelligence that have been mentioned here: academic, practical, social, and emotional. Of course, being highly intelligent is never a drawback in life. Furthermore, as we discuss in a later section of this chapter, certain kinds of intelligence may play a particularly important role in PN. Nevertheless, being a successful navigator of life is about much more than raw intellectual power.

Planning

Cognitive researchers have also recognized the importance of direction—of being able to set a course and then follow it. The rubric under which PN has most commonly been pursued in the cognitive literature is that of planning.

Some of the seminal theories of cognitive psychology emphasized the importance of planning. A well-known theory that emphasized the importance of planning was the theory of Luria (1973, 1980), which viewed this skill as coordinate with simultaneous information processing, successive information processing, and attention (see also Das, Kirby, & Jarman, 1979).

Of equal importance was the theory of Miller, Galanter, and Pribram (1960), which viewed planning as central to all information processing. According to these authors, "A Plan is, for an organism, essentially the same as a program for a computer, especially if the program has the sort of hierarchical character" that the authors view as essential to higher cognition (p. 16). Although the authors view a Plan as a "rough sketch of some course of action" (p. 17), their conceptualization of the Plan more applies to specific cognitive tasks than to life in general.

Perhaps one exception to the tendency of cognitive investigators to focus on planning as it applies to specific problems rather than to life in general involves the research on social and practical intelligence that we mentioned in the previous section, often as manifested in theory and empirical work on the interface between personality and intelligence (see, e.g., Cantor & Harlow, 1994; Chiu, Hong, & Dweck, 1994; Haslam & Baron, 1994). However, as we discussed, although practical intelligence (and other types of intelligence) probably contribute to PN, they are not identical to PN. Intelligence involves having mental power, whereas PN involves actually using that power effectively in one's life. Similarly, planning in one's life certainly is an aspect of PN, but clearly, not the whole thing. Some people are capable of making great plans for their lives, but then never actually implement the plans, or implement them badly.

Wisdom

Perhaps we come closer to PN when we talk about wisdom rather than some form of intelligence or planning. Wisdom has been defined in many different ways (see Sternberg, 1990). For example, Csikszentmihalyi and Rathunde (1990) have defined it as a growth-oriented way to mediate between conflicting types of information; Labouvie-Vief (1990) sees it as an integration of reflection and critical thinking; Baltes and Smith (1990) have defined it in terms of factual and procedural knowledge integrated with contextualism, relativism, and uncertainty. Kitchener and Brenner (1990) speak of synthesizing knowledge from opposing views, and Orwoll and Perlmutter (1990) emphasize the integration of cognition and affect.

A wise person will almost certainly be a better personal navigator than will be one who is unwise. But wisdom appears, in most definitions, to refer to a set of skills, acquired over the course of a lifetime, for making good judgments of many kinds, perhaps within a certain domain. Many people show great wisdom in their work, but not in their personal lives. Or they may show wisdom in their advice to others, but not necessarily in their advice to themselves. Someone, such as a psychotherapist, may give wise advice at the same time that he or she has failed in various respects in his or her own personal life. Wisdom thus appears to be, at least in part, domain specific. In the later years of one's life, wisdom may be involved in reflecting on the successes and fortunes of the PN of one's life, as well as in continuing one's PN into old age. So wisdom seems to be related to PN, but again, not the same thing.

Motivation and Self-Efficacy

There are a number of different streams of research on motivation. One line of research has been done under the rubric of achievement motivation, which, we believe, is quite distinct from PN. David McClelland and his colleagues, for example, have been particularly interested in need for achievement (McClelland, 1961; McClelland, Atkinson, Clark, & Lowell, 1978; McClelland & Winter, 1969). The achievement motive, which involves competition with an internalized standard of excellence, is present in every culture and so has been the focus of dozens of cross-cultural studies (see Markus & Kitayama, 1991). Because increases in the achievement motive may be linked to increases in productivity, several projects in various cultures have tried to increase levels of this motive among workers and managers. In one such project, investigators assessed the effectiveness of the attempts by Indian business owners to encourage their employees to emulate the achievement motive shown by many Western businessmen and women. Toward this end, the Indian employees were subjected to an intense series of seminars designed to get them to think, talk, and act like achievement-oriented business people. The project was modestly successful (McClelland & Winter, 1969).

One of the numerous studies of the achievement motive in China found that Chinese parents place great emphasis on achievement, but their focus is different from that of American parents (Ho, 1986). Whereas American children are motivated to achieve for the purpose of being independent, Chinese children are motivated to please the family and the community.

The more one examines the literature on achievement motivation, the less it looks like the kind of construct we have in mind for PN. Achievement motivation may well be a component of PN, but only for some people; even for them it is certainly not the whole thing. Some people may decide that high levels of societally sanctioned achievement, such as in a career, are not really what they want out of life. They may have less career-oriented personal goals that are more important to them, for instance, raising a family. Or they may desire societally sanctioned achievement, but believe they are unlikely to attain it and so do not even bother trying.

Consider, for example, Jay MacLeod's (1995) ethnographic study of a primarily white gang of low-income teenage boys called the Hallway Hangers, most of whom had poor academic achievement and had dropped out of school. These boys did not subscribe to what MacLeod terms the "achievement ideology"—the belief that anyone can succeed in life with hard work. The Hallway Hangers did not view high school achievement as relating to improved chances of success on the job market; rather, they felt that even with great effort and personal sacrifice, they were not likely to be able to go to college, have satisfying careers, or move up the socioeconomic ladder. Furthermore, these feelings were not irrational, given their life circumstances. However, many of the boys did show considerable motivation and initiative outside of school. For instance, they were highly motivated to find work, and most had held a series of jobs, although, unfortunately, the legitimate work available to them tended to be poorly paid and without the opportunity for advancement. (Some of the boys also pursued illicit "careers," such as drug dealing, one building a thriving business until, in the words of another Hallway Hanger, "he became his own best customer," p. 157).

Most assuredly, some of their teachers would describe the Hallway Hangers as lacking in achievement motivation. However, in our view, the construct of PN may provide a better way of thinking about the Hallway Hangers' problems. In terms of PN, these boys could not find a fruitful direction in life; as one sympathetic teacher put it, they could not "project themselves into the future" (p. 97). This lack of direction seemed not so much the result of low motivation as of at least two other factors. First, the Hallway Hangers had trouble planning for their futures in part because they were consumed by the local processes of navigation, by the day-to-day exigencies of getting by. For instance, they and their families depended on the income they could generate from work, limited as that income was, in a way that would be unknown among middle- or upper-income teenagers. And second, they viewed success (at least societally sanctioned success) as being largely unattainable—somewhat like the person who contemplates trying to round the horn of South America in a canoe. Society does not give everyone the same kind of "boat" to navigate. At the same time, we need to realize that some people may furnish their own boat; it just may not be one of which society approves.

Other motivational work also touches upon certain aspects of PN. For example, some theorists have dealt with action controls, which are self-regulatory mechanisms used to accomplish goals. One theory, called "action control theory," has produced a theory of action versus state orientation (Kuhl & Kraska, 1989).

According to this theory, action-oriented individuals tend to take immediate action to achieve their goals. They are able to attend either successively or simultaneously to the present, the future, and discrepancies between the two. They seek to transform the present state into the desired future state (Kuhl, 1987). In a sense, they may be viewed as experts in the heuristic defined by Newell and Simon (1972) as "means–ends analysis," which involves computing the shortest path from where one is to where one wants to go. In contrast, state-oriented individuals cannot deal effectively with the transformation of the present into the future. They tend to dwell on past difficulties they have encountered, as well as intentions that are not appropriate to the situations in which they find themselves. They tend toward "fixation on past, present, or future states, for example, on a past failure to attain a goal, on the present emotional consequences of that failure, or on the desired goal state itself" (Kuhl & Kraska, 1989, p. 366).

The action orientation bears some resemblance to at least one aspect of PN, but seems not to be equivalent to it. For one thing, not everyone chooses an orientation toward action for his or her life. In fact, some taking a more Eastern orientation toward life might view the preoccupation with action rather than reflection and contemplation as a distinctively Western bias. For another thing, an action orientation can be toward a desired future state that, in fact, proves to have been ill-chosen. One of us has an acquaintance who, because of her love of animals, spent years training to be a veterinarian; having attained her goal, she finds that she hates the job, largely because she cannot stand having to euthanize animals on a regular basis. Almost all of us have, at one time or another, acted assertively toward the attainment of certain ends, only to find, when we achieved them, that they were not what we wanted, or should have wanted.

Yet another line of relevant research is that on self-efficacy (e.g., Bandura, 1986, 1991, 1997), which involves the belief that one has control over the events in one's life and the ability to overcome failure or adversity. People are most self-efficacious when they set realistic, fairly proximal goals that they have some hope of attaining. Self-efficacy and goal setting are in turn moderators of self-regulation. People who lack a sense of self-efficacy tend to believe that events are beyond their control and are likely to give up easily in the face of failure. The Hallway Hangers clearly lacked a sense of self-efficacy with regard to conventional success; they did not believe that they were capable of advancing in a traditional career or substantially improving their economic lot. And these beliefs certainly influenced their ability to find direction in life. However, the Hallway Hangers did not seem to lack self-efficacy in a global sense, for instance, with regard to their ability to get by "on the streets" or to be successful in an illicit business such as drug dealing. Furthermore, one might have a very healthy sense of self-efficacy, but still pursue the wrong direction in life or fail to find any direction at all; for example, we think it unlikely that the problems of Sidis or De Mello stemmed from a lack of self-efficacy, and neither did the problems of the veterinarian mentioned in the previous paragraph. Thus, in our view, self-efficacy is likely related, but not identical, to PN.

The stream of motivational research that may be most relevant to PN involves self-actualization (e.g., Maslow, 1943, 1970). Some psychological theorists, most notably, humanistic psychologists, have viewed people as in a constant striving for self-actualization. These theorists, including Rogers (1961, 1980) as well as Maslow, have attempted to specify the characteristics of self-actualized people. Perhaps the difference in our emphasis is that we believe in focusing on the process of getting to the destination as much as in focusing on the preconditions for PN (e.g., having basic food or shelter needs met).

From the standpoint of PN, life is a journey, but not always one with a clear destination. Furthermore, in life, one often must navigate by dead reckoning—through estimation and guesswork, without explicit guidance from modern navigational aids. Both having direction, on the one hand, and the actual process of navigation, on the other, are important in PN. The constructs of motivation and self-efficacy capture some (but only some) of what we mean by "direction," such as having goals and the belief that one can achieve them. They capture much less of the equally important, day-to-day process of trying to reach one's destination that consumes so much of one's time and energy.

Personality

One line of research in the area of personality has examined individual differences in temperament that are evident very early in life, such as temperamentally "easy" and "difficult" infants (e.g., Thomas & Chess, 1977). Among other characteristics, "easy" infants adapt fairly readily to new situations, smile often and otherwise evidence a positive mood, and show a high degree of regularity in basic biological functions, such as eating and sleeping. "Difficult" infants are relatively slow to adapt to new situations, cry often and loudly, and demonstrate considerable irregularity in basic biological functions. Not surprisingly, children with an easy temperament appear to elicit more positive responses from the adults around them, including their own parents, than do difficult children. Having an easy temperament also appears to provide a protective factor for children in highly adverse situations, such as children raised in extreme poverty and family instability (e.g., Werner & Smith, 1992), like Gregory Howard Williams, whom we mentioned in our opening. It is probably protective for adults as well. Those individuals who do well in life—despite being raised in a variety of adverse circumstances—are sometimes termed "resilient."

We believe that an easy temperament may contribute to PN, especially in childhood, when individuals are relatively powerless in their abilities to control their own life voyages, and eliciting sympathy from adults is particularly important. An easygoing temperament may also facilitate PN in adulthood, simply because people may treat those with easy temperaments in a way that better enables those individuals to have the flexibility to navigate through life as they wish. Once more, however, we do not think that this is all that there is to PN. For instance, we have known many highly successful adults who would hardly be described as having easygoing temperaments. Thus, a certain type of temperament may be an asset in life, especially in childhood and especially under adverse conditions. Nevertheless, PN involves much more than an easygoing temperament, and people who are excellent navigators in life may have a variety of temperaments.

Finally, we should say something about possible sources of individual differences in what we have categorized as PN skills. The abilities and personality traits that contribute to PN probably are strongly shaped by one's experiences and environment, as well as by one's inborn tendencies. For example, as MacLeod (1995) points out, the Hallway Hangers' dim view of their chances for success was quite consistent with their experiences (and furthermore, in MacLeod's view, with objective reality). And Williams (1995) attributes the sharply divergent paths in life taken by himself and his younger brother partly to the influence of their father, who strongly supported and encouraged young Gregory's dream of an education. Unfortunately, the father did not encourage similar dreams in Gregory's brother, whom he appeared to view as a ne'er-do-well like himself. To use the navigational metaphor, Gregory was strongly encouraged to find a direction in life; his brother was not. Navigation is directly under one's control, but what are not usually under one's control are the forces acting on one as one attempts to navigate.

CONCLUSION

In this chapter, we have introduced the construct of personal navigation, or PN. PN involves both having direction in life (as well as changing that direction when appropriate), and the process of successfully navigating events in one's life. Existing tests undoubtedly measure aspects of the skills needed for PN. But, as we noted at the outset of the chapter, none of them seems to come anywhere close to enabling us to predict who will navigate life successfully and who will not. Ultimately, we will need to decide whether PN is a legitimate and important construct, and if so, whether to try to assess it, either through a combination of existing tests, or through the creation of new tests. Perhaps the reason we have so often encountered barriers in predicting real-world performance from various kinds of tests is that existing tests measure just minor aspects of the PN skills that are necessary for achieving life goals. Moreover, PN probably represents not just a set of skills, but also their integration. It is also interactive—the interaction between individual and context.

Psychology has a lot of constructs. Does it need another? We suggest that it may. Dating back to the structuralist movement, the tendency of psychologists has been to try to figure out the elements that lead to one kind of process (e.g., sensation and perception in the case of the structuralists) or another behind overall adaptation. We believe that dealing with the construct of PN, under whatever name, requires us to be more holistic and to ask what it is that people genuinely seek in their lives. They may seek fame and fortune, outstanding professional achievement, satisfaction in their personal lives, or whatever. But we suggest that all these goals are subordinated to an overarching goal: setting a direction and then doing what needs to be done to get there, changing direction along the way, after realizing at times that the sought-after direction was not the right one. These are the things that, for many people, give meaning to their lives, and that might possibly even give some new meaning to the psychological study of people.

ACKNOWLEDGMENTS

The work reported herein was supported under the Javits Act program (Grant No. R206R50001) as administered by the Office of Educational Research and Improvement, U.S. Department of Education. The findings and opinions expressed in this chapter do not reflect the positions or policies of the Office of Educational Research and Improvement or the U.S. Department of Education. We are grateful to Michel Ferrari for his comments on this work.

REFERENCES

American heritage dictionary. (1985). Boston, MA: Houghton Mifflin.

Anthony, E. J. (1987). Children at high risk for psychosis growing up successfully. In E. J. Anthony & B. J. Cohler (Eds.), *The invulnerable child* (pp. 147–184). New York: Guilford Press.

Baltes, P. B., & Smith, J. (1990). Toward a psychology of wisdom and its ontogenesis. In R. J. Sternberg (Ed.), *Wisdom: Its nature, origins, and development* (pp. 87–120). New York: Cambridge University Press.

Bandura, A. (1986). *Social foundations of thought and action: A social-cognitive theory.* Englewood Cliffs, NJ: Prentice-Hall.

Bandura, A. (1991). Social cognitive theory of self-regulation. *Organizational Behavior and Human Decision Processes, 50,* 248–287.

Bandura, A. (1997). *Self-efficacy: The exercise of control.* New York: Freeman.

Binet, A., & Simon, T. (1916). *The development of intelligence in children.* Baltimore: Williams & Wilson. (Original work published 1905)

Buckingham, B. R. (1921). Intelligence and its measurement. *Journal of Educational Psychology, 12,* 271–275.

Cantor, N., & Harlow, R. (1994). Social intelligence and personality: Flexible life-task pursuit. In R. J. Sternberg & P. Ruzgis (Eds.), *Personality and intelligence* (pp. 137–168). New York: Cambridge University Press.

Cantor, N., & Kihlstrom, J. F. (1987). Social intelligence: The cognitive basis of personality. In P. Shaver (Ed.), *Review of personality and social psychology* (Vol. 6, pp. 15–34). Beverly Hills, CA: Sage.

Chiu, C.-Y., Hong, Y.-Y., & Dweck, C. S. (1994). Toward an integrative model of personality and intelligence: A general framework and some preliminary steps. In R. J. Sternberg & P. Ruzgis (Eds.), *Personality and intelligence* (pp. 104–134). New York: Cambridge University Press.

Cronbach, L. J., & Snow, R. E. (1977). *Aptitudes and instructional methods.* New York: Irvington.

Csikszentmihalyi, M. (1990). *Flow: The psychology of optimal experience.* New York: Harper & Row.

Csikszentmihalyi, M., & Rathunde, K. (1990). The psychology of wisdom: An evolutionary interpretation. In R. J. Sternberg (Ed.), *Wisdom: Its nature, origins, and development* (pp. 25–51). New York: Cambridge University Press.

Das, J. P., Kirby, J., & Jarman, R. (1979). *Simultaneous and successive cognitive processes.* New York: Academic Press.

Deary, I. J., & Stough, C. (1996). Intelligence and inspection time: Achievements, prospects, and problems. *American Psychologist, 51,* 599–608.

Elder, G. H. (1974). *Children of the Great Depression.* Chicago: University of Chicago Press.

Gardner, H. (1983). *Frames of mind: The theory of multiple intelligences.* New York: Basic Books.

Goleman, D. (1995). *Emotional intelligence.* New York: Bantam.

Gruber, H. (1981). *Darwin on man: A psychological study of scientific creativity* (2nd ed.). Chicago: University of Chicago Press. (Original work published 1974)

Haslam, N., & Baron, J. (1994). Intelligence, personality, and prudence. In R. J. Sternberg & P. Ruzgis (Eds.), *Personality and intelligence* (pp. 32–58). New York: Cambridge University Press.

Ho, D. Y. F. (1986). Chinese patterns of socialization. In M. H. Bond (Ed.), *The psychology of the Chinese people* (pp. 1–37). Hong Kong: Oxford University Press.

Hunt, E. B. (1980). Intelligence as an information-processing concept. *British Journal of Psychology, 71,* 449–474.

Kitchener, K. S., & Brenner, H. G. (1990). Wisdom and reflective judgment: Knowing in the face of uncertainty. In R. J. Sternberg (Ed.), *Wisdom: Its nature, origins, and development* (pp. 212–229). New York: Cambridge University Press.

Kuhl, J. (1987). Feeling versus being helpless: Metacognitive mediation of failure-induced performance deficits. In F. Weinert & R. Kluwe (Eds.), *Metacognition, motivation, and understanding* (pp. 217–235). Hillsdale, NJ: Erlbaum.

Kuhl, J., & Kraska, K. (1989). Self-regulation and metamotivation: Computational mechanisms, development and assessment. In R. Kanfer, P. L. Ackerman, & R. Cudeck (Eds.), *Abilities, motivation and methodology* (pp. 343–374). Hillsdale, NJ: Erlbaum.

Labouvie-Vief, G. (1990). Wisdom as integrated thought: Historical and developmental perspectives. In R. J. Sternberg (Ed.), *Wisdom: Its nature, origins, and development* (pp. 52–83). New York: Cambridge University Press.

Luria, A. R. (1973). *The working brain.* London: Penguin.

Luria, A. R. (1980). *Higher cortical functions in man* (2nd ed., rev. & expanded). New York: Basic Books.

MacLeod, J. (1995). *Ain't no makin' it: Aspirations and attainment in a low-income neighborhood.* Boulder, CO: Westview Press.

Markus, H. R., & Kitayama, S. (1991). Culture and the self: Implications for cognition, emotion, and motivation. *Psychological Review, 98,* 224–253.

Maslow, A. H. (1943). A theory of human motivation. *Psychological Review, 50,* 370–396.

Maslow, A. H. (1970). *Motivation and personality* (2nd ed.). New York: Harper.

McClelland, D. C. (1961). *The achieving society.* Princeton, NJ: Van Nostrand.

McClelland, D. C. (1973). Testing for competence rather than for "intelligence." *American Psychologist, 28,* 1–14.

McClelland, D. C., Atkinson, J. W., Clark, R. A., & Lowell, E. L. (1976). *The achievement motive.* New York: Irvington.

McClelland, D. C., & Winter, D. G. (1969). *Motivating economic achievement.* New York: Free Press.

Miller, G., Galanter, E., & Pribram, K. (1960). *Plans and the structure of behavior.* New York: Holt.

Nelson, T. O. (1996). Consciousness and metacognition. *American Psychologist*, *51*, 102–116.

Newell, A., & Simon, H. A. (1972). *Human problem solving*. Englewood Cliffs, NJ: Prentice-Hall.

Orwoll, L., & Perlmutter, M. (1990). The study of wise persons: Integrating a personality perspective. In R. J. Sternberg (Ed.), *Wisdom: Its nature, origins, and development* (pp. 160–177). New York: Cambridge University Press.

Pellegrino, C. (1991). *Unearthing Atlantis: An archeological odyssey*. New York: Random House.

Ree, M. J., & Earles, J. A. (1993). g is to psychology what carbon is to chemistry: A reply to Sternberg and Wagner, McClelland, and Calfee. *Current Directions in Psychological Science*, *1*, 11–12.

Rogers, C. R. (1961). *On becoming a person: A client's view of psychotherapy*. Boston: Houghton Mifflin.

Rogers, C. R. (1980). *A way of being*. Boston: Houghton Mifflin.

Rutter, M. (1987). Psychosocial resilience and protective mechanisms. *American Journal of Orthopsychiatry*, *57*, 316–331.

Rutter, M. (1989). Pathways from childhood to adult life. *Journal of Child Psychology and Psychiatry*, *30*, 23–51.

Rutter, M. (1996, April). *Profound early deprivation and later social relationship in early adoptees from Rumanian orphanages followed at age 4*. Paper presented at the 10th Biennial International Conference on Infant Studies, Providence, RI.

Salovey, P., & Mayer, J. D. (1990). Emotional intelligence. *Imagination, Cognition, and Personality*, *9*, 185–211.

Schmidt, F. L., & Hunter, J. E. (1981). Employment testing: Old theories and new research findings. *American Psychologist*, *36*, 1128–1137.

Sheldon, K. M., & Emmons, R. A. (1995). Comparing differentiation and integration within personal goal systems. *Personality and Individual Differences*, *18*, 39–46.

Stanley, J. C. (1974). Intellectual precocity. In J. S. Stanley, D. P. Keating, & L. H. Fox (Eds.), *Mathematical talent: Discovery, description, and development* (pp. 1–22). Baltimore: Johns Hopkins University Press.

Stanley, J. C. (1976). The case for extreme educational acceleration of intellectually brilliant youths. *Gifted Child Quarterly*, *20*, 66–75.

Stanovich, K. E. (1996). *Thinking straight about psychology* (4th ed.). New York: HarperCollins.

Sternberg, R. J. (1985). Implicit theories of intelligence, creativity, and wisdom. *Journal of Personality and Social Psychology, 49*, 607–627.

Sternberg, R. J. (Ed.). (1990). *Wisdom: Its nature, origins, and development*. New York: Cambridge University Press.

Sternberg, R. J. (1996). *Successful intelligence*. New York: Simon & Schuster.

Sternberg, R. J., Conway, B. E., Ketron, J. L., & Bernstein, M. (1981). People's conception of intelligence. *Journal of Personality and Social Psychology, 41*, 37–55.

Sternberg, R. J., & Detterman, D. K. (Eds.). (1986). *What is intelligence? Contemporary viewpoints on its nature and definition*. Norwood, NJ: Ablex.

Sternberg, R. J., & Lubart, T. I. (1995). *Defying the crowd: Cultivating creativity in a culture of conformity*. New York: Free Press.

Sternberg, R. J., & Wagner, R. K. (1993). The g-ocentric view of intelligence and job performance is wrong. *Current Directions in Psychological Science, 2*, 1–4.

Sternberg, R. J., Wagner, R. K., Williams, W. M., & Horvath, J. A. (1995). Testing common sense. *American Psychologist, 50*, 912–927.

Stevenson, H. W., & Stigler, J. W. (1992). *The learning gap*. New York: Summit Books.

Thomas, A., & Chess, S. (1977). *Temperament and development*. New York: Brunner/Mazel.

Wechsler, D. (1939). *The measurement of adult intelligence*. Baltimore: Williams & Wilkins.

Weissbourd, R. (1996). *The vulnerable child: What really hurts America's children and what we can do about it*. Reading, MA: Addison-Wesley.

Werner, E. E., & Smith, R. S. (1992). *Overcoming the odds: High risk children from birth to adulthood*. Ithaca, NY: Cornell University Press.

Williams, G. H. (1995). *Life on the color line*. New York: Penguin Books.

Winner, E. (1996). *Gifted children*. New York: Basic Books.

CHAPTER NINE

The Developing Self-System and Self-Regulation of Primary School Children

✧

THÉRÈSE BOUFFARD
CAROLE VEZEAU

An important concern in contemporary cognitive psychology is to identify those factors implicated in a person's use and active control of his or her own cognitive processes. This concern is also shared by researchers in education who argue that a key issue for educational research is to determine how students become self-regulated—that is how they come to actively use cognitive, motivational, and volitional resources in school (Zimmerman, 1990). A domain in which the study of self-regulation has led to considerable research is metacognition, which refers both to (1) one's knowledge of one's own cognitive processes and (2) the active monitoring and regulation of those processes during an ongoing cognitive enterprise (Flavell, 1976). Self-regulation is assumed to play a crucial role in metacognitive development not only because it allows people to be more efficient in using their metacognitive knowledge to solve a given task, but because it also allows them to enrich their metacognitive knowledge base (Lefebvre-Pinard & Pinard, 1985). According to Pinard (1986), Piaget's notions of reflective and reflected abstractions come into play in a reverse way from on-line self-regulation of available metacognitive knowledge. Through repeated experiences and on-line self-regulation, persons become able to extract information about the positive and negative factors that affect the quality of their intellectual functioning. This information is then incorporated into the person's existing metacognitive knowledge, knowledge that can be activated to solve further cognitive enterprises.

Many authors now admit that, while an available repertoire of cognitive resources is a prerequisite for self-regulation, it is insufficient to warrant it (Bandura, 1989; Borkowski, Johnston, & Reid, 1987; Bouffard-Bouchard & Pinard, 1988; Paris, Lipson, & Wixson, 1983). Because students' initiatives are central to self-regulation, and because self-regulation processes are time-consuming and require students to invest effort, suggests that motivation (and factors that serve to establish and maintain motivation) is extremely important. As Weinert (1987) reminds us, good learning requires not only knowing how to learn, but also being motivated to learn.

Authors from diverse theoretical perspectives agree that motivation is rooted in students' beliefs about themselves as active cognitive agents (Bandura, 1986; Borkowski et al., 1987; Flavell, 1981, 1987; McCombs, 1989; Paris et al., 1983; Pressley, Borkowski, & O'Sullivan, 1985). Therefore, unlike earlier models of metacognition, most current models explicitly include affective and motivational variables—included among the beliefs in students' self-system. The self-system is seen as instrumental for academic achievement, and for cognitive and metacognitive development, in that it is both an agent of motivation and regulates people's behaviors (Markus, Cross, & Wurk, 1990; Markus, Niedenthal, & Nurius, 1986; McCombs & Marzano, 1990).

The first part of this chapter addresses some issues relevant to the self-system, particularly the socialization involved in its construction. The second part focuses on developmental trends in the self-system, examining both age-related variations in self-perceptions and changes in the relation between self-perceptions and achievement. This part also discusses theoretical and empirical studies that link self-beliefs to self-regulation. We then present a longitudinal study of primary school children that examines the development of general and task-specific beliefs and how these are related to self-regulation. Concluding remarks focus on future research directions.

SOCIALIZATION CONTEXTS AND CONSTRUCTION OF THE SELF-SYSTEM

Susan Harter, whose main ideas derived from the seminal work of William James, is certainly one of the most influential theorists currently exploring the role of the self in cognitive functioning. For Harter (1986, 1987, 1992), self-evaluations of competence are important determinants of motivation and achievement. She considers the self-system a multidimensional construct comprising global and domain-specific beliefs about one's own competence (Harter, 1983, 1986); a view also shared by many researchers who suggest that individuals come to organize their self-beliefs into self-structures that represent their view of self-attributes, self-worth, self-competence, self-efficacy, and other similar constructs (Eccles, Wigfield, Harold, & Blumenfeld, 1993; Marsh, Byrne, & Shavelson, 1988; McCombs, 1986; Wigfield & Karpathian, 1991). Self-evaluation of competence is the key process involved in constructing self-structures and in exercising personal control over efficient action in meaningful situations (Bandura, 1986; McCombs, 1989); however, personal and environmental characteristics influence self-evaluation processes.

Social Contexts and Self-System Development

Individual characteristics such as gender, temperament, and learning disabilities have been found to be related to the self-system (Clever, Bear, & Juvonen, 1992; Eccles et al., 1993; Klein, 1992; Leondari, 1993; Lintunen, Leskinen, Oinonen, & Marjo, 1995; Marsh, 1993; Meece & Courtney, 1992; Orosan, Weine, Jason, & Johnson, 1992; Priel & Leshem, 1990; Schneewind, 1995; Short, 1992; Stetsenko, Little, Oettingen, & Baltes, 1995). However, it is through socialization within family and school environments that such individual characteristics help define the self-system.

Family as a Socialization Context

Moretti and Higgins (1990) present a framework through which parental socialization practices can influence the development of infant self-perceptions. This framework is based on how parents (1) seek to identify and respond to the child's characteristics that do or do not correspond to their expectations and (2) are oriented toward positive or negative outcomes for their child. Very briefly, the development of a positive self-system is promoted in children whose parents have congruent expectancies for them, based on a realistic appraisal of their needs and abilities. Such parents provide clear self-guides to their children, allowing them to experience positive feelings and self-perceptions when they behave as parents expect. On the other hand, parents who set standards and expectancies for their children that do not match with their children's needs and capacity will tend to undermine construction of a positive self-system as their children experience repeated feedback of dissatisfaction from them, and thus come to develop negative self-perceptions. According to Moretti and Higgins (1990), consequences of parental socialization practices on the child's self-system increase as the child's capacity for mental representation increases. Thus, positive or negative self-perceptions become increasingly consolidated as children develop cognitively.

Other authors also suggest that parental practices are important to how children's self-perceptions develop. Pulkkinen (1982) showed that high parental involvement was associated with children's higher competence and achievement motivation. Alexander and Entwistle (1988) showed that parents' beliefs about children's ability are more important predictors of children's performance than is their actual ability. Eccles, Adler, and Kaczala (1982) reported that parents' perceptions of, and expectations for, their children were related to children's perception of parents' beliefs and to children's self and task perceptions. Furthermore, parents' attitudes about their children abilities had more impact on children's attitudes than had children's own past performance. Grolnick, Ryan, and Deci (1991) verified a structural model of relations between parental context, children's inner resources, and academic achievement. They showed that paternal and maternal autonomy support and involvement predicted children's perceptions of control and of competence, as well as children's relative autonomy; these inner resources in turn predicted achievement. Other authors reported evidence for the influence of parents' attitudes and beliefs on children's self-perceptions (Barnes & Austin, 1995; Grolnick & Slowiaczek, 1994; Jacobs & Weisz, 1994; Killeen, 1993). But, as children get older, they are enrolled in school, which becomes another important context for socialization likely to influence their self-perceptions.

School as a Socialization Context

Many studies have examined how features of the school environment (e.g., classroom climate, interaction patterns promoted in the classroom, teacher's own self-efficacy, teacher's behavior and feedback) may influence children's construction of a self-system (Ashton, 1985; Brattesani, Weinstein, & Marshall, 1984; Butler & Marinov-Glassman, 1994; Cairns, 1990; Deci, Schwartz, Sheinman, & Ryan, 1981; Hoge, Smit, & Hanson, 1990; Johnson & Johnson, 1985; Marshall & Weinstein, 1986; Ryan & Grolnick, 1986; Seidman, Allen, Aber, Mitchell, & Feinman, 1994; Weinstein, 1989). Social comparison is another important issue relevant to school socialization, one that plays a major role in developing self-perceptions. This process of social comparison is examined in more detail in the following paragraphs.

In a study comparing self-perceptions of competence in normal, learning-disabled, and mentally retarded children, Renick and Harter (1984; see Harter, 1986) have found that normal and mentally retarded children reported equivalent self-perceptions, whereas learning-disabled children reported lowered self-perceptions. The authors suggest that these surprising results were probably due to the social comparison group children used to evaluate themselves. If mentally retarded children compared themselves to their mentally retarded peers, they did not experience negative self-perceptions. If learning-disabled children compared themselves to their normal-achieving peers, logically, they would report lower self-perceptions. Renick and Harter (1989) documented this process by showing that learning-disabled children's self-perceptions were significantly higher when explicitly they were asked to evaluate themselves in relation to their learning-disabled peers.

Bandura (1986) also argued that social comparison is one of the most influential sources for one's own judgment of self-efficacy. Observing a peer dealing more or less successfully with a task affected observers' self-efficacy regarding their own competence relative to that peer (Schunk, Hanson, & Cox, 1987). For example, Schunk (1983) increased children's self-efficacy for solving math problems by providing them with comparative information about the number of problems completed by similar children. Social comparison was equally able to experimentally induce high or low self-efficacy in older students (Bouffard-Bouchard & Pinard, 1988).

Ruble (1987) proposed that children's self-perceptions develop through a self-socialization process in which social comparison serves as a significant source of information. The self-socialization process emphasizes the child's active role in assessing his or her own competence, that is, what information is sought and used in self-evaluation, and developmental changes in the interest and use of social comparison.

Some evidence suggests that, unlike older children, kindergartners and children beginning elementary school do not use social comparison in self-evaluation, although they do use it when evaluating other children's competence (Ruble, 1987). Ruble, Boggiano, Feldman, and Loebl (1980) examined age-related increases in use of social comparison for self-evaluation. In an initial study involving first and second graders, they showed than even when children were provided with specific information about how well they performed compared to peers, they did not use this information when evaluating themselves or the task. A second study used kindergarten, second- and fourth-grade children in a situation that provided strong incentive to use social comparison of abilities. Again, younger children did not use the comparative information, but the fourth graders did so. According to Ruble, Grosouvsky, Frey, and Cohen (1992), young children probably pay more attention to their own actions than to those of others; thus, direct experience with the task provides them with more salient information for self-evaluation than does social comparison. However, in later school grades, the importance of evaluation is increasingly emphasized, and more opportunities are provided for social comparisons among students. Such changes may explain the shift from focusing on their own performance to social comparison as a significant source for self-evaluation as children grow older. Besides studying the processes involved in constructing self-perceptions, many studies have examined age-related changes in self-perceptions, and it is to this literature that we now turn.

DEVELOPMENTAL TRENDS IN SELF-EVALUATION AND THEIR RELATION TO SELF-REGULATION AND ACHIEVEMENT

Age-related changes in self-evaluation bear on different but related issues to those mentioned earlier, including the stability of self-perceptions across elementary school years and changes in the relation between self-perceptions, self-regulation, and achievement with age.

Stability of Self-Perceptions during the Elementary School Years

According to Paris and Byrnes (1989), children under 8 years old usually report exaggerated appraisals of their abilities. Many other authors hold the view that children start school with very optimistic self-perceptions that rapidly and systematically decline throughout the elementary school years (Licht, 1992; Miller, 1987; Stipek, 1981). For example, Marsh (1989) reported linear decreases in children self-perceptions of competence in various academic domains across elementary school years. Eccles et al. (1993) reported similar results in a study of first- to fourth-grade elementary school children (see also Stipek & MacIver, 1989, for a review). However, researchers show no consensus about why children have such initial unrealistic optimism. Ruble et al. (1992) suggests that perhaps young children tend to make judgments consistent with their desires. For others, the generality of the phenomenon suggests that changes in cognitive processing abilities and changes in school environment as children progress are probably involved in this decline (Stipek & MacIver, 1989). Nicholls (1978, 1979) suggests that young children tend to equate effort with ability; so having tried hard, or having mastered the task, could both lead young children to feel smart and to positively evaluate their own abilities. Also, as Flammer (1989) pointed out, the teacher's feedback in kindergarten and preschool is relatively arbitrary and positive, and aims to encourage children.

Other authors also observed the inflation in young children's self-perceptions but disagree with the frequently reported rapid decline in early elementary school years. In a study about children's self-concept for reading, Chapman and Tunmer (1995) observed that while children's attitude about reading began to decline from grade 4 to 5, self-perceptions of competence in reading remained very stable across the first five elementary school years. Harter (1982) and Harter and Pike (1984) reported no decline in perceived competence between preschool and second grade, or between third and ninth grade, when children were questioned about specific skills. Even though changes in school environment, when going from elementary school to junior high, may provoke a reevaluation of self-competence, Harter (1992) showed that there is no unique direction to this change; self-perceptions remained intact for some students, whereas the self-perceptions of others either increased or decreased.

The controversial findings just reported concerning how self-perceptions change during elementary school years clearly indicate the need for more research on this topic. A study examining this question is presented later in the chapter. However, independent of whether children's self-perceptions decline, there remains the crucial question of how self-perceptions relate with achievement and self-regulated learning. This question is the focus of our next section.

Relationship between Self-Perceptions and Achievement and Self-Regulation

Research with beginning elementary school children has provided contradictory evidence as to the role of the beliefs self-system in learning. Young children's self-appraisals tend to be unstable because they are heavily dependent on salience and immediate outcomes (Bandura, 1983). Most authors argue that it is only around the third or fourth grade that moderate correlations exist between beliefs about oneself on the one hand, and between these beliefs and school achievement on the other hand. Substantial relations between beliefs about one's competence and performance should not expected before the late elementary school years (Harter, 1986; Harter & Pike, 1984; Nicholls, 1978, 1979). For older children, numerous empirical studies have found that self-perceptions are important determinants of achievement motivation, cognitive engagement in tasks, and school performance (see Assor & Connell, 1992, for a review). Some authors have even suggested that with age and experience, children bring their self-perceptions into line with their actual performance (Chapman & Tunmer, 1995; Saarnio, Oka, & Paris, 1990). Thus, initial perceptions would become more positive for those who experience success, whereas they would become more negative for those who experience failure or poor performance. But relationship between self-perceptions and academic achievement is probably not direct and is likely due to variations in self-regulation related to levels of self-perceptions. Before discussing this issue, the next section briefly reviews works that link self-regulation to achievement.

Self-Regulation and Achievement

Studies have shown that there are broad individual differences in the use of general cognitive strategies or metacognitive skills in achieving skilled performance (Bransford et al., 1982; Rabinowitz & Glaser, 1985; Sternberg, 1981a, 1986). An important line of research on this issue compares gifted and average students. Assuming that self-regulation is a hallmark of intelligence, some authors argued that gifted people may be more apt than others to employ such strategies to acquire, organize, and use their knowledge (Sternberg & Davidson, 1983), as well as to better self-monitor their ongoing cognitive enterprises (Shore & Dover, 1987). Instructional and correlational studies in this perspective have provided consistent evidence as to the importance of self-regulation in skilled performance (Bouffard-Bouchard, Parent, & Larivée, 1993; Bransford et al., 1982; Peck & Borkowski, 1983; Scruggs, Mastropieri, Jorgensen, & Monson,1986; Schofield & Ashman, 1987; Sternberg, 1981b).

There is also substantial evidence attesting to the importance of self-regulated processes in academic achievement for students of various levels of cognitive ability. For example, studies by Zimmerman and Martinez-Pons (1986, 1988) showed moderate to strong relations between academic achievement and students' reported use of active learning strategies as well as teachers' rating of their students' strategic approach for learning (see also Wang, Haertel, & Walberg, 1990, for a review). Other studies have showed that underachievers and learning-disabled children often have good learning strategies in their repertoire but fail to use them (Wong, 1991). Therefore, underachievement is sometimes more a matter of deficit of production of self-regulatory processes than deficit of knowledge of these processes.

Therefore, questions often raised by both researchers and educators are why, even when students possess the necessary requisite skills, some do engage in active self-regulation and others do not, and why even the same student engages in self-regulation in some situations but not in others.

Relationship between Self-Perceptions and Self-Regulation

Bandura (1986) argues that personal agency is a requisite for self-regulation. And, as Zimmerman and Martinez-Pons (1992) remind us, "It is one thing to possess self-regulatory skills but another thing to be able to get oneself to apply them persistently in the face of difficulties, stressors or competing attractions" (p. 219). There is a growing number of researchers who also agree that beliefs about own abilities are key factors to motivating self-regulated learning (McCombs, 1986, 1989; Pokay & Blumenfeld, 1990; Skinner, Wellborn, & Connell, 1990).

Studies with learning-disabled children have consistently shown that their pessimistic views about themselves lead them to perform even more poorly than they might otherwise perform (Licht, 1992). However, detrimental effects of low self-perceptions on self-regulation and performance are not restricted to learning-disabled children. Pintrich and De Groot (1990) found that self-efficacy beliefs influenced the self-regulation of learning in regular students. Phillips (1984, 1987) has clearly shown that because beliefs about one's own abilities are not necessarily realistic, even very capable students often underestimate themselves. But more important, such underestimation leads to low engagement, poor self-regulation, and lower performance than should be expected given their real competence (Bouffard-Bouchard, Parent, & Larivée, 1991).

Until now, studies on the relation between self-perceptions and self-regulation have not measured self-regulation directly but have assessed it through self-report questionnaires or teachers' rating. And, although such studies have examined students of various grade levels, they commonly use a cross-sectional design. The study we present used a longitudinal approach to examine whether a developmental pattern exists in the relations between self-perceptions and self-regulation. Primary school children participated in the study over a 3-year period. Each year, we assessed various types of ability beliefs and examined children's on-line self-regulation when solving a reading task.

A LONGITUDINAL STUDY OF CHILDREN'S DEVELOPING SELF-PERCEPTIONS AND HOW THEY RELATE TO SELF-REGULATION AND PERFORMANCE

Our study of 178 elementary school children (an almost equal number of boys and girls), began when children were in fourth grade (mean age = 10.0, SD = 0.77). Children were recruited from 10 regular classes in five Montreal public elementary schools.

One objective of the study was to examine whether children's beliefs self-system—as well as the relations between these beliefs, acquisition of metacognitive knowledge, and academic achievement—would change with development. For practical reasons, we decided to restrict the study to the area of reading. Furthermore, although the self-system clearly comprises many beliefs, we focused on three types of beliefs that we believed to be crucial for achievement: (1) children's beliefs about their ability as cognitive agents; (2) children's beliefs about their ability to exercise control to achieve success; and (3) children's beliefs about their ability as readers. Accordingly, each spring, over the course of the study, we asked children to fill out a set of questionnaires in a group session during one of their regular class periods.

The Cognitive subscale of the Perceived Competence Scale for Children (Harter, 1982) was used to assess children's beliefs about their ability as cognitive agents. Children's beliefs about their ability to exercise control to achieve success were assessed using the Control Beliefs subscale of the Control, Agency, and Means–Ends Interview for Children (Skinner, Chapman, & Baltes, 1988). Children's beliefs about their ability as readers were assessed with the instrument developed by Paris and Oka (1986).

In order to examine the relations between these three types of beliefs and metacognitive knowledge, children were also asked to fill out the Index of Reading Awareness (Jacobs & Paris, 1987), designed to assess metacognitive knowledge about reading strategies. Children's reading ability was also assessed using a standardized instrument for French Canadian readers in grades 3 to 6. Finally, reading achievement was measured through children's final marks in reading, obtained at the end of each school year.

A second objective of the study was to examine whether these various components of the self-system might account for the variation in children's on-line self-regulation during reading. To this end, within a few weeks after the group session, each child was seen individually in a quiet room of the school, during which time we observed their self-regulation during a reading error detection task. The error detection task was chosen for three main reasons: (1) It allowed us to record several self-regulatory behaviors; (2) it allowed us to use operational criteria to distinguish self-regulation from performance (Zimmerman, 1994); and (3) it provided subjects with the possibility to make choices and exert control during the reading task.

The task was designed to be executed individually on a Macintosh computer and to automatically and unobtrusively record several indicators of on-line self-regulation. This was accomplished by using a HyperCard[1] that allowed sentence-by-sentence presentation of the reading passages, and allowed children unconstrained forward and backward movements through the text, permitting revisions, changes of mind, and so on. Children were told they would read some stories with problems in them that they should try to solve. We then explained the three types of problems: lexical error, external (prior-knowledge violations), and internal inconsistencies. Children were also told how many errors of each type were in each of the passages and the importance of limiting error detection to the types described. Finally, they were told that they had 30 minutes to complete the task and were given a sample passage to familiarize them with the task and the computer. Because many authors argue that self-efficacy is a more proximal determinant of behavior than are other components of the self-system, children were asked to specify their *self-efficacy* in finding the errors in the subsequent passages.

In addition to the computer-recorded information about self-regulation, direct observations were made of childrens' monitoring of their working time and their use of additional resources (dictionary, grammar book) made available. Finally, while many covert, self-regulatory processes are inaccessible to direct observation, the ability to accurately evaluate the quality of one's responses was taken to be a good indicator of how actively children self-regulated their performance of a task (Baker & Brown, 1984). Accordingly, children were asked to indicate their confidence in having truly detected an error for each sentence they indicated as containing an error.

Developmental Changes in the Self-System

The first two questions addressed in this study were whether children's beliefs self-system changed with development and how these changes, if any, were related to metacognitive development and reading achievement.

As mentioned earlier, some controversy exists in the literature over whether children's self-perceptions change during the elementary school years. In our study, the multivariate analysis of variance revealed that only girls' self-perceptions as readers decreased, whereas those of boys remained the same from grades 4 to 6. For both genders, children's beliefs about their ability as cognitive agents remained stable across the 3 years, as did girls' beliefs about their capability to exercise control. In this latter case, boys' beliefs in their abilities even increased.

As clearly shown in Table 9.1, there were significant between-year correlations for each type of beliefs. In addition, we observed no change linked to school level in the intercorrelations between the three types of beliefs (see Table 9.2).

Our findings, therefore, do not support those of authors who report a general decrement in children's self-beliefs throughout the elementary school years (Marsh, 1989; see also Stipek & MacIver, 1989, for a review). One possible explanation for our finding may be that, unlike most other studies that compared different children, we adopted a longitudinal approach. This approach may have minimized intersubject variability and thus produced a clearer picture of how a child's different self-perceptions change with age. However, recall that some researchers also reported no decline in self-perceptions of reading competence from fourth to fifth grade, even using a transversal approach (Chapman & Tunmer, 1995).

As another possible explanation, most studies report a decline in self-perceptions involved children in earlier elementary school grades (Wigfield & Harold, 1992) or students in transition to junior high school (Seidman et al., 1994; Wigfield, Eccles, MacIver, Reuman, & Midgley, 1991); with few exceptions, studies show that elementary school children tend to be very optimistic in their self-perceptions of competence, and that their perceptions become more realistic with cognitive development, school experience, and a better understanding of the relation between effort and ability. The transition from elementary to secondary school appears to be another period marked by a general decline in students' self-perceptions, one usually attributed to changes in the school and classroom environment that em-

TABLE 9.1. Between-Year Coefficients of Correlation for the Beliefs Self-System According to Gender

		1	2	3
		Grade 4		
Grade 5				
1. Beliefs about	B	.54**		
cognitive ability	G	.58**		
2. Beliefs about	B		.42**	
reading ability	G		.48**	
3. Beliefs about	B			.39**
control	G			.43**
		Grade 5		
Grade 6				
1. Beliefs about	B	.49**		
cognitive ability	G	.51**		
2. Beliefs about	B		.50**	
reading ability	G		.53**	
3. Beliefs about	B			.34**
control	G			.45**

$*p < .05; **p < .01.$

TABLE 9.2. Intercorrelation of the Beliefs Self-System According to School Grade and Gender

	1	2	3
Grade 4			
1. Beliefs about cognitive ability		.39**	.49**
2. Beliefs about reading ability	.51**		.54**
3. Beliefs about control	.65**	.59**	
Grade 5			
1. Beliefs about cognitive ability		.52**	.43**
2. Beliefs about reading ability	.53**		.39**
3. Beliefs about control	.46**	.29*	
Grade 6			
1. Beliefs about cognitive ability		.46**	.44**
2. Beliefs about reading ability	.51**		.32**
3. Beliefs about control	.52**	.58**	

Note. Correlation coefficients above the diagonal are for boys and those below the diagonal are for girls.
$*p < .05; **p < .01.$

phasize evaluation and that provides more opportunity for social comparisons among students. All in all, our study's findings suggest that, unlike younger and older children, late elementary school children's beliefs about themselves as learners organized in their self-system show some stability, consistency, and coherence. Since we intend to follow the students in our study through the first 3 years of high school, we will soon be able to examine the developmental pattern of changes in the self-system over a 6-year period that includes the transition to another school environment.

Metacognitive knowledge about reading naturally increased each year for both boys and girls. However, its relation to the various self-system components, and between the self-system and reading achievement, hardly changed from year to year (r's about .22 and .38, respectively). These findings converge with those of other authors (Harter & Pike, 1984; Paris & Oka, 1986), who affirm that moderate relations are established between self-beliefs and school achievement by grade 4. However, contrary to some authors (Saarnio et al., 1990), these relations did not appear to become stronger with age and school experience. In the present study, similar patterns of moderate relation between the self-system and metacognitive knowledge and reading achievement were observed across years, despite a slight decline in fifth grade. This suggests, following Assor and Connell (1992), that children of this age can make valid self-appraisals of their capability.

Developmental Pattern of Relations between the Self-System and Self-Regulation

Another question we sought to answer was how various components of the self-system accounted for variations in children's on-line self-regulation during reading.

By definition, self-regulation cannot be reduced to single or isolated behaviors. Assessing self-regulation requires considering several behaviors that provide information about what and how much control is exerted when performing a given task (Bandura, 1991; Schunk, 1994; Zimmerman, 1994). Furthermore, it appears necessary to distinguish between "appropriate" and "inappropriate" self-regulation. Although we have never come across this distinction in the literature, it allows us to distinguish behaviors that reflect decisions that help students succeed at the task from behaviors that are detrimental to so doing. Finally, as mentioned earlier, many self-regulatory processes cannot be directly observed, but accurately evaluating the quality of one's responses can serve as an indicator of the amount of self-regulation during performance of the task.

Scores of appropriate, inappropriate, and covert self-regulation were computed. *Appropriate self-regulatory behaviors* included behaviors such as moving backward and forward through the text, reviewing the entire text or previous sentences, using the dictionary and the grammar book, and verifying work time remaining: Subjects' score for appropriate self-regulation was computed by summing over all instances of these behaviors. *Inappropriate self-regulatory behaviors* included behaviors such as lowering confidence about correct detection or increasing confidence about false detection, and indicating more errors than the number specified for the passages: Subjects' score for inappropriate self-regulation was computed by summing over all instances of these behaviors. *Covert self-regulation* refers to children's sensitivity to response correctness and reflects the extent to which they could distinguish between correct and incorrect responses. This ability was measured by the discrepancy between their confidence in having detected real and false errors.

Finally, performance was assessed by computing the number of errors correctly detected across the passages.

We used path-analysis techniques to examine the direct and indirect effects of self-system–belief variables on self-regulation and performance for each year of the study. Because reading is sometimes considered as a sex-stereotyped domain, gender was also included in the models. Figures 9.1, 9.2, and 9.3 illustrate the path-analysis models found for each year of the study.

Overall, the models were relatively consistent from year to year and, each year, they accounted for a large portion of the variance in the task performance. Reading ability was systematically related to beliefs about ability as a cognitive agent and to beliefs about capability to exert control and about ability as a reader, suggesting that children of this age can make more or less realistic self-appraisals of their abilities (Assor & Connell, 1992). But except in grade 6, reading ability had no direct impact on self-efficacy. Only beliefs about ability as a cognitive agent had some influence on self-efficacy. These finding support Bandura's (1986) contention about the subjective nature of self-efficacy judgment and about its relative specificity to a given task.

The models also showed that, each year, over and above the effects of reading ability on task performance, covert and inappropriate self-regulation directly affected performance—whereas appropriate self-regulation affected performance only in grade 5. The systematic and important impact of covert self-regulation on performance supports the idea that self-evaluation of one's own performance is a good indicator of how well the person controlled the ongoing endeavor. These findings also suggest the validity of distinguishing self-regulatory decisions that may have positive or negative impact on performance. As Miller (1994) reminds us, strategic behaviors are not necessarily useful, and some may even worsen performance, a phenomenon she calls "utilization deficiency." This worsening was clearly evident in this study for indicators of what we call "inappropriate self-regulation"; these behaviors had substantial negative influence on task performance at each school level.

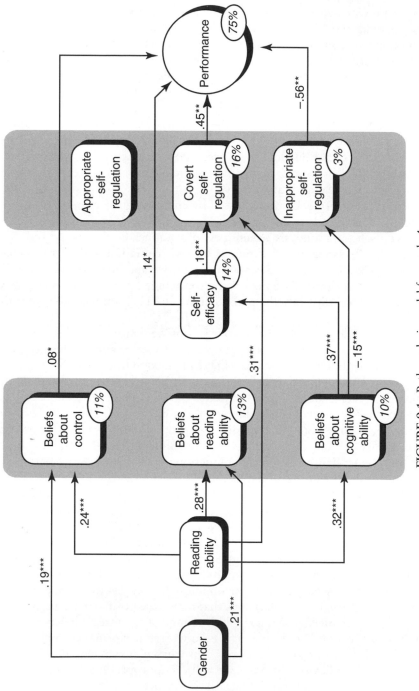

FIGURE 9.1. Path analysis model for grade 4.

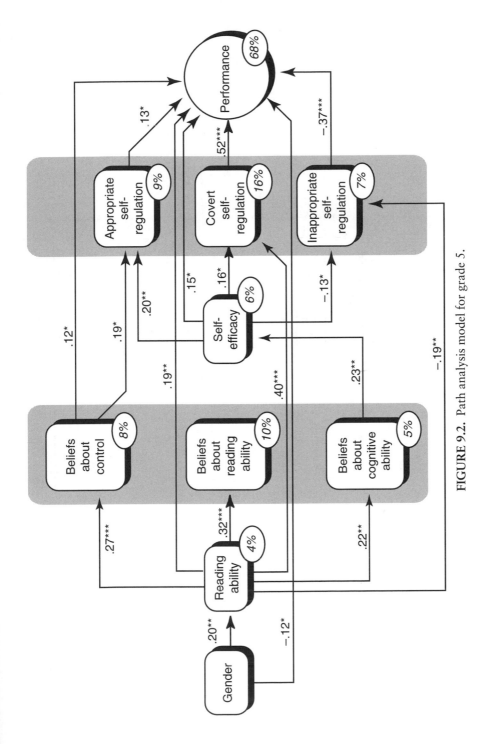

FIGURE 9.2. Path analysis model for grade 5.

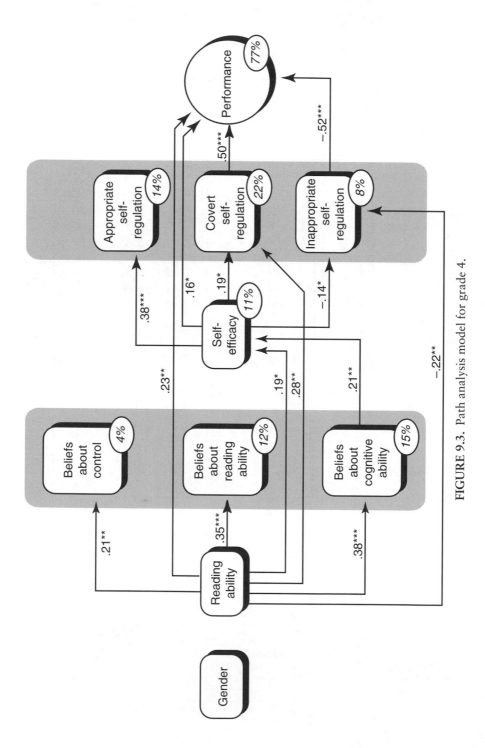

FIGURE 9.3. Path analysis model for grade 4.

Regarding the effect of the self-system on self-regulation and performance, beliefs about control had a slight effect on performance at grade 4 and 5, but not at grade 6. Beliefs about cognitive ability did not affect performance directly, but indirectly through self-efficacy. However, self-efficacy did affect performance both directly and indirectly, through its effect on (1) covert self-regulation at each school grade, (2) appropriate self-regulation at grade 5, and (3) inappropriate self-regulation at grades 5 and 6. These findings agree with those recently reported by Pajares and Miller (1994) in the domain of mathematics. Using path-analysis procedures, these authors showed that math self-efficacy was more predictive of problem solving than was math self-concept. Together with our findings, this finding support the claim made by Bandura (1989) that self-efficacy is probably a more proximal determinant of behaviors and achievement than is self-concept.

Some people might criticize the ecological validity of our study by arguing that our reading task was not necessarily representative of standard school reading tasks for children at this age. Even if this were true, in our view, this argument in no way invalidates the relations found between the self-system and self-regulation. But more to the point, we believe that children's self-regulation in the present task does, at least to some extent, portray what they usually do during reading. If so, and if self-regulation encourages metacognitive development and academic excellence, we should expect self-regulation in the preceding year to be related to the development of metacognitive knowledge about reading strategies and reading achievement in the following year. Table 9.3 presents between-year correlations for each measure of self-regulation with both reading strategic

TABLE 9.3. Between-Years Correlations between Self-Regulation and Strategic Knowledge and Reading Achievement

	Self-regulation		
	Appropriate	Inappropriate	Covert
	Grade 4		
Grade 5			
Strategic knowledge	.09	−.17*	.25***
Reading achievement	.15*	−.17*	.29***
	Grade 5		
Grade 6			
Strategic knowledge	.19**	−.15*	.34***
Reading achievement	.20**	−.23***	.44***

*$p < .05$; **$p < .01$; ***$p < .001$.

knowledge and reading achievement. This table shows that with few exceptions, measures of self-regulation in a given year relate to strategic knowledge and reading achievement at the following year.

CONCLUSION

This chapter discusses the importance of considering children's self-beliefs about their own ability to self-regulate reading. The longitudinal study presented shows that by grade 4, children have already developed a stable and consistent set of beliefs about themselves, and that their general beliefs about cognitive ability, and about their ability to exert control over performances, influence self-efficacy about reading comprehension. But more important, this study highlighted the importance of having children develop positive self-beliefs by showing the influence of self-efficacy on self-regulation which, in turn, influences the acquisition of metacognitive knowledge and academic achievement.

Most previous research has examined how self-perceptions come to affect cognitive functioning, with some effort devoted to understanding how personal characteristics and socialization contexts affect the development of self-beliefs. However, the role of cognitive skills in self-appraisals of capability needs more clarification.

In fact, from our perspective, self-appraisals result from cognitive processing in which individuals coordinate and weigh the importance of various sources of information (e.g., past performances, task difficulty, effort expended, task value, social feedback). Such analysis requires a general ability to select and classify information, to understand how some of them may compensate the effect of the others; and to resolve apparent contradictions. Classification has two important features. First, classification criteria are not uniform, but instead depend on children's experiences and environment. Particularly for young children, what constitutes a good performance is often determined by the children's social milieu. Such variability is also found in selecting activities that are considered important enough to be explicitly categorized. In this sense, self-perceptions are sort of "ad hoc" categories whose use develops between 5 and 10 years of age (Markman, 1989), and whose construction implies some rules of reasoning.

To date, few studies have explored the cognitive skills needed for children to appropriately reason about their own competence. Surber (1980) used Piaget's cognitive developmental model to examine the development of reversible thinking in causal schemas about ability, effort, and performance at four age levels, from kindergarten to college. Results showed that younger children (up until fifth grade) found it difficult to consistently use inverse compensation to evaluate the role of ability and effort. Nicholls

(1978, 1979) also used Piaget's theory to qualify the developmental pattern of children's (ages 5 to 13) ability to differentiate the role of effort and ability when inferring competence. Nicholls distinguished four levels of reasoning that revealed that understanding of how information about effort and ability combine to infer that competence increased with age. He argues that both age trends in level of reasoning and nature of explanations provided by children at each level represented different levels of thinking in Piaget's model. However, neither Surber nor Nicholls included any independent measures to allow an objective evaluation of cognitive development. And more important, we cannot know whether children had the requisite cognitive skills for making accurate self-evaluation of competence, because no specific cognitive skill was clearly identified and examined (except for inverse compensation in the Surber's study).

For all of these reasons, we believe that given the importance of the self-system beliefs for cognitive and metacognitive functioning and development, a theoretically and practically promising, and valuable, area of research is that of investigating the cognitive skills and processes needed to construct a positive self-system.

NOTE

1. Special thanks are addressed to Claude Barron, who specifically developed the program for the purposes of this study.

REFERENCES

Alexander, K. L., & Entwistle, D. R. (1988). Achievement in the first two years of school: Patterns and processes. *Monographs of the Society for Research in Child Development*, 53(2, Serial No. 218).

Ashton, P. (1985). Motivation and the teacher's sense of efficacy. In C. Ames & R. Ames (Eds.), *Research on motivation in education: Vol. 2. The classroom milieu* (pp. 141–171). Orlando, FL: Academic Press.

Assor, A., & Connell, J. P. (1992). The validity of students' self-reports as measures of performance affecting self-appraisals. In D. H. Schunk & J. L. Meece (Eds.), *Student perceptions in the classroom* (pp. 25–47). Hillsdale, NJ: Erlbaum.

Baker, L., & Brown, A. L. (1984). Metacognitive skills and reading. In T. D. Pearson (Ed.), *Handbook of reading research* (pp. 353–394). New York: Longman.

Bandura, A. (1983). Self-referent thought: A developmental analysis of self-efficacy. In J. H. Flavell & L. Ross (Eds.), *Social cognitive development* (pp. 200–239). Cambridge, England: Cambridge University Press.

Bandura, A. (1986). *Social foundations of thought and action: A social cognitive theory*. Englewood Cliffs, NJ: Prentice-Hall.

Bandura, A. (1989). Human agency in social cognitive theory. *American Psychologist*, *44*, 1175–1184.

Bandura, A. (1991). Self-regulation of motivation through anticipatory and self-regulatory mechanisms. In R. A. Dienstbier (Ed.), *Perspectives on motivation: Nebraska Symposium on Motivation* (Vol. 38, pp. 69–164). Lincoln: University of Nebraska Press.

Barnes, T. P., & Austin, A. M. (1995). The influence of parents and siblings on the development of a personal premise system in middle childhood. *Journal of Genetic Psychology*, *156*, 73–85.

Borkowski, J. G., Johnston, M. B., & Reid, M. K. (1987). Metacognition, motivation, and controlled performance. In S. J. Ceci (Ed.), *Handbook of cognitive, social and neuropsychological aspects of learning disabilities* (pp. 147–173). Hillsdale, NJ: Erlbaum.

Bouffard-Bouchard, T., Parent, S., & Larivée, S. (1991). Influence of self-efficacy on self-regulation and performance among junior and senior high school age students. *International Journal of Behavioral Development*, *14*, 153–164.

Bouffard-Bouchard, T., Parent, S., & Larivée, S. (1993) Self-regulation on a concept-formation task among average and gifted students. *Journal of Experimental Child Psychology*, *56*, 115–134.

Bouffard-Bouchard, T., & Pinard, A. (1988). Sentiment d'auto-efficacité et exercice des processus d'autorégulation chez des étudiants de niveau collégial. *Journal International de Psychologie*, *23*, 409–431.

Bransford, J. D., Stein, B. S., Vye, N. J., Franks, J. J., Auble, P. M., Mezynski, K. J., & Perfetto, G. A. (1982). Differences in approaches to learning: An overview. *Journal of Experimental Psychology: General*, *111*, 390–398.

Brattesani, S. K., Weinstein, R. S., & Marshall, H. H. (1984). Student perceptions of differential teacher treatments as moderators of teacher expectation effects. *Journal of Educational Psychology*, *76*, 236–247.

Butler, R., & Marinov-Glassman, D. (1994). The effects of educational placement and grade level on the self-perceptions of low achievers and students with learning disabilities. *Journal of Learning Disabilities*, *27*, 325–334.

Cairns, E. (1990). The relationship between adolescent perceived self-competence and attendance at single-sex secondary school. *British Journal of Educational Psychology*, *60*, 207–211.

Chapman, J. W., & Tunmer, W. E. (1995). Development of young children's reading self-concepts: An examination of emerging subcomponents and their relationship with reading achievement. *Journal of Educational Psychology*, *87*, 154–167.

Clever, A., Bear, G. G., & Juvonen, J. (1992). Discrepancies between competence and importance in self-perceptions of children in integrated classes. *Journal of Special Education*, *26*, 125–138.

Deci, E. L., Schwartz, A. J., Sheinman, L., & Ryan, R. M. (1981). An instrument to assess adults' orientation toward control versus autonomy with children: Reflections on intrinsic motivation and perceived competence. *Journal of Educational Psychology*, *73*, 642–650.

Eccles, J., Adler, T. F., & Kaczala, C. M. (1982). Socialization of achievement attitudes and beliefs: Parental influences. *Child Development*, *53*, 310–321.

Eccles, J., Wigfield, A., Harold, R. D., & Blumenfeld, P. (1993). Age and gender differences in children's self and task perceptions during elementary school. *Child Development*, 64, 830–847.

Flammer, A. (1989). Developmental analysis of control beliefs. In A. Bandura (Ed.), *Self-efficacy in changing societies* (pp. 69–113). New York: Cambridge University Press.

Flavell, J. H. (1976). Metacognitive aspects of problem solving. In L. B. Resnick (Ed.), *The nature of intelligence* (pp. 231–235). Hillsdale, NJ: Erlbaum.

Flavell, J. H. (1981). Cognitive monitoring. In W. P. Dickson (Ed.), *Children's oral communication skills* (pp. 35–60). New York: Academic Press.

Flavell, J. H. (1987). Speculations about the nature and development of metacognition. In R. H. Kluwe & F. E. Weinert (Eds.), *Metacognition, motivation and understanding* (pp. 21–29). Hillsdale, NJ: Erlbaum.

Grolnick, W. S., Ryan, R. M., & Deci, E. L. (1991). Inner resources for school achievement: Motivational mediators of children's perceptions of their parents. *Journal of Educational Psychology*, 83, 508–517.

Grolnick, W. S., & Slowiaczek, M. L. (1994). Parent's involvement in children's schooling: A multidimensional conceptualization and motivational model. *Child Development*, 65, 237–262.

Harter, S. (1982). The perceived competence scale for children. *Child Development*, 53, 87–97.

Harter, S. (1983). Developmental perspectives on the self-system. In P. H. Mussen (Ed.), *Handbook of child psychology* (Vol. 4, pp. 275–386). New York: Wiley.

Harter, S. (1986). Processes underlying the construction, maintenance and enhancement of the self-concept in children. In J. Suls & A. C. Greenwald (Eds.), *Psychological perspectives on the self* (Vol. 3, pp. 137–181). Hillsdale, NJ: Erlbaum.

Harter, S. (1987). The determinants and mediational role of global self-worth in children. In N. Eisenberg (Ed.), *Contemporary topics in developmental psychology* (pp. 219–242). New York: Wiley.

Harter, S. (1992). The relationship between perceived competence, affect, and motivational orientation within the classroom: Processes and patterns of change. In A. K. Boggiano & T. S. Pittman (Eds.), *Achievement and motivation: A social-developmental perspective* (pp. 77–114). New York: Cambridge University Press.

Harter, S., & Pike, R. (1984). The pictorial scale of perceived competence and social acceptance for young children. *Child Development*, 55, 1969–1982.

Hoge, D. R., Smit, E. K., & Hanson, S. L. (1990). School experiences predicting changes in self-esteem of sixth- and seventh-grade students. *Journal of Educational Psychology*, 82, 117–127.

Jacobs, J. E., & Paris, S.G. (1987). Children's metacognition about reading: Issues and definition, measurement, and instruction. *Educational Psychologist*, 22, 255–278.

Jacobs, J. E., & Weisz, V. (1994). Gender stereotypes: Implication for gifted education. *Roeper-Review*, 16, 152–155.

Johnson, D. W., & Johnson, R. T. (1985). Motivational processes in cooperative, competitive, and individualistic learning situations. In C. Ames & R. Ames

(Eds.), *Research on motivation in education: Vol. 2. The classroom milieu* (pp. 249–286). Orlando, FL: Academic Press.

Killeen, M. R. (1993). Parent influences on children's self-esteem in economically disadvantaged families [Special issue: Socially vulnerable populations]. *Issue in Mental Health Nursing, 14,* 323–336.

Klein, H. A. (1992). Individual temperament and emerging self-perception. *Journal of Research in Childhood Education, 6,* 113–120.

Lefebvre-Pinard, M., & Pinard, A. (1985). Taking charge of one's cognitive activity: A moderator of competence. In E. D. Neimark, R. De Lisi, & J. L. Newman (Eds.), *Moderators of competence* (pp. 191–211). Hillsdale, NJ: Erlbaum.

Leondari, A. (1993). Comparability of self-concept among normal achievers, low achievers and children with learning difficulties. *Educational Studies, 19,* 357–371.

Licht, B. G. (1992). The achievement-related perceptions of children with learning problems: A developmental analysis. In D. H. Schunk & J. L. Meece (Eds.), *Student perceptions in the classroom* (pp. 247–264). Hillsdale, NJ: Erlbaum.

Lintunen, T., Leskinen, E., Oinonen, M., & Marjo, S. (1995). Change, reliability and stability in self-perceptions in early adolescence: A four-year follow-up study. *International Journal of Behavioral Development, 18,* 351–364.

Markman, E. M. (1989). *Categorization and naming in children: Problems of induction.* Cambridge, MA: MIT Press.

Markus, H., Cross, S., & Wurk, E. (1990). The role of the self-system in competence. In R. J. Sternberg & J. Kolligian, Jr. (Eds.), *Competence considered* (pp. 205–225). New Haven, CT: Yale University Press.

Markus, H., Niedenthal, P., & Nurius, P. (1986). On motivation and the self-concept. In R. M. Sorrentino & E. T. Higgins (Eds.), *Handbook of motivation and cognition: Foundations of social behavior.* (pp. 96–121). New York: Guilford Press.

Marsh, H. W. (1989). Age and sex effects in multiple dimensions of self-concept: Preadolescence to early adulthood. *Journal of Educational Psychology, 81,* 417–430.

Marsh, H. W. (1993). Physical fitness self-concept: Relations of physical fitness to field and technical indicators for boys and girls aged 9–15. *Journal of Sport and Exercise Psychology, 15,* 184–206.

Marsh, H. W., Byrne, B. M., & Shavelson, R. J. (1988). A multifaceted academic self-concept: Its hierarchical structure and its relation to academic achievement. *Journal of Educational Psychology, 80,* 366–380.

Marshall, H. W., & Weinstein, R. S. (1986). Classroom context of student-perceived differential teacher treatment. *Journal of Educational Psychology, 78,* 441–453.

McCombs, B. L. (1986). The role of the self-system in self-regulated learning. *Contemporary Educational Psychology, 11,* 314–332.

McCombs, B. L. (1989). Self-regulated learning and academic achievement: A phenomenological view. In B. J. Zimmerman & D. H. Schunk (Eds.), *Self-regulated learning and academic achievement* (pp. 51–82). New York: Springer-Verlag.

McCombs, B. L., & Marzano, R. J. (1990). Putting the self in self-regulated learning: The self as agent in integrating will and skill. *Educational Psychologist*, 25, 51–69.

Meece, J. L., & Courtney, D. P. (1992). Gender differences in students' perceptions: Consequences for achievement-related choices. In D. H. Schunk & J. L. Meece (Eds.), *Student perceptions in the classroom* (pp. 209–228). Hillsdale, NJ: Erlbaum.

Miller, A. (1987). Changes in academic self-concept in early school years: The role of conceptions of ability. *Journal of Social Behavior and Personality*, 2, 551–558.

Miller, P. H. (1994). Individual differences in children's strategic behaviors: Utilization deficiencies. *Learning and Individual Differences*, 6, 285–307.

Moretti, M. M., & Higgins, E. T. (1990). The development of self-system vulnerabilities: Social and cognitive factors in developmental psychopathology. In R. J. Sternberg & J. Kolligian (Eds.), *Competence considered* (pp. 286–314). New Haven, CT: Yale University Press.

Nicholls, J. (1978). The development of the concepts of effort and ability, perception of academic attainment, and the understanding that difficult tasks require more ability. *Child Development*, 49, 800–814.

Nicholls, J. (1979). Development of perception of own attainment and causal attributions for success and failure in reading. *Journal of Educational Psychology*, 71, 94–99.

Orosan, P. G., Weine, A. W., Jason, L. A., & Johnson, J. H. (1992). Gender differences in academic and social behavior of elementary school transfer students. *Psychology in the Schools*, 29, 394–402.

Pajares, F., & Miller, M. D. (1994). Role of self-efficacy and self-concept beliefs in mathematical problem solving: A path analysis. *Journal of Educational Psychology*, 86, 193–203.

Paris, S. G., & Byrnes, J. P. (1989). The constructivist approach to self-regulation and learning in the classroom. In B. J. Zimmerman & D. H. Schunk (Eds.), *Self-regulated learning and academic achievement* (pp. 169–200). New York: Springer-Verlag.

Paris, S. G., Lipson, M. Y., & Wixson, K. (1983). Becoming a strategic reader. *Contemporary Educational Psychology*, 8, 293–316.

Paris, S. G., & Oka, E. R. (1986). Children's reading strategies, metacognition, and motivation. *Developmental Review*, 6, 25–56.

Peck, V. A., & Borkowski, J. G. (1983, April). *The emergence of strategic behavior in the gifted*. Paper presented at the Biennial Meeting of the Society for Research in Child Development, Detroit, MI.

Phillips, D. (1984). Socialization of perceived academic competence among highly competent children. *Child Development*, 55, 2000–2016.

Phillips, D. (1987). The illusion of incompetence among academically competent children. *Child Development*, 58, 1308–1320.

Pinard, A. (1986). "Prise de conscience" and taking charge of one's own cognitive functioning. *Human Development*, 29, 341–354.

Pintrich, P. R., & De Groot, E. V. (1990). Motivational and self-regulated learning components of classroom academic performance. *Journal of Educational Psychology*, 82, 33–40.

Pokay, P., & Blumenfeld, P. C. (1990). Predicting achievement early and late in the semester: The role of motivation and use of learning strategies. *Journal of Educational Psychology, 82,* 41–50.

Pressley, M., Borkowski, J. G., & O'Sullivan, J. T. (1985). Children's metamemory and the teaching of memory strategies. In D. L. Forrest-Pressley, G. E. MacKennon, & T. G. Waller (Eds.), *Metacognition, cognition and human performance* (pp. 111–153). San Diego: Academic Press.

Priel, B., & Leshem, T. (1990). Self-perceptions of first- and second-grade children with learning disabilities. *Journal of Learning Disabilities, 23,* 637–642.

Pulkkinen, L. (1982). Self-control and continuity from childhood to adolescence. In P. B. Baltes & O. G. Brim (Eds.), *Life span development and behavior* (Vol. 4, pp. 63–105). San Diego: Academic Press.

Rabinowitz, M., & Glaser, R. (1985). Cognitive structure and process in highly competent performance. In F. D. Horowitz & M. O'Brien (Eds.), *The gifted and talented: Developmental perspectives* (pp. 75–98). Washington, DC: American Psychological Association.

Renick, M. J., & Harter, S. (1989). Impact of social comparisons on the developing self-perceptions of learning disabled students. *Journal of Educational Psychology, 81,* 631–638.

Renick, M. J., & Harter, S. (1984). *A developmental study of the perceived competence of learning and disabled children.* Unpublished manuscript, University of Denver, Denver, CO.

Ruble, D. N. (1987). The acquisition of self-knowledge: A self-socialization perspective. In N. Eisenberg (Ed.), *Contemporary topics in developmental psychology* (pp. 243–270). New York: Wiley.

Ruble, D. N., Boggiano, A. K., Feldman, N. S., & Loebl, J. H. (1980). Developmental analysis of the role of social comparison in self-evaluation. *Developmental Psychology, 16,* 105–115.

Ruble, D. N., Grosouvsky, E. H., Frey, K. S., & Cohen, R. (1992). Developmental changes in competence assessment. In A. K. Boggiano & T. S. Pittman (Eds.), *Achievement and motivation: A social-developmental perspective* (pp. 138–164). New York: Cambridge University Press.

Ryan, R. M., & Grolnick, W. S. (1986). Origins and pawns in the classroom: Self-report and projective assessments of individual differences in children's perceptions. *Journal of Personality and Social Psychology, 50,* 550–558.

Saarnio, D. A., Oka, E. R., & Paris, S. G. (1990). Developmental predictors of children's reading comprehension. In T. H. Carr & B. A. Levy (Eds.), *Reading and its development: Components skills approaches* (pp. 57–79). Hillsdale, NJ: Erlbaum.

Schneewind, K. A. (1995). Impact of family processes on control beliefs. In A. Bandura (Ed.), *Self-efficacy in changing societies* (pp. 114–148). New York: Cambridge University Press.

Schofield, N. J., & Ashman, A. F. (1987). The cognitive processing of gifted, high average, and low average ability students. *British Journal of Educational Psychology, 57,* 9–20.

Schunk, D. H. (1994). Self-regulation of self-efficacy and attributions in academic settings. In D. H. Schunk & B. J. Zimmerman (Eds.), *Self-regulation of learn-

ing and performance: Issues and educational applications (pp. 75–99). Hillsdale, NJ: Erlbaum.

Schunk, D. H. (1983). Developing children's self-efficacy and skills: The roles of social comparative information and goal setting. *Contemporary Educational Psychology, 8*, 76–86.

Schunk, D. H., Hanson, A. R., & Cox, P. D. (1987). Peer model attributes and children's achievement behaviors. *Journal of Educational Psychology, 79*, 54–61.

Scruggs, T. E., Mastropieri, M. A., Jorgensen, C., & Monson, J. (1986). Effective mnemonic strategies for gifted learners. *Journal for the Education of the Gifted, 9*, 105–121.

Seidman, E., Allen, L., Aber, J. L., Mitchell, C., & Feinman, L. (1994). The impact of school transitions in early adolescence on the self-system and perceived social context of poor urban youth. *Child Development, 65*, 507–522.

Shore, B. M., & Dover, A. C. (1987). Metacognition, intelligence and giftedness. *Gifted Child Quarterly, 31*, 37–39.

Short, E. J. (1992). Cognitive, metacognitive, motivational, and affective differences among normally achieving, learning-disabled, and developmentally handicapped students. *Journal of Clinical Child Psychology, 21*, 229–239.

Skinner, E. A., Chapman, M., & Baltes, P. B. (1988). Control, means–ends, and agency beliefs: A new conceptualization and its measurement during childhood. *Journal of Personality and Social Psychology, 54*, 117–133.

Skinner, E. A., Wellborn, J. G., & Connell, J. P. (1990). What it takes to do well in school and whether I've got it: A process model of perceived control and children's engagement and achievement in school. *Journal of Educational Psychology, 82*, 22–32.

Sternberg, R. J. (1981a). A componential theory of intellectual giftedness. *Gifted Child Quarterly, 25*, 86–93.

Sternberg, R. J. (1981b). Intelligence and nonentrenchment. *Journal of Educational Psychology, 73*, 1–16.

Sternberg, R. J. (1986). A triarchic theory of intellectual giftedness. In R. J. Sternberg, & J. E. Davidson (Eds.), *Conceptions of giftedness* (pp. 223–243). Cambridge, England: Cambridge University Press.

Sternberg, R. J., & Davidson, J. E. (1983). Insight in the gifted. *Educational Psychologist, 18*, 51–57.

Stetsenko, A., Little, T. D., Oettingen, G., & Baltes, P. B. (1995). Agency, control, and means–ends beliefs about school performance in Moscow children: How similar are they to beliefs of Western children? *Developmental Psychology, 31*, 285–299.

Stipek, D. (1981). Children's perceptions of their own and their classmates' ability. *Journal of Educational Psychology, 73*, 404–410.

Stipek, D., & MacIver, D. (1989). Developmental change in children's assessment of intellectual competence. *Child Development, 60*, 521–538.

Surber, C. F. (1980). The development of reversible operations in judgments of ability, effort, and performance. *Child Development, 51*, 1018–1029.

Wang, M. C., Haertel, G. D., & Walberg, H. J. (1990). What influences learning? A content analysis of review literature. *Journal of Educational Research, 84*, 30–43.

Weinert, F. E. (1987). Introduction and overview: Metacognition and motivation as determinants of effective learning and understanding. In F. E. Weinert & R. H. Kluwe (Eds.), *Metacognition, motivation and understanding* (pp. 1–16). Hillsdale, NJ: Erlbaum.

Weinstein, R. S. (1989). Perceptions of classroom processes and student motivation: Children's views of self-fulfilling prophecies. In C. Ames & R. Ames (Eds.), *Research on motivation in education* (Vol. 3, pp. 187–221). San Diego: Academic Press.

Wigfield, A., Eccles, J. S., MacIver, D., Reuman, D. A., & Midgley, C. (1991). Transitions during early adolescence: Changes in children's domain-specific self-perceptions and general self-esteem across the transition to junior high school. *Developmental Psychology, 27*, 552–565.

Wigfield, A., & Harold, R. D. (1992). Teacher beliefs and children's achievement self-perceptions. In D. H. Schunk & J. L. Meece (Eds.), *Student perceptions in the classroom* (pp. 95–121). Hillsdale, NJ: Erlbaum.

Wigfield, A., & Karpathian, M. (1991). Who am I and what can I do? Children's self-concepts and motivation in achievement situations. *Educational Psychologist, 26*, 233–261.

Wong, B. Y. L. (1991). The relevance of metacognition to learning disabilities. In B. Y. L. Wong (Ed.), *Learning about learning disabilities* (pp. 213–258). New York: Academic Press

Zimmerman, B. J. (1990). A social cognitive view of self-regulated academic learning. *Journal of Educational Psychology, 81*, 329–339.

Zimmerman, B. J. (1994). Dimensions of academic self-regulation: A conceptual framework for education. In D. H. Schunk & B. J. Zimmerman (Eds.), *Self-regulation of learning and performance: Issues and educational applications* (pp. 3–21). Hillsdale, NJ: Erlbaum.

Zimmerman, B. J., & Martinez-Pons, M. (1986). Development of a structured interview for assessing student use of self-regulated learning strategies. *American Educational Research Journal, 23*, 614–628.

Zimmerman, B. J., & Martinez-Pons, M. (1988). Construct validation of a strategy model of student self-regulated learning. *Journal of Educational Psychology, 80*, 284–290.

Zimmerman, B. J., & Martinez-Pons, M. (1992). Perceptions of efficacy and strategy use in self-regulation of learning. In D. H. Schunk & J. L. Meece (Eds.), *Student perceptions in the classroom* (pp. 185–207). Hillsdale, NJ: Erlbaum.

CHAPTER TEN

The Brain, the Me, and the I

✧

KARL H. PRIBRAM
RAYMOND BRADLEY

> We have inherited from our forefathers the keen
> longing for unified, all embracing knowledge. . . .
> But the spread, both in width and depth, of the
> multifarious branches of knowledge during the last
> hundred odd years has confronted us with a queer
> dilemma. We feel clearly that we are only now
> beginning to acquire reliable material for welding
> together the sum total of all that is known into a
> whole: but, on the other hand, it has become next
> to impossible for a single mind fully to command
> more than a small specialized portion of it.
>
> I see no other escape from this dilemma (lest
> our true aim be lost forever) than that some of us
> should venture to embark on a synthesis of facts
> and theories, albeit with second hand and
> incomplete knowledge of some of them—and at
> the risk of making fools of ourselves.
> —SCHRÖDINGER (1944, p. 178)

PREAMBLE

When Raymond Bradley (1987) had finished a book relating his longitudinal studies of communes, Lois Erickson, a mutual friend of Bradley and Karl Pribram, asked Pribram to read it, hoping a recommendation for publication might develop. Pribram read the manuscript and was impressed with what appeared to him some basic commonality between the functional structures of social and neuronal collectives, between collectives of different levels or scales of system organization. Over the years, this commonality has become more and more striking—and, at the same time, more challenging.

The challenge is the following: General systems theory is based on the finding that often collectives of different scales can be shown to operate according to the same—or, at least, very similar—principles of organization. This is an intriguing finding but does not tell us *how—the process by which*—such operations come about.

In order to determine the "how" of processing, it becomes necessary to identify the transfer functions that make it possible for operations at one scale to influence those at the adjacent higher- and lower-order scales. A "scale" or "level" is defined as a "description" of the organization of the elements in a system that is simpler than a description of the elements themselves.

The recognition of multiple levels of organization in a system, is not, as Robert Hinde (1992, p. 1019) rightly points out, an argument for reductionism, because each level "must be thought of not as an entity but rather in terms of processes continually influenced by the dialectical relations between levels." Such an approach, which requires scientists to "cross and recross" the boundaries between levels (Hinde's phrase, quoted in Bateson, 1991, p. 14), thus demands integration between adjoining disciplines.

With respect to the challenge posed by relating social collectives to neural collectives, it thus became imperative, as a first step, to delineate scales intermediate between those describing the operations of communes and those describing the operations of the brains of the persons composing the communes. This chapter attempts such a delineation.

In this attempt, we were immediately faced with the problem of a plethora of different terms used in different scientific enterprises. It became critical to try to develop a uniform vocabulary for processes that seemed to us to characterize the same operations.[1] However, in doing this we may be treading close to what in philosophy is called theoretical reduction. But we are *not* reductionists in the sense that if we just knew everything there is to know at a particular level of inquiry, we would be able to explain everything at the next higher level. Rather, as Lévi-Strauss (1963) pointed out, this type of reduction may work for very simple systems (systems controlled by few variables). But for more diverse (complex) systems, a structural approach becomes necessary. The structural method places emphasis on the arrangement of *relations* among elements rather than on the elements per se. The classical example of this emphasis on relation is the periodic table in chemistry. Compounds—such as water—are composed of atoms joined in particular relations called bonds. The bonding is largely accounted for by the structure of—the number of protons in—the atoms composing the compound. But chemical science is basically a science of relations —of the nature of the bonding structures that compose molecules.

Another example is language, which is composed of some 20–30 phonemes. It is not these phonemes per se, however, that characterized a language. Rather, it is the relational structure, the ordering, and contextual embedding that make possible linguistic meaning.

Our approach, therefore, is based on the structural premise of *transposable invariance*: that "structure indicates an ordered arrangement of parts, which can be treated as transposable, being relatively invariant, while the parts themselves are variable" (Nadel, 1957, p. 8; see also Piaget, 1970). This is what makes it possible, for example, to describe the "structure of a tetrahedron without mentioning whether it is a crystal, a wooden block, or a soup cube" (Nadel, 1957, p. 7). Here, we are pursuing just such a structural approach. At the same time, we are nonetheless sympathetic to and uneasy with theoretical "reduction," which for us might better be called theoretical translation. In short, what we aim for is a translation of the concepts essential to understanding relations at one level of inquiry in order to articulate the meaning of concepts at an adjacent level.

INTRODUCTION

A few years back, one of us (KHP), during a seminar, noted that the left arm of a graduate student was moving somewhat awkwardly in arranging papers on the table in front of her. KHP asked the student, Ms. C., if she was all right, while pointing to her left arm. She replied, "Oh, that's just Alice; she doesn't live here anymore." At the end of the semester, Ms. C. presented a detailed account of her experiences with Alice, which is presented under "Some Case Histories."

Shortly thereafter, KHP was asked to supervise a graduate student at another university who was examining a boy *unable* to recount his experiences.

The two case histories provided the seed that provoked us to crystalize some latent ideas based on a wide range of data obtained on brain-lesioned monkeys, on brain electrical activity and the development of self in humans, and on factors leading to stability in social relationships and stability in social collectives. Our presentation is tentative but intrigues us sufficiently to venture it here.

The case histories describe two distinct modes of coping that are disrupted by brain injury: one articulates the organism—in egocentric space—and *locates* it—allocentrically—in its environment; the other *evaluates* and *monitors* experience. We proceed to suggest that these two dimensions of coping are embodiments of Piaget's processes of assimilation and accommodation. As accommodation is "the source of changes [that] bind the organism to successive constraints in the environment" (Piaget 1936/1954, p.

352), we explore the source in the consequences of behavior and the resulting changes in the organism's competence in assimilating (coping with) its
environment.

Next, we note that this development of competence in assimilation entails three additional dimensions, dimensions that also have been shown to
involve distinguishable brain systems: arousal–familiarization; activation–
selective readiness; and effort–comfort. We go on to suggest that these dimensions are the same as those obtained by Sternberg (1986) in his analysis
of the dimensions that lead to competence and stability in loving relationships: passion (arousal); commitment (selective readiness); and intimacy (ef-
fort-comfort). Finally, we relate these dimensions to those that have been
shown to operate in producing stability in competent social collectives: flux
(passion); control (commitment); and collaboration (intimacy).

SOME CASE HISTORIES

From Ms. C.:

> I was doing laundry about midmorning when I had a migraine. I felt a
> sharp pain in my left temple, and my left arm felt funny. I finished my
> laundry toward midafternoon and called my neurologist. He told me to
> go to the emergency room. I packed a few things and drove about 85
> miles to the hospital where he is on staff (the nearest was 15 minutes
> away). In the ER, the same thing happened again. And again, the next
> morning after I was hospitalized, only it was worse. The diagnosis of a
> stroke came as a complete surprise to me because I felt fine, and I didn't
> notice anything different about myself. I remember having no emotional
> response to the news. I felt annoyed and more concerned about getting
> home, because I was in the process of moving.
>
> Not until several days later, while I was in rehabilitation, did I no-
> tice strange things happening to me. I was not frightened, angry, or an-
> noyed. I didn't feel anything—nothing at all. Fourteen days after I was
> admitted to the hospital, I became extremely dizzy, and I felt I was falling
> out of my wheelchair. The floor was tilting to my left, and the wheelchair
> was sliding off the floor. Any stimulus on my left side or repetitive move-
> ment with my left arm caused a disturbance in my relationship with my
> environment. For instance, the room would tilt down to the left, and I
> felt my wheelchair sliding downhill on the floor, and I was falling out of
> my chair. I would become disoriented, could hardly speak, and my whole
> being seemed to enter a new dimension. When my left side was placed
> next to a wall or away from any stimuli, this disturbance would gradual-
> ly disappear. During this period, the left hand would contract, and the
> arm would draw up next to my body. It didn't feel or look like it be-
> longed to me. Harrison moved the left arm repeatedly with the same

movement, and a similar behavior occurred, except I started crying. He asked me what was I feeling, and I said anger. In another test he started giving me a hard time until the same episode began to occur, and I began to cry. He asked me what I was feeling, and I said anger. Actually, I didn't feel the anger inside, but in my head when I began to cry. Not until I went back to school did I become aware of having no internal physical feelings.

I call that arm Alice (Alice doesn't live here anymore)—the arm I don't like. It doesn't look like my arm and doesn't feel like my arm. I think it's ugly, and I wish it would go away. Whenever things go wrong, I'll slap it and say, "Bad Alice" or "It's Alice's fault." I never know what it's doing or where it is in space unless I am looking at it. I can use it, but I never do consciously, because I'm unaware of having a left arm. I don't neglect my left side, just Alice. Whatever it does, it does on its own, and most of the time, I don't know it's doing it. I'll be doing homework and then I'll take a sip of coffee. The cup will be empty. I was drinking coffee with that hand and didn't know it. Yet I take classical guitar lessons. I don't feel the strings or frets. I don't know where my fingers are, or what they are doing, but still I play.

How do I live with an illness I'm not aware of having? How do I function when I'm not aware that I have deficits? How do I stay safe when I'm not aware of being in danger?

Ms. C. is obviously intelligent, attending lecture material, asking interesting questions. She is a widowed lady in her mid-50s, enrolled in adult education, majoring in clinical psychology. She gets around splendidly despite Alice and despite a history of a temporary left hemiparesis. The diagnosis was damage of the right temporal–parietal cortex confirmed by an abnormal electroencephalogram (EEG) recorded from that location. The damage was not sufficiently extensive to show in a positron-emission tomography (PET) scan.

Contrast Ms. C.'s story with the following observations made on an 8-year-old boy by Chuck Ahern as a part of a thesis program that KHP supervised:

TJ had an agenesis of the corpus callosum with a midline cyst at birth. During the first six months of his life, two surgical procedures were carried out to drain the cyst. Recently performed Magnetic Resonance Imaging (MRI) showed considerable enlargement of the frontal horns of the lateral ventricle—somewhat more pronounced on the right. The orbital part of the frontal lobes appeared shrunken as did the medial surface of the temporal pole.

TJ appears to have no ability for quantifying the passage of time [what Bergson (1922/65) called durée] and no experiential appreciation of the meaning of time units. For example, a few minutes after tutoring begins, he cannot say—even remotely—how long it has been since the session started. He is as apt to answer this question in years as in minutes. He does always use one of seven terms of time quantification (seconds, minutes, hours, days, weeks, months or years) when asked to estimate the duration of an episode but uses them randomly. He can put these terms in order, but does not have any sense of their meaning or their numerical relationships to one another.

When TJ returned from a trip to the Bahamas he did recall that he had been on the trip; however, the details he could recount about the trip numbered fewer than 5. His estimates of how long it had been since his trip were typical in that they were inaccurate and wildly inconsistent on repeated trials. Also, the first five times back at tutoring he stated that he had not been at tutoring since his trip. It appears that he is unable to place in sequence those few past events that he can recall. Nonetheless, he can answer questions correctly based on his application of general knowledge about development, e.g., he knows he was a baby before he could talk because "everyone starts as a baby." But, one day he asked his tutor if he knew him when he was a kid, indicating, I think, his incomprehension of the duration of each of these developmental periods and his unawareness of what events constituted such a period for him.

TJ is aware that he has a past, that events have happened to him but he cannot recollect those events. He also spontaneously speaks of events in his future such as driving an automobile and dating and growing a beard. He has play-acted on separate occasions his own old age and death. TJ is capable of excitement about the immediate future. On the very day that he was going to the Bahamas he was very excited as he exclaimed repeatedly: "I'm going to the Bahamas." But when his tutor asked him when, he said blankly: "I don't know." He also displayed keen anticipation when one day he saw a helicopter preparing to take off from the hospital. The helicopter engines revved approximately 13 minutes before it took off and TJ become increasingly more vocal and motorically active, laughing as he repeated, "When's it going to take off?" He also anticipates future punishment when he is "bad." He is aware, on some level, of the immediate future in his constant question "What's next," which he asks his mother at the end of each activity.

There are a variety of other occasions on which he demonstrated this capacity regarding tempo (as opposed to evaluating the duration of an experience). There have been several breaks in his usual thrice weekly tutoring schedule. Each of four times this schedule has been interrupted, he has run to meet his tutor when he approached rather than waiting inside as he usually does. Also, on these occasions he has typically asked if his tutor missed him. However he states he does not know how long it has been since his last session, and there was no evidence that he knew it had been longer than usual.

TJ compares who walks faster or who draws faster. He has at least a basic sense of sequencing as when he says, "I'll take a turn and then you take a turn." He also uses terms like "soon" and "quick" correctly in conversation. For example, when he wanted to do a drawing at the beginning of a session, and his tutor said that we needed to begin to work and he countered, "This will be quick." Unsurprisingly, he finished his drawing at his normal pace. He somehow seems to use such terms correctly without any experiential appreciation of them. (Modified from letter written by Chuck Ahern on March 19, 1995, addressed to Karl H. Pribram)

These two case histories illuminate two very important dimensions of self. One dimension, portrayed by Ms. C., *locates* us in the world and also with respect to our body's configural integrity. The other dimension, highlighted by TJ, *monitors* our experience. Without such monitoring, the events comprising the experience fail to become evaluated and encoded into memory. Location is kin to—but more primitive than—a spatial dimension[2]; monitoring is kin to—but more basic than—a temporal dimension.[3]

After I (KHP) read Ms. C.'s paper and gave her a well-deserved A+ and showed her Ahern's description of TJ, we discussed Dan Dennett's views—expressed in his book modestly entitled *Consciousness Explained* (1991)—on "the Cartesian Theater" and on "Narrative Consciousness." She agreed that the paper she wrote was indeed a narrative, but it had taken her a good while to piece together the various episodes that she had experienced into a coherent story. Meanwhile, and this was still the case, her experience was and is dramatically locational, that is, theatrical. What could provide better theater than "Alice doesn't live here anymore" with the emphasis on the "live *here*"? Despite the current vogue to trash Descartes (see, e.g., Antonio Damasio's *Descartes' Error* [1994][4]), many persons do, in fact, experience them*selves* as actors on the stage of life, as Shakespeare so eloquently expressed it.

Ms. C. experienced devastations to her locational *integrity*. Other patients, after injuries to their occipital lobes, demonstrate "blindsight," the ability to visually identify objects in the "blind" field despite the fact that they fail to be consciously aware of these objects. Patients such as those who are blindsighted and Ms. C., who might be considered to have a tactile and kinesthetic blindsight, both have damage to the cortex of the posterior convexity of their brains. Thus, they suffer disruption of their egocentric (essentially tactile and kinesthetic) and allocentric (essentially visual and auditory) organization. I (KHP) have called this a disruption of "objective" awareness because it relates the patient to his or her impairment as if it were a relationship among objects. The relationship is "intentional" in Brentano's (1874/1914) sense of an ability to differentiate the perceiver from the perceived (Pribram, 1976). Note that in such patients, narrative abilities do not suffer—as you can judge for yourself from Ms. C.'s report.

Contrast this with the boy's disability: This boy's episodic memory is severely deficient. But, as compared to Ms. C., he has no problem with his egocentric space, nor is he blindsighted —he has no difficulty in experiencing his allocentric whereabouts. Despite his disability in monitoring, he continually defines his location both in space and in clock-time. However, his narrative self is severely limited by his inability to monitor events and place them into sequences of episodes. The narrative self is composed of such sequences of episodes. TJ's attempts to do so are contrived and depend on his intact ability to deal with egocentric and allocentric experience.

To summarize this part of the chapter: Two dimensions of self can be distinguished on the basis of selective damage to different parts of the brain. These dimensions concern an objective "me" and a narrative "I."[5]

The *objective "me"* is characterized as spatiotemporally articulated (egocentrically) and located in the world (allocentrically). The *narrative "I,"* by contrast, is constituted by a hermeneutic monitor of episodes and events that themselves are the consequences of the monitoring, and thus self-organizing. Next, let us look into how these two dimensions of the self develop.

THE DEVELOPMENT OF A STABLE SELF

Piaget formulated two complementary processes that guide cognitive growth. One process he labels "accommodation"; the other, "assimilation." "In their initial directions, assimilation and accommodation are obviously opposed to one another, since assimilation is conservative and tends to subordinate the environment to the *organism as it is*, whereas accommodation is the source of changes and *bends the organism* to the successive constraints of the environment" (Piaget 1954, p. 352) Thus,

> the nursling's psychic activity is at first only simple assimilation of the external environment to the functioning of the organs. Through the medium of assimilatory schemata, at first fixed, then mobile, the child proceeds from this elementary assimilation to putting means and ends into relationships such that the assimilation of things to personal activity and the accommodation of schemata to the external environment find an increasingly stable balance. The undifferentiated and chaotic assimilation and accommodation which characterize the first months of life are superseded by assimilation and accommodation simultaneously dissociated and complementary. (Piaget 1954, p. 219)

With respect to the brain functions described here, assimilation appears to be affected by the process of locating the child in egocentric and allocentric space–time schemata; accommodation resembles the effect that a self-organizing, neurologically based monitoring hermeneutic process would be expected to exert.

Assimilation and accommodation are ordinarily complementary, and Piaget's developmental stages reflect this complementarity. Nonetheless, as Piaget points out in the preceding quotation, one can discern differences in balance: The sensory–motor stage is more assimilative than accommodative; the various operational stages employ both processes in a complementary fashion and more or less equally; and the postoperational stage(s) are primarily accommodative. In both assimilation and accommodation, biological and experiential factors interact, and the maturation of brain electrical activity reflects the resultant of this interaction. During infancy, maturation centers on the sensory–motor regions of the convexal cortex. In childhood and adolescence, two stages of "growth spurts," as they are called, can be discerned, each stage beginning posteriorly and ending frontally. Of great interest, and surprise, to us (Hudspeth & Pribram, 1992) was the discovery of a major increment in the maturation of the frontal cortex during the ages 17–21! This is the age at which we send our children to college (or to the armed services in case of war). Their accommodation is critically vulnerable, and how they mould their frontolimbic hermeneutic monitoring systems during these years will determine their citizenship for the rest of their lives.

This nice sequence of consequential events (events: Latin, *ex venire:* out-come) is, on occasion, interrupted. Interruption leads to a different process: The organism becomes aroused, demarcating an episode, but if the interruption is repeated, the organism habituates to it; the marked episode becomes familiar. When we remove one of the components of the frontolimbic forebrain (the amygdala), the organism's clock-like oscillations flatten out and familiarization fails to occur (reviewed by Pribram & McGuinness, 1975, 1992). Episodes become recurrent, distracting from the progressive development of competence through commitment; the organism has become hung up, treating each occurrence and reoccurrence of the distracting event as arousing. Interestingly, a kitten raised for 8 weeks in darkness and in an otherwise also restricted environment will, in the first few weeks of release, show symptoms identical to those of amygdalectomized cats: failure of visceroautonomic responses, continual and repetitious investigation of the environment (behavioral orienting), and a low threshold to startle (Konrad & Bagshaw, 1970).

Schore (1994), in an aptly titled volume, *Affect Regulation and the Origin of Self*, reviews additional evidence for the importance of the amygdala and related orbitofrontal systems in the development of the infant. Schore bases his analysis on the importance of the dyadic relationship between an infant and the primary caregiver. Through social bonding with the infant (touching, holding, feeding and especially mutual [eye] gazing, etc.), the primary caregiver creates a relationship with the infant that, according to Schore, involves interaction along two dimensions. The first is the stimulation of positive affect by the primary caregiver; this dimension seems to us to correspond to our arousal–familiarization dimension. Arousal involves establishing contexts to facilitate the dyadic mutual positive affect that becomes the conduit through which, according to Schore, the infant's worldly experience is channeled. The second dimension of this interaction is the regulation by the primary caregiver of the infant's behavior: This corresponds to what we have called a "readiness to selectively respond" system, which is centered on the basal ganglia of the brain. Schore describes interactions of the brain structures that become involved in these two processes:

> In line with the principle that information is processed in stages, the entire temporal sequence involves sensory–perceptual encoding of a socioaffective facial stimulus in the posterior cortex, cognitive appraisal in the frontolimbic anterior cortex, amplification of the signal during cortico–subcortical transmission along the visuolimbic pathway, activation of hypothalamic motivational systems which influence internal states and generate emotion–specific action tendencies, activation of the ascending ventral tegmental mesocortical dopamine circuit, and finally activation of prefrontal areas and frontal cortical motor regions responsible for facial and body emotion response expression. (pp. 195–196)[6]

Schore extensively reviews the evidence that "environmental stimuli regulate the anatomical, cellular, and even the molecular organization of the developing nervous system" (p. 161). With regard to the (dopaminergic) functions of the postnatal infant frontal cortex, the stimuli are primarily social. Thus, in situations where the primary caregiver's nurturing love has broken down during a critical period (approximately the first year of life) and the infant is subjected to prolonged exposure of heightened negative affect, the growth and organization of the infant's developing frontal cortex can be affected with enduring pathological consequences. This results in structurally defective neurobiological organization, which, in turn, produces disturbances in social attachments. These functional impairments of the neural circuitries result in a persisting susceptibility to further patterns of pathophysiological growth that are associated with later forming psychiatric disorders (see the review of the evidence by Schore, pp. 159–167).

To translate into the terms of this chapter, the requisite neurobiological organization for the development of a stable self is prompted by the interactions in the mother–infant dyad, which entail affect stimulation and modulation/regulation. Affect stimulation entails *accommodation by the infant to the egocentric and allocentric assimilation of his or her relation with the external world* and corresponds to arousal–familiarization. Accommodative modulation/regulation via the mother produces this egocentric and allocentric assimilation of the environment and corresponds to selective readiness. Often, it takes effort to disengage from the disposition toward arousal, that is, to return to commitment. Still another part of the forebrain (the hippocampal formation) is critical to determining the duration of engagement (Pribram & McGuinness, 1975, 1993). What is involved is the matching of the cycles of the clockwork to registrations in memopad. A model has been developed to detail how such a matching process might be articulated (Pribram, 1996). Thus, the infant's development of competence for (eventual) autonomous socio-emotional attachment corresponds to an effort–comfort dimension. More on this shortly.

Two further commonalities with Schore's findings should be noted. The first is that there is some evidence that the significance of these two dimensions of interaction continues beyond infant development to also be important in the development of the young child. Drawing from a study of aggression among 4-year-olds in preschools, Robert Hinde (1992) reports [on the basis of "three replications"] that aggression was found to lower when "maternal warmth" and "maternal control" in the mother–child relationship were "more or less in balance" (see pp. 1025–1026, especially Figure 5). The second is a commonality with Piaget's ideas on the role of affect in intellectual development. In his final lectures, Piaget (1983) links the development of the individual's intelligence and creativity as an adult to a nurturing fabric of loving relationships as a child.

THE ACCOMMODATIVE PROCESS

We have learned a good deal about the way accommodation is achieved. Reinforcement theory has made solid contributions to understanding, as have the results of neuropsychological experiments. In experimental psychology, reinforcements (and deterrents) are conceived as those consequences of behavior that influence an organism to increase (or decrease) the recurrence of the behavior that generated the consequence. Two separate processes are apparently involved: One is a biologically based, clocklike oscillator, the other a register or "memopad" that keeps track of the consequences per se. Many of the oscillations display an appetitive and a consummatory phase and are critical to organizing the accommodative process. The memopad registers an environmental event when it matches, that is, resonates with, the appropriate moment on the clock cycle. Recurrence of the consequential behavior is determined by the density of registrations (see Killeen, 1994, for review and details).

A ready way to conceptualize the reinforcement (or deterrent) process is as follows: (1) initially *consequences* are those sequences of behavior that fit into the context of prior registrations (thereby becoming consequential); (2) at a certain point, the density of such consequences attain a sufficient probability that the recurrence of the behavior will produce a reliable match, so that the organism becomes *confident* in its *competence* to perform in this context. At this point, the laws of learning are replaced by those that characterize performance (see Pribram, 1971, Chap. 16). The density of consequences engenders "affluence," which leads to such a means–ends reversal:

> What happens when a man, or for that matter an animal, has no need to work for a living? . . . The simplest case is that of the domesticated cat— a paradigm of affluent living more extreme than that of the horse or the cow. All basic needs of the domesticated cat are provided for almost before they are expressed. It is protected against danger and inclement weather. Its food is there before it is hungry or thirsty. What then does it do? How does it pass its time?
>
> We might expect that having taken its food in a perfunctory way it would curl up on its cushion and sleep until faint internal stimulation gave some information of the need for another perfunctory meal. But no, it does not just sleep. It prowls the garden and the woods killing young birds and mice. It *enjoys* life in its own way. The fact that life can be enjoyed, and is most enjoyed, by many living beings in the state of affluence (as defined) draws attention to the dramatic change that occurs in the working of the organic machinery at a certain stage of the evolutionary process. *This is the reversal of the means–end relation in behaviour.* In the state of nature the cat must kill to live. In the state of affluence it lives to kill. This happens with men. When men have no need to work for a living there are broadly only two things left to them to do. They can "play" and they can cultivate the arts. These are their two ways of enjoying life. It is true that many men work because they enjoy it, but in this case "work" has changed its meaning. It has become a form of "play." "Play" is characteristically an activity which is engaged in for its own sake—without concern for utility or any further end. "Work" is characteristically an activity in which effort is directed to the production of some utility in the simplest and easiest way. Hence the importance of ergonomics and work study—the objective of which is to reduce difficulty and save time. In play the activity is often directed to attaining a pointless objective in a difficult way, as when a golfer, using curious instruments, guides a small ball into a not much larger hole from remote distances and in the face of obstructions deliberately designed to make the operation as difficult as may be. This involves the reversal of the means–end relation. The "end"—getting the ball into the hole—is set up as a *means* to the new end, the real end, the enjoyment of difficult activity for its own sake. (Mace, 1962, pp. 10–11)

A somewhat similar statement has been presented by Robert W. White (1960). He emphasizes the role played by the progressive achievement of competence in the maintenance of behavior and makes a strong case that the "feeling of efficacy" is an important guide to behavior.

> Effectance is to be conceived as a neurogenic motive, in contrast to a viscerogenic one. It can be informally described as what the sensory–neuro–muscular system wants to do when it is not occupied with homeostatic business. Its adaptive significance lies in its promotion of spare-time behavior that leads to an extensive growth of competence, well beyond what could be learned in connection with drive-reduction. (White, 1960, p. 103)

White is concerned with the implications of effectance in clinical psychology; here, our concern is with what the sensory–neuromuscular system "wants."

According to the foregoing analysis, the common problem for means–end theory and effectance theory is that activities of a certain type appear to be self-maintaining. The consequences of the actions must provide their own set within which a subsequent event will be consequent, that is, reinforcing.

In many respects, what has been discussed is the development of behavior differentiation, that is, skill. Effectance and competence, play and gamemanship, demand precise timing of actions within larger sequences of actions, so that consequences—sequences in context—will form a harmonious production. And a great deal is known about the neurology of skill. Here, perhaps, more than anywhere else, the model of "sequence in context" can be realized in tissue—and in fact, the model was originally devised to handle some new neurological facts in this area (Miller, Galanter, & Pribram, 1960).

At the reflex level, control of muscular contraction can no longer be conceived simply in terms of the reflex arc (some excitation of receptors, transmission of the signal aroused by such excitation to the central nervous system, and back again to the muscle in question). The change in conception is necessitated by the discovery that the activity of the γ efferent fibers, fibers that transmit signals from the central nervous system to the receptors in the muscle (muscle spindles), acts as a feedback, that is, controls the amount of activity recordable from the afferents that signal the state of the receptor to the central nervous system. The presence of this feedback loop makes it difficult at any moment in time to assess the origin of a particular amount of activity in the afferent nerves, and thus the state of the receptor. That state could reflect the state of contraction (isometric or isotonic) of its muscle group or it could reflect the amount of activity of the γ efferent system, or both. Only a comparison between states at successive moments, in the context of γ efferent activity, will give a signal of the state of contraction of the muscle group. The γ efferent activity provides the setting, the context, the bias on the muscle receptor. (On occasion, the reverse may well be the case. The bias may be set by the muscle contraction and changes in γ efferent activity computed.)

Sherrington, in his classic lectures, *The Integrative Action of the Nervous System* (1911/1947), was not unaware of the problem, and his statement of it is worth repeating (his solution is cast in simple, associative terms—reinforcement for Sherrington occurs through immediate spinal induction [summation through increased intensity and coextensity of convergent inputs]):

> We note an orderly sequence of actions in the movement of animals, even in cases where every observer admits that the coordination is merely reflex. We see one act succeed another without confusion. Yet, tracing this sequence to its external causes, we recognize that the usual thing in nature is not for one exciting stimulus to begin immediately after another ceases, but for an array of environmental agents acting concurrently on the animal at any moment to exhibit correlative change in regard to it, so that one or other group of them becomes—generally by increase in intensity—temporarily prepotent. Thus there dominates now this group, now that group in turn. It may happen that one stimulus ceases coincidentally as another begins, but as a rule one stimulus overlaps another in regard to time. *Thus each reflex breaks in upon a condition of relative equilibrium, which latter is itself reflex.* In the simultaneous correlation of reflexes some reflexes combine harmoniously, being reactions that mutually reinforce [Sherrington, 1911/1947, p. 120; emphasis added].

At the cerebral level, also, neurology has a great deal to say about skill. Removals of the precentral "motor" cortex of primates (including man) certainly results in awkward performance (Pribram et al., 1955–56). It follows that in the cerebral mechanisms in control of action excitation "breaks in upon a condition of relative equilibrium." . . . This has been shown by John Lilly (1959). Prolonged trains of excitation (subliminal to those that would produce movement) were delivered to the precentral motor cortex whenever the lever was depressed by the subject (a monkey). Lever pressing had to be paced so the on–off nature of the excitation could be maintained. The monkey learned to perform such paced lever pressing behavior and spent many (may I say "happy"?) hours at this occupation.

This has been a long way from means–end reversal to effectance to skill. The point is simply that these areas of interest pose a common problem: how is it that selective behavior is maintained in the absence of guides from drive stimuli—or, in the extreme, when behavior apparently goes in a direction contrary to one plausibly related to drive stimuli? The suggestion made in this section is that the consequences of actions are truly stimulus events that occur in sequence and that, once some order has been initiated in this sequence of stimuli, this order per se can provide the set or context for the occurrence of the next or sub-sequent event. Actions have consequence and the consequences of actions are reinforcers. Behavior, thus, becomes its own guide. (Pribram, 1963, pp. 137–141)

The means–ends reversal based on confidence in one's competence can thus be defined in terms of the *density of consonant consequences* generated in a particular context. Other things being equal, the organism becomes committed to organize its actions not just to reduce dissonance but to actively produce consonance. The production of consonance is no haphazard affair—consequences will be registered when they match the organism's repertoire of behaviors relevant to its biologically based "interests" such as drinking or running. The more densely consonant become nested within the context, the framework, of the less densely consonant. (For a detailed review of the data upon which these ideas are based, see Pribram, 1980; 1995.) This nested hierarchical arrangement commits the organism to a Plan or program. The importance of Plans to the organization of behavior has been discussed fully in Miller et al. (1960). In addition, however, as noted here, consequential behavior is self-organizing, a theme that has been stressed by Konrad Lorenz (1969) and by Bandura (1991).

To summarize: Development concerns primarily the effect of changes occurring in the accommodative process as they alter assimilation. Accommodation is achieved by a process in which *con-sequences* (sequences of behavior that are consonant within the context created by prior consequences) become densely registered by virtue of their consonance. This process leads to *confidence*, which is based on the development of a behavioral *competence* in that context. The means–ends reversal *commits* the organism to pursuing the development of this competence. Commitment, therefore, is not to the status quo of current egocentric and allocentric, intentional competence but to a selective readiness to modify competence in accord with continuing assimilation of experience.

MODIFICATION OF ASSIMILATION
BY ACCOMMODATION

In order for an assimilated content to be influenced by and to influence an accommodating context, sensory channels must be flexible in their organization. Evidence (from event-related brain electrical recordings, reviewed by Pribram, 1991, Lecture 10) indicates that updating occurs and thus calls into question the notion that processing capacity is inflexible. Additional evidence makes it unlikely that limitations in processing span are due to limitations imposed by some fixed channel capacity. Briefly, the evidence runs as follows:

Two million input nerve fibers converge by way of the central nervous system onto 350 thousand output fibers. Sherrington (1911/1947) conceptualized a restriction on performance due to this convergence in his doctrine of a "limited final common path." However, Donald Broadbent (1974) showed that with regard to cognitive operations such as attention, limited span is not so much a function of the final common path as it is a function of the central processing mechanisms in the brain. Broadbent reviewed this aspect of his work in the Neurosciences Study Program III (1974), and Pribram devoted a whole section of this program to the topic entitled "How Is It That Sensing So Much We Can Do So Little."

The issue of limited span is usually discussed in terms of a fixed channel capacity. But as reviewed by Pribram and McGuinness (1975, 1992), a considerable volume of work has shown that the central processing span is not fixed. Thus Miller (1956), Garner (1962), and Simon (1974, 1986), among others, have clearly shown that information-processing span can be enhanced by reorganization such as that provided by "chunking." In fact, the limitations in processing can be overcome to such an extent that one is hard put to defining any "ultimate" limit. These data and others have led us to conceptual limitations in processing span as limitations in channel competence rather than in channel capacity (Pribram & McGuinness, 1975), a view also expressed by Maffei (1985). Thus, the conception of a limited capacity depending on some fixed "exoskeleton" constraining channels becomes untenable. An increase in processing capability, in competence, becomes possible by way of challenges to a flexible "endoskeleton" supporting processing channels. Chunking has been shown, by neurobehavior experiments, to be influenced by resections of the far frontal cortex. Furthermore, electrical excitation of the frontal cortex changes receptive field properties of neurons in the sensory channels of the primary visual cortex. These changes are directly related to the ability to parse or chunk the input.

The particular experiments that demonstrate top-down neurophysiological processing—processing that implements changes in channel structure and, therefore, in egocentric and allocentric processing capability—were performed on the receptive field organization of single neurons in the lateral geniculate nucleus and the primary visual cortex of cats and monkeys. Receptive fields of visual cortex neurons were mapped by displaying a small moving dot on a contrasting background. The location and motion of the dot were computer controlled. Thus, the computer could sum (in a matrix of bins representing the range over which the dot was moved) the number of impulses generated by the neuron whose receptive field was being mapped. This was done for each position of the dot, because the computer "knew" where the dot was located (Lassonde, Ptito, & Pribram, 1981; Spinelli & Pribram, 1967).

The map obtained for the lateral geniculate nucleus (the halfway house between retina and cortex) is usually called the "Mexican hat" function for obvious reasons. The brim of the hat represents the spontaneous background of impulse activity of the neuron. The crown of the hat represents the excitation of the cell by the dot of light shown to the animal when the cell is located at the center of the visual field. Where the crown meets the brim, there is a depression indicating that the output of the cell has been inhibited.

It is this inhibitory surround that can be augmented or diminished by electrical excitation of other parts of the forebrain. Stimulation of the far frontal cortex or the head of the caudate nucleus diminishes the inhibitory surround; stimulation of the posterior intrinsic (association) cortex (specifically, in this case, the inferotemporal portion of this cortex) or of another of the basal ganglia, the putamen, produces an augmentation of the inhibitory surround.

Receptive fields of adjacent neurons overlap to a considerable extent. Thus, when the excitatory portion of the receptive fields becomes enlarged, the dendritic fields essentially merge into a more or less continuous functional field. By contrast, when the excitatory portion of the receptive fields shrinks, each neuron becomes functionally isolated from its neighbor.

This modifiability of the competence of the primary visual system was supported by testing the effects of the same electrical stimulations on the recovery cycles of the system as recorded with small macroelectrodes. Far frontal stimulations produce a slowing of recovery, whereas posterior stimulations result in a more rapid recovery as compared with an unstimulated baseline. Slow recovery indicates that the system is acting in unison; rapid recovery indicates that the system is "multiplexed"—that its channels are separated and not encumbered by a more extensively interconnected system with consequent greater inertia.

Further confirmation of the importance of frontal accommodative processing on posterior assimilatory functions has been provided by Tucker, Potts, and Posner (1995), who, recording on a 128-electrode geodesic array, used visual event-related brain electrical activity to study the course of this electrical activity in the brain. After an initial activity produced in the primary visual cortex in the occipital lobe, the frontal lobes became activated only to have the occipital cortex once more involved. This "reprise," as Tucker calls it, takes about 200 msec to occur. As yet, we do not know the pathways by which this occipital–frontal–occipital interaction takes place, nor whether the shape of the occipital reprise is altered by the frontal activity—but these issues are currently being investigated.

In summary, commitment entails accommodation of the competence of processing channels to assimilate progressively greater differentiation of the organism's egocentric and allocentric relationship to its environment.

STABILITY IN LOVE RELATIONSHIPS

So much for a top-down "deconstruction" of the accommodation–assimilation processes. We now take a bottom-up look to see how these processes influence social structures. Drawing on Sternberg's work on love relationships (1986; Sternberg & Barnes, 1985; Sternberg & Grajek, 1984), we now consider how the three dimensions we have just distinguished in brain systems–arousal–familiarization, activation–selective readiness, and effort–comfort—are, in essence, the same as the three dimensions that Sternberg has identified in his social-psychological research.

Grounded in the results of a series of empirical studies (Levinger, Rands, & Talaber, 1977; Rubin, 1970; Sternberg & Grajek, 1984; Sternberg & Barnes, 1985; and Swensen, 1972, among others), Sternberg (1986) has proposed a triarchical dimensionality to account for stability in love relationships. According to his "triangular theory," love has three components or dimensions. One is *passion*, which leads to romantic and physical attraction, and sexual intercourse. A second is *decision/commitment*, a cognitive component that involves two temporal considerations: a short-term decision that one loves someone else, and a long-term commitment to maintain that love. The third is *intimacy*, which refers to feelings of closeness, connectedness, and bondedness, feelings that give rise to the experience of warmth in a loving relationship. While the passion component of a close relationship is "relatively unstable," commitment and intimacy are "relatively stable" (Sternberg, 1986, pp. 119–122).

Constructing a taxonomy of eight kinds of love from the different (logical) combinations of the presence or absence of the three components in a relationship, Sternberg characterizes "consummate love," the category that involves the "full combination" of all three, as the ideal toward which most people strive in romantic relationships (1986, p. 124). And while noting that it may be easier to achieve consummate love than to maintain it on a long-term basis, he points out that the results from his empirical research (Sternberg & Barnes, 1985) suggest that the "ability to *communicate effectively* is almost a *sine qua non* of a successful loving relationship" (p. 134; emphasis added). The significance of this finding will become clear in the next section.

Translating Sternberg's dimensionality into our terms, there is a match between the "I" that becomes totally accommodative (the arousal–familiarization dimension), as developed here, and the "passion" dimension as portrayed in Sternberg's analysis. In a like manner, we would propose that commitment (by "me" to egocentrically and allocentrically assimilate—via selective readiness—the situation) as outlined in this chapter is equivalent to "decision/commitment" as developed in his analysis. Sternberg (1986, p. 134) indicates that this component is "most subject to conscious control" by an individual. Note that this type of conscious control corresponds to Brentano's (1973) intentionality and Searle's (1983) intention, as discussed earlier. Finally, there appears to be a close correspondence between the effort–comfort dimension, worked out on the basis of neuropsychological analysis, and the "intimacy" dimension in Sternberg's work. Intimacy often entails effort and can lead to comfort. Intimacy involves matching passion to commitment.

By way of summary, we have used Piaget's concepts of accommodation and assimilation as a framework of common terminology to translate concepts describing the operations of a particular scale or level of organization into the terms of those at an adjacent level. Thus, in describing, so far, operations at the neurobiological, neuropsychological, and social-psychological levels, we have endeavored to show correspondences in the organization of behavior that produce a stable, competent, social self. As we move to the final level, the sociological level of collective organization, it is worth noting that our approach is consistent with the strategy that Robert Hinde has often advocated (e.g., 1979, 1987, 1992; see Bateson, 1991, for examples), and our effort can be seen as an attempt to describe, in *substantive* terms, linkages between physiological, psychological, and sociological organization that Hinde (1992) identifies in his "levels of social complexity" framework.

In the next section, data from a longitudinal analysis of the factors leading to a stable order of collaboration in communes offer support for the correspondences we have established earlier. Interestingly enough, the data are also consistent with the notion of cooperation that Piaget develops in his essay, "Logical Operations and Social Life" (Piaget, 1965/1995a). Piaget's concept of cooperation, a "system of operations carried out in common" (p. 153) that is based on reciprocal interactions within the context of a common system of language, values, and social norms, is analogous to our concept of collaboration.[7]

STABILITY IN SOCIAL COLLECTIVES

Following up on previous analyses of the communes that Bradley studied (1987; Bradley & Roberts, 1989a, 1989b; Carlton-Ford, 1993; Zablocki,

1980), we (i.e., Bradley and Pribram) have developed an empirically based model of the endogenous processes of communication by which stability is generated in (small) social collectives (Bradley & Pribram, 1995, 1996). By communication, we mean a process by which information about the collective's internal organization is gathered, processed, and distributed throughout the collective as a whole. A "social collective" is defined as a durable arrangement of individuals distinguished by shared membership (a boundary) and collaboration in relation to a common purpose or goal. Stability is the degree to which structural integrity and functional viability are sustained by a collective over time.

The results from this earlier body of work have shown that two patterns of social relations form the communicative structure in stable communes. As shown for the stable communes (groups surviving at least 24 months beyond measurement of their social structure; see Figure 10.1),[8] one pattern is a dense web of mutual relations of positive affect interconnecting virtually all members. This web is organized as a *field*, a distributed, massively parallel order of symmetrical monitoring processes in which individuals are essentially interchangeable. The second pattern is a

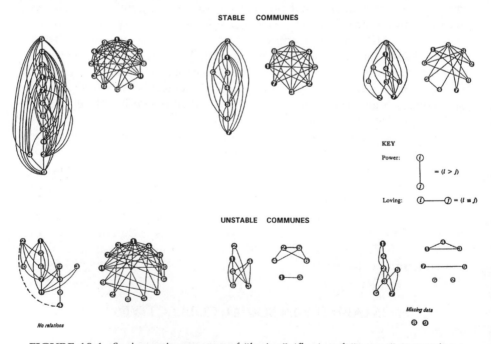

FIGURE 10.1. Sociometric structure of "loving" (flux) and "power" (control) relations for selected stable and unstable communes.

densely interlocking order of power relations that also extends to connect virtually all individuals. This is a *hierarchy*, a highly stratified system of asymmetrical, transitively ordered relations that define, for each individual, a location that is spatially and temporally identified and, therefore, unique. The relationship between the two orders was found to be associated with group survival (see Bradley 1987, Chap. 7; Bradley & Roberts, 1989a).[9] In following up on these earlier findings that describe *what* the communicative structure was composed of, we sought to understand how the interaction between field and hierarchy operates as an information processing system that *in*-forms (gives shape to) the production of stable collective order.

The *collaborations*[10] within the collective that form the communication system are formed by the interpenetration of networks of endogenous monitoring processes organized along two dimensions in which the values allocated in each dimension define points within a social field (Bradley & Roberts, 1989a). The values ascribed to the horizontal dimension represent *flux*, the amount of activation of potential energy (*passion*) in a social collective. The values ascribed to the vertical dimension represent the amount of *control* (*commitment*)—the degree to which individuals are interconnected by a transitively ordered network of relations—exercised at that location (see Figure 10.2). The coordinates representing the dimensions bound a phase space within which each value represents an amount of information—a *quantum* of information in Gabor's (1946) terms (see Appendix B)—that characterizes the communicative structure and *in*-forms the collective's expenditure of energy. Thus, each quantum of information, a configuration of flux and control, is associated with a potential for successful collaboration and, hence, stability of organization.

Using data from Bradley's (1987) nationwide longitudinal study of urban communes, we found strong correspondences between observed patterns of communication and group stability (see Figure 10.2). Sociometric data mapping all possible dyadic ties among adult members within each commune were used to construct structural measures (following Bradley & Roberts, 1989b) of flux (triads of mutual ties of positive affect: "loving," or "exciting," or "improving" relations) and control (transitively ordered triads of "power" relations); group stability (the minimum functional requirement of successful collaboration) was measured by survival status 24 months beyond the measurement of sociometric structure (for details, see Bradley & Pribram, 1995, 1996).

The scatterplot for the 46 communes in Figure 10.2 (the measure of control is plotted on the vertical ordinate, and flux is plotted on the horizontal ordinate; unstable, nonsurviving groups are shown as hollow dots) shows that the communes form a triangular pattern, and that those located in the peripheral areas are more likely to be unstable. This triangular re-

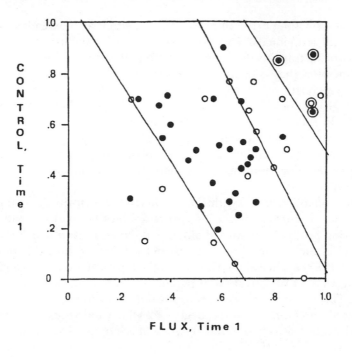

FIGURE 10.2. Scatterplot of communes on flux and control at Time 1. Stability (survival status) at Time 3 (24 months later) shows partitions for stability.

gion appears to be divided into two stable subregions, separated by zone of high instability. Beyond the zone of instability, in the apex of the triangle, are five communes, four of which had a charismatic leader in residence (shown with a circle in Figure 10.2) and were intent on achieving a radical restructuring of social order; the fifth group is a noncharismatic commune whose members expressed a strong desire for charismatic leadership as a means to facilitate their efforts at social change. The differences between these four groupings of communes in Figure 10.2, in terms of their patterns of flux, control, and stability, were found to be statistically significant (Bradley & Pribram, 1996).

A second, striking finding is that flux and control are predictive of stable collective organization (successful collaboration): that the relationship between flux and control at a given point in time was found to predict the survival status of communes 24 months in the future. This also is evident from the pattern of data in Figure 10.2. It is clearly apparent that the communes tend to cluster in the midregion of the field formed by flux and control, the area theoretically expected to be associated with efficient information processing. This region of optimal stability is consistent with thermodynamically inspired connectionist models of neural networks (e.g., Hopfield, 1982; Hinton & Sejnowski, 1986). In such models, *efficient pattern matching is found to occur in a region between total randomness and total organization*—in our terms, between rapid flux and rigid control. Moreover, it can be seen that location in the midregion (the area characterizing successful collaboration) is associated with a high probability of survival in the future.[11] However, this finding should *not* be taken to mean that the data generated by flux and control, at a given moment, necessarily enfold long-term information about collective order many months in the future. Rather, it is more likely that the efficiency of information processing in the midregion operates as an attractor.

To summarize: Our data concern the communicative structure formed by endogenous networks of interaction that monitor the activation and expenditure of energy by the collective. These endogenous processes are conceived to be based on the biological potential of the individuals composing the collective to engage in physical work, measured as energy. When activated by the collective, this biological energy is made available for the accommodation necessary for collaboration as a field of potential (passionate) energy. We have labeled this dimension of the endogenous order, flux.

In the other dimension, individuals are connected hierarchically. We have labeled this dimension control (commitment), because it appears to direct and assimilate the activation of the energy to the needs of the collective. Controls over the activation and distribution of flux result in social communication by way of quantum-like units of information (logons)— moment-by-moment descriptions, in terms of space-time and spectral coordinates, of the collective's endogenous organization.

The efficiency of the internal dynamics, and its relationship to collaboration, was found to display an optimal (energy conserving) combination of flux (passion) and control (commitment) that is associated with stable collective organization. Our empirical results thus show that for the group to survive as an effective collaborative unit, an efficient, self-maintaining communicative structure was required. Only those configurations of flux and control that produce a path of least action—one that entailed the smallest amount of turbulence—resulted in successful collaboration, and therefore, a stable, effective collective.

Despite a difference in focus—we focus primarily on the movements of energy and information in collaborative systems, Piaget (1965/1995b) focuses on the "operatory logic" in cooperative systems that underlies the development of thought and reason—a basic commonality is apparent. In Piaget's system of cooperation, reciprocity in "interindividual relations"— the free distribution and movement of information back and forth between individuals—is one of the two conditions for "equilibrium." This idea of a free exchange among individuals is analogous to our concept of flux. The second condition is a "common system of signs and references" (p. 148), namely, common language, values, and social norms. According to Piaget, the system of common signs and references acts as a constraint system, in our terms, control: It functions to "conserve" the operatory logic and outputs (what Piaget refers to as "propositions") of prior interactions, thereby acting to in-form the development and evolution of subsequent "cooperations." Thus, when reciprocal interactions are coupled with the information-conserving system of common sign and references, "mobile equilibrium" results, that is, stability in cooperative interindividual relations (pp. 145–153). In a personal communication from Inhelder, she stated that Piaget, after becoming acquainted with the ideas proposed by Prigogine (Prigogine & Stengers, 1984), agreed completely with a change in terminology for his (Piaget's) concepts from "equilibrium" to "stability far from equilibrium."

We have analyzed our data with regard to efficient communication within a collective but have, as yet, not addressed the effectiveness of social collectives with regard to the larger community within which they operate. This fascinating topic can now be studied within the purview of open, nonlinear systems dynamics. We plan to address this important question in a subsequent investigation.

CONCLUSION

At a time when cognitive science was just beginning to make a stir, Bill Estes and one of us (KHP) were asked to summarize a conference in Prague (then in Czechoslovakia). We looked at each other in dismay: Industrial, social, clinical, and a variety of breeds of experimental psychologists had presented a dense program that seemed to have little internal cohesion. As we puzzled and read our notes, it suddenly occurred to us that however diverse the presentations seemed, they did deal with issues in psychology, and that if we could define these issues, we might make a real contribution in our summaries. It turned out that there were only a half-dozen issues, and that we could rationally organize the Babel of discipline-specific terminologies around them.

In this chapter we have made a similar attempt. For years, we have felt that an affinity must exist between what the brain is doing to organize the behavior of organisms and the organization of social interactions among these organisms (Bradley & Pribram, 1988). Each discipline develops its own terminology; thus, the affinities must be sought in the issues the disciplines' data address, not in the terms used in the addressing. The danger is, of course, that surface similarities will obscure a search for deeper meanings, but, as in ordinary communication, a good place to *start* is to take the communication at face value.

Here, we have started with two case histories—exemplars of many clinical and experimental primate observations—and given our interpretation of the basic processes delineated by them. Our interpretation is couched in terms familiar to cognitive neuroscience. Because of our interest in development, we related this interpretation to that made by Piaget, whose work with Inhelder we have followed for many decades (see Bradley, 1987; McGuinness, Pribram, & Pirnazar [including the "endnote" by Inhelder], 1990; Pribram & Hudspeth, 1992). The connection between accommodation and the work on motivation and emotional controls on attention (Pribram & McGuinness, 1975, 1992) did not occur to us until we were preparing this chapter. Having made this connection, it was but a step to incorporating the data on love relationships and collaboration in social collectives (Figure 10.3)—a step we had been contemplating for several years (Bradley & Pribram, 1991).

To summarize: Neuropsychological observation and experimentation have distinguished two dimensions of self: on the one hand, an objective "me" that egocentrically articulates and allocentrically locates us in our environment, and on the other, a narrative "I" that monitors and evaluates that articulation. We suggest that these dimensions embody Piaget's assimilative and accommodative developmental processes. As it is accommodation that changes the organism's competence in assimilating the environment, we inquired as to how the accommodative process works, how competence becomes updated. We found the answers in our own neuropsychological data and those reviewed by Schore: three additional dimensions were discerned. (1) arousal–familiarization, (2) activation–selective readiness, and (3) effort–comfort. We further claimed that arousal leading to familiarization is the essence of accommodation; that activation of selective readiness is the essence of assimilation; and that effort–comfort mediates between accommodation and assimilation. Finally, we became intrigued with the similarity between these dimensions and those developed by Sternberg in investigations on the stability of love relationships that were relevant to our investigations on stability in social collectives. We believe that the connection between the neuropsychological dimensions and those

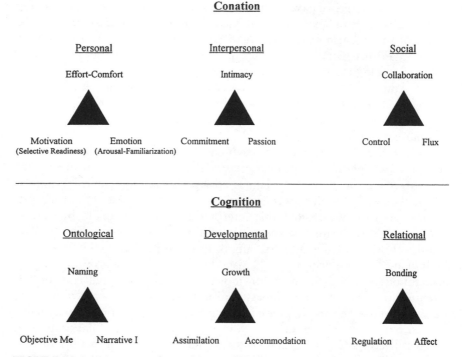

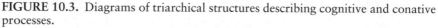

FIGURE 10.3. Diagrams of triarchical structures describing cognitive and conative processes.

obtained on the social psychology of love suggest the following correspondences: arousal with passion; selective readiness with commitment; and effort–comfort with intimacy. Moreover, based on the relational predictors of stability in social collectives, we further believe that these correspondences can be extended to include sociological dimensions of collective organization—to wit: arousal–familiarization to accommodation, to passion, and to flux; selective readiness to assimilation, to commitment, and to control; and effort–comfort to stability, to intimacy, and to communicative collaboration (see Figure 10.4).

Though the techniques of data collection are to some extent different, the relevance of the Sternberg data to those derived from social collectives, because both were obtained from dyads, makes this comparison robust. The connection between brain science and social science through neuropsychological and cognitive-developmental investigations, such as we make here, are more speculative, but we believe that they will prove to hold explanations that cannot be arrived at in any other way.

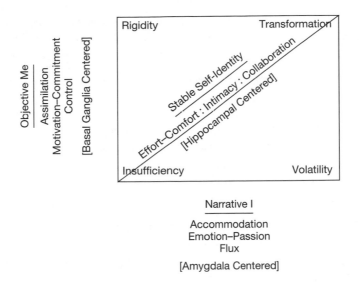

FIGURE 10.4. Speculative diagram of relations between triarchical structures at various scales (levels) of inquiry and the resultant effects on stability

APPENDIX A[12]

The ubiquitous role of quantum mechanics in neurobiology is obvious. Every time a photon enters the eye, an energy transaction obeying the law $E = h\nu$ takes place. The last 50 years have increasingly revealed that the role played by quantum mechanics in all of biology is significant. As Schrödinger wrote in 1946, "The mechanism of heredity is closely related to, nay founded on the very basis of quantum theory" (Schrödinger, 1946, p. 47).

The possibility that processes in the brain's connection web may result in the transmission of photons of frequencies in the 10^{13} Hz, i.e., far-infrared range, was suggested by Fernandez–Moran in 1951, and has been increasingly borne out. In 1960 Wiener wrote: ". . . the active bearer of the specificity of a molecule may lie in the frequency pattern of its molecular radiation, an important part of which may lie in the infrared electromagnetic range or even lower" [p. 52].

Since the late 1960's Fröhlich's researches have confirmed collective behavior of assemblies of biomolecules resulting in long-range coherent radiation in the 10^{11} Hz range. And in the 1980's Adey suggested that coherent infrared radiation could be the basis of intracellular signaling and energy transfer over short distances. The quantum role is further addressed in the papers of Jibu, Yasue, and Pribram (1993, 1996).

We must, however note that the *phase* of the undulations is conspicuously absent from the Planck–Einstein equations. To an extent this missing phase is restored in the researches of Fröhlich by virtue of his emphasis on coherence. But the tendency in quantum mechanical circles has been to ignore the phase of the de Broglie wave, and attribute significance only to its frequency, wavelength, and the non-negative square of its amplitude. Aside from a few great physicists such as Born, Dirac, and Feynman, the common attitude has been to dismiss rather naively the de Broglie wave as a mathematical "tool."

It was the pioneering thought of Haldane (1934) that *the full-fledged de Broglie wave (with frequency, wavelength, amplitude and phase) is involved in all phenomena in the universe*, and thus the phase of the de Broglie wave is most germane to the understanding of cognitive processes. More precisely, what Haldane proposed was that the resonances of the de Broglie wave systems of highly organized material systems constitute their potential for "mental" prowess. His pithy paper (Haldane, 1934) therefore merits a brief digression on our part.

The central theme of Haldane's paper is that the wave-mechanics of de Broglie and Schrödinger can explain the phenomena of both life and mind. He admits as limiting extremes the billiard-ball atomism of Lucretius and Newton on the one hand, and, on the other, the ideal world of Plato, these limits being attained as the mass–energy of the system is allowed to tend to zero or infinity, respectively. The fact that the universe is in-between these extremes is what makes life and mind possible. Recall in this regard, that posterior and frontal lobe stimulation of the brain can bring about tendencies toward these extremes of conscious processing (Pribram, 1991; Lecture 10).

APPENDIX B[13]

Gabor's discovery was based on the fact that the Fourier theorem opposes two different orders, two different ways in which signals become organized. In Lecture 3, we became acquainted with these two domains as characterizing the input to and output from a lens that performs a Fourier transform. On one side of the transform lies the space–time order we ordinarily perceive. On the other side lies a distributed enfolded holographic-like order referred to as the frequency or spectral domain.

Gabor (1946), as had Heisenberg and Hilbert before him, chose to represent the spectral and space–time orders by orthogonal coordinates, thus forming a phase space. Gabor was intrigued by the fact that in psychophysics, as in quantum physics, one could accurately determine either frequency (e.g., of a tone) or time (e.g., of its occurrence) but not both. Thus an uncertainty principle holds for psychophysics as well as for quantum physics:

In Gabor's own words published in 1946:

> Fourier's theorem makes of description in time and description by the spectrum, two mutually exclusive methods. If the term "frequency" is used in the strict

mathematical sense which applies only to infinite wave-trains, a "changing frequency" becomes a contradiction in terms, as it is a statement involving both time and frequency.

The terminology of physics has never completely adapted itself to this rigorous mathematical definition of "frequency." For instance, speech and music have a definite "time pattern" as well as a frequency pattern. It is possible to leave the time pattern unchanged, and double what we generally call "frequencies" by playing a musical piece on the piano an octave higher, or conversely, it can be played in the same key, but in different time.

Let us now tentatively adopt the view that both time and frequency are legitimate references for describing a signal and illustrate this—by taking them as orthogonal coordinates. In this diagram harmonic oscillation is represented by a vertical line. Its frequency is exactly defined while its epoch is entirely undefined. A sudden surge or "delta function" (also called "unit impulse function"), on the other hand, has a sharply defined epoch, but its energy is distributed over the whole frequency spectrum. This signal is therefore represented by a horizontal line. (p. 431)

Changing from a function of either [Space]time *or* frequency [as in the Fourier relation]—[to] a function of two variables—[space]time *and* frequency—[we compose a phase space]. [Thus] we have the strange feature that, although we can carry out the analysis with any degree of accuracy in the [space]time direction or the frequency direction, we cannot carry it out simultaneously in both beyond a certain limit. In fact, the mathematical apparatus adequate for treating this diagram in a quantitative way has become available only fairly recently to physicists, thanks to the development of quantum physics.

The linkage between the uncertainties in the definition of "[space]time" and "frequency" has never passed entirely unnoticed by physicists. It is the key to the problem of the "coherence length" of wave trains. . . . But these problems came into the focus of physical interest only with the discovery of wave mechanics, and especially by the formulation of Heisenberg's principle of indeterminacy in 1927. This discovery led to a great simplification in the mathematical apparatus of quantum theory, which was recast in a form of which use will be made in the present paper. (p. 432)

Gabor defined his elementary function, as a *logon* or quantum of information.

NOTES

1. Recently there has been a growing recognition among scientists (e.g., Barkow, Cosmides, & Tooby, 1992; Goldsmith, 1991; Kauffman, 1995) and philosophers (e.g. Searle, 1995) of the importance of developing a common language and set of common principles of organization by which studies in the natural and social sciences can be integrated.

2. This locational dimension includes clock-time, what the Greeks called chronos and Minkowski and Einstein related to space. Location for a moving organism is always in space-time.

3. Monitoring entails not only the experiencing of duration but also the decisive moment; what the Greeks referred to as *Kairos*. The dictionary defines "moni-

302 DEVELOPMENT OF SELF-AWARENESS ACROSS THE LIFESPAN

tor" as follows: a device to record or control a process; to check for significant content.

4. Descartes's "cogito ergo sum" is currently taken to mean *cogito* as a purely cognitive dimension. But *cogito* is not *cognito*. As Freud described in his "Project for a Scientific Psychology," there are several types of thinking that are purely emotive: for example, circular ruminations that give evidence of a "hang up" (see Pribram & Gill, 1976). So, despite the sharp separation between feeling and reason during the enlightenment, this distinction need not have carried over to the production of thought. It is only in the late 20th century that *cogito,* thought, has been exclusively identified with logic and cognition (see also Weiskrantz, 1988).

As to Cartesian dualism, its origins in propositional utterances is traced in Pribram (1963, 1971). On the issue of elemental nonphysical "substance," our modern understanding of electricity has totally undermined earlier views on "spirit." With regard to this, and the intimate place of body in mind, Damasio's illumination of Descartes's error is, of course, correct. But we may be equally wrong in going to the opposite extreme—when we interpret radiant energy, massless bosons, to be material substance (see Pribram, 1996). Should current speculation on the role of Einstein–Bose condensation in promoting superconductivity in neural membranes (especially in dendrites) be correct, a bosonic, soft photon, "nonmaterial" aspect of neural functioning in the generation of thought may yet have to be seriously considered in the philosophy of mind (Jibu, Hagan, Hameroff, Pribram, & Yasue, 1994; Jibue, Pribam, & Yasue, in press; see also Appendix A).

5. An excellent review of the history of differentiating an objective "me" from a hermeneutic (interpretive) "I" can be found in Hermans, Kempen, and Van Loon (1992).

6. We quote Schore's paragraph in full so that the reader can see that we are not the only authors who construct unintelligible sentences to describe the intricacies of the impossibly complicated interactions among brain structures that underlie every psychological process.

7. Searle's (1995) recent account of the construction of social reality is also based on a relational logic he finds in cooperative interactions. Searle argues that "genuine cooperative behavior" is the basis for a nonreductionist order of social life that he calls "collective intentionality": "The crucial element in collective intentionality is a sense of doing (wanting, believing, etc.) something together, and the individual intentionality that each person has is derived *from* the collective intentionality that they share" (1995, pp. 24–25; emphasis in original).

8. The sociograms in Figure 10.1 were constructed from sociometric enumeration of all possible pairwise relations (dyads) in which each adult member was asked a set of standardized questions about his or her relationship with each other member. See Bradley (1987) or Bradley and Roberts (1989b) for further details.

9. A similar finding, documenting the importance of both reciprocity (monitoring processes) and transitivity (location in social space) in communication, was made by Rice (1982) in a study of networking in computer conferencing systems.

10. The term collaboration is derived from the French verb *collaborer* and means working (*laborer*) together (*col*) to produce (Fowler & Fowler, 1964, p. 234). See Roberts and Bradley (1991) for a full discussion of collaboration as a sociological concept.

11. The three lines shown marking the boundaries of the regions in Figure 10.2 were established by dividing the full sample of 46 communes into stable and unstable sets such that the probability of survival for the former was maximized, while being minimized for the latter. Discriminant analysis, comparing the four grouping of communes separated by the lines, provided a strong statistical confirmation of these results as 45 (98%) of the 46 communes were correctly classified by two canonical discriminant functions constructed from the measures of flux and control. It is worth noting that *none* of the other nine sociological variables (measuring aspects of ideological orientation, normative regulation, formal organization, structural characteristics, and member commitment) investigated in this analysis met the statistical criteria for inclusion in the multivariate stepwise procedure. A split-sample reliability analysis confirmed the generalizability of these results (see Bradley & Pribram, 1996).

12. From Pribram (1997, pp. 317–318). Copyright 1996 by American Mathematical Society. Reprinted by permission.

13. From Pribram (1991, pp. 70–71). Copyright 1991 by K. H. Pribram. Reprinted by permission.

REFERENCES

Barkow, J., Cosmides, L., & Tooby, J. (1992). *The adapted mind*. New York: Oxford University Press.

Bateson, P. (1991). Levels and processes. In P. Bateson (Ed.), *The development and integration of Behavior: Essays in honour of Robert Hinde* (pp. 3–17). Cambridge, England: Cambridge University Press.

Bergson (1965). *Duration and simultaneity*. Indianapolis: Bobbs-Merrill. (Original work published 1922)

Bradley, R. T. (1987). *Charisma and social structure: A study of love and power, wholeness and transformation*. New York: Paragon House.

Bradley, R. T., & Roberts, N. C. (1989a). Relational dynamics of charismatic organization: The complementarity of love and power. *World Futures, 27,* 87–123.

Bradley, R. T., & Roberts, N. C.(1989b). Network structure from relational data: Measurement and inference in four operational models. *Social Networks, 11,* 89–134.

Bradley, R. T., & Pribram, K. H. (1988). *Some correspondences in the organization of brain function and group function*. Unpublished presentation to the Third Holonomic Processes in Social Systems Conference, Esalen Institute, Big Sur, CA.

Bradley, R. T., & Pribram, K. H. (1991). *Communication and the stability of social collectives*. Unpublished manuscript, Institute for Whole Social Science, Carmel, CA.

Bradley, R. T., & Pribram, K. H. (1995; in press). Communication and optimality in biosocial collectives. In D. S. Levine & W. R. Elsberry (Eds.), *Optimality in biological and artificial networks*. Hillsdale, NJ: Erlbaum.

Bradley, R. T., & Pribram, K. H. (1996). Self-organization and the social collective.

In K. H. Pribram & J. King (Eds.), *Learning as self-organization*. Mahwah, NJ: Erlbaum.

Brentano, F. W. (1973). *Psychology from an empirical standpoint*. New York: Routledge.

Broadbent, D.E. (1974). Divisions of function and integration. *Neurosciences Study Program III*. New York: MIT Press.

Carlton-Ford, S. (1993). *The effects of ritual and charisma*. New York: Garland.

Damasio, A. R. (1994). *Descartes' error*. New York: Putnam's Sons.

Dennett, D. C. (1991). *Consciousness explained*. Boston: Little, Brown.

Fowler, H. W., & Fowler, F. G. (Eds.). (1964). *The Concise Oxford Dictionary of Current English* (5th ed.) London: Oxford University Press.

Gabor, D. (1946). Theory of communication. *Journal of the Institute of Electrical Engineers, 93,* 429–457.

Garner, W. R. (1962). *Uncertainty and structure as psychological concepts*. New York: Wiley.

Goldsmith, T. (1991). *The biological roots of human nature*. New York: Oxford University Press.

Hermans, H. J. M., Kempen, H. J. G., & Van Loon, R. J. P. (1992). The dialogical self: Beyond individualism and rationalism. *American Psychologist, 47(1),* 23–33.

Hinde, R. A. (1979). *Towards understanding relationships*. London: Academic Press.

Hinde, R. A. (1987). *Individuals, relationships and culture*. Cambridge, England: Cambridge University Press.

Hinde, R. A. (1992). Developmental psychology in the context of other behavioral sciences. *Developmental Psychology, 28(6),* 1018–1029.

Hinton, G. E., & Sejnowski, T. J. (1986). Learning and relearning in Boltzmann machines. In D. E. Rumelhart & J. L. McClelland (Eds.), *Parallel distributed processing: Explorations in the microstructure of cognition* (Vol. 1, pp. 282–317). Cambridge, MA: MIT Press.

Hopfield, J. J. (1982). Neural networks and physical systems with emergent collective computational abilities. *Proceedings of the National Academy of Sciences, 79,* 2554–2558.

Hudspeth, W. J., & Pribram, K. H. (1992). Psychophysiological indices of cognitive maturation. *International Journal of Psychophysiology, 12,* 19–29.

Jantsch, E. (1980). *The self-organizing universe: Scientific and human implications of the emerging paradigm of evolution*. Oxford, England: Pergamon Press.

Jibu, M., Hagan, S., Hameroff, S. R., Pribram, K. H., & Yasue, K. (1994). Quantum optical coherence in cytoskeletal microtubules: Implications for brain function. *BioSystems, 32,* 195–209.

Jibu, M., Pribram, K. H., & Yasue, K. (in Press). From conscious experience to memory storage and retrieval: The role of quantum brain dynamics and boson condensation of evanescent photons. *International Journal of Modern Physics B Memorial Issue* dedicated.

Kauffman, S. (1995). *At home in the universe: The search for the laws of self-organization and complexity*. New York: Oxford University Press.

Killeen, P. R. (1994) Mathematical principles of reinforcement. *Behavioral and Brain Sciences, 17*(1), 105–172.

Konrad, K. W., & Bagshaw, M. H. (1970). Effect of novel stimuli on cats reared in a restricted environment. *Journal of Comparative Physiology and Psychology, 70,* 157–164.

Lassonde, M. C., Ptito, M., & Pribram, K. H. (1981). Intracerebral influences on the microstructure of visual cortex. *Experimental Brain Research, 43,* 131–144.

Levinger, G., Rands, M., & Talaber, R. (1977). *The assessment of involvement and rewardiness in close and casual pair relationships* (NSF Tech. Report). Amherst: University of Massachusetts.

Lévi-Strauss, C. (1963). *Structural anthropology.* New York: Basic Books.

Lorenz, K. (1969). Innate bases of learning. In K. H. Pribram (Ed.), *On the biology of learning* (pp. 13–94). New York: Harcourt Brace & World.

Lorenz, K. (1965). *Evolution and modification of behavior.* Lansing, MI: Phoenix Books.

Mace, G. A. (1962). Psychology and aesthetics. *British Journal of Aesthetics, 2,* 3–16.

Maffei, L. (1985). Complex cells control simple cells. In D. Rose & V. G. Dobson (Eds.), *Models of the visual cortex* (pp. 334–340). New York: Wiley.

McGuinness, D., Pribram, K. H., & Pirnazar, M. (1990). Upstaging the stage model. In C. N. Alexander & E. Langer (Eds.), *Higher stages of human development* (pp. 97–113). New York: Oxford University Press.

Miller, G. A. (1956). The magical number seven, plus or minus two, or some limits on our capacity for processing information. *Psychological Review, 63,* 81–97.

Miller, G. A., Galanter, E., & Pribram, K. H. (1960). *Plans and the structure of behavior.* New York: Henry Holt. (Russian trans.; also in Japanese, German, Spanish, Italian)

Nadel, S. F. (1957). *The theory of social structure.* London: Cohen & West.

Nevin, J. A., & Grace, R. C. (Submitted). *The Law of Effect reconsidered. Was Thorndike right?*

Piaget, J. (1983). *On affectivity and intelligence.* New York: Academic Press.

Piaget, J. (1970). *Structuralism* (C. Maschler, trans. and ed.) New York: Basic Books.

Piaget, Jean (1965/1995a). Logical operations and social life. In L. Smith (Ed.), *Sociological studies* (pp. 134–157). London and New York: Routledge. (Original work published 1965)

Piaget, Jean (1965/1995b). Genetic logic and sociology. In L. Smith (Ed.), *Sociological studies* (pp. 184–214). London and New York: Routledge. (Original work published 1965)

Piaget, J. (1936/1954). *The construction of reality in the child.* New York: Ballentine Books.

Pribram, K. H. (1963). Reinforcement revisited: A structural view. In M. Jones (Ed.), *Nebraska Symposium on Motivation* (pp. 113–159). Lincoln: University of Nebraska Press.

Pribram, K. H. (1971). *Languages of the brain: Experimental paradoxes and principles in neuropsychology.* Englewood Cliffs, NJ: Prentice-Hall; Monterey, CA:

Brooks/Cole, 1977; New York: Brandon House, 1982. (Trans. in Russian, Japanese, Italian, Spanish)

Pribram, K. H. (1976). Self-consciousness and intentionality: A model based on an experimental analysis of the brain mechanisms involved in the Jamesian theory of motivation and emotion. In G. S. Schwartz & D. Shapiro (Eds.), *Consciousness and self-regulation: Advances in research* (Vol. 1, pp. 51–100). New York: Plenum.

Pribram, K. H. (1980). Cognition and performance: The relation to neural mechanisms of consequence, confidence and competence. In A. Routtenberg (Ed.), *Biology of reinforcement: Facets of brain stimulation reward* (pp. 11–36). New York: Academic Press.

Pribram, K. H. (1986). The cognitive revolution and mind/brain issues. *American Psychologist, 41*(5), 507–520.

Pribram, K. H. (1991). *Brain and perception: Holonomy and structure in figural processing.* Hillsdale, NJ: Erlbaum.

Pribram, K. H. (1995). The enigma of reinforcement. *Neurobehavioral plasticity: Learning, development and response to brain insults.* Proceedings of the Bob Isaacson Symposium in Clearwater, FL.

Pribram, K. H. (1997). What is mind that the brain may order it?. In V. Mandrekar & P. R. Masani (Eds.), *Proceedings of Symposia in Applied Mathematics, Vol. 2: Proceedings of the Norbert Wiener Centenary Congress, 1994* (pp. 301–329). Providence, RI: American Mathematical Society.

Pribram, K. H. (in press). The composition of conscious experience. *Journal of Consciousness Studies.*

Pribram, K. H., & Gill, M. M. (1976). *Freud's 'project' re-assessed: Preface to contemporary cognitive theory and Neurophychology.* New York: Basic Books.

Pribram, K. H., & McGuinness, D. (1992). Attention and para-attentional processing: Event-related brain potentials as tests of a model. In: D. Friedman & G. Bruder (Eds.), *Annals of the New York Academy of Sciences, 658,* 65–92. New York: New York Academy of Sciences.

Pribram, K. H., & McGuinness, D. (1975). Arousal, activation and effort in the control of attention. *Psychological Review, 82*(2), 116–149.

Prigogine, I., & Stengers, I. (1984). *Order out of chaos: Man's new dialogue with nature.* New York: Bantam Books.

Rice, R. E. (1982). Communication networking in computer-conferencing systems: A longitudinal study of group roles and system structure. In M. Burgoon (Ed.), *Communication Yearbook, 6,* 925–944. Beverly Hills, CA: Sage.

Rubin, Z. (1970). Measurement of romantic love. *Journal of Personality and Social Psychology, 16,* 265–273.

Schore, A. N. (1994). *Affect regulation and the origin of the self: The neurobiology of emotional development.* Hillsdale, NJ: Erlbaum.

Schrödinger, E. (1944). *What is life?: Mind and matter.* Cambridge, England: Cambridge University Press.

Searle, J. R. (1983). *Intentionality: An essay on the philosophy of mind.* Cambridge, England: Cambridge University Press.

Searle, J. R. (1995). *The construction of social reality.* New York: Free Press.

Sherrington, C. (1947). *The integrative action of the nervous system.* New Haven, CT: Yale University Press. (Original work published 1911)

Simon, H. (1974). How big is a chunk? *Science, 183,* 482–488.

Simon, H. (1986). The parameters of human memory. In F. Klix & H. Hagendorf (Eds.), *Human memory and cognitive capabilities: Mechanisms and performances* (pp. 299–309). Amsterdam, Holland: Elsevier.

Spinelli, D. N., & Pribram, K. H. (1967). Changes in visual recovery function and unit activity produced by frontal cortex stimulation. *Electroencephalography and Clinical Neurophysiology, 22,* 143–149.

Sternberg, R. J. (1986). A triangular theory of love. *Psychological Review, 93*(2), 119–135.

Sternberg, R. J., & Barnes, M. L. (1985). Real and ideal others in romantic relationships: Is four a crowd? *Journal of Personality and Social Psychology, 49*(6), 1586–1608.

Sternberg, R. J., & S. Grajek (1984). Real and ideal others in romantic relationships: Is four a crowd? *Journal of Personality and Social Psychology, 47*(2), 312–329.

Swensen, C. H. (1972). The behavior of love. In H. A. Otto (Ed.), *Love today* (pp. 86–101). New York: Association Press.

Tucker, D., Potts, G. F., & Posner, M. I. (1995). *Recursive event-related potentials over human visual cortex.* Publication of the Brain Electrophysiology Laboratory, Department of Psychology, University of Oregon, Eugene, OR 97403.

Weiskrantz, L. (Ed.). (1988). *Thought without language.* Oxford: Clarendon Press.

White, R. W. (1960). Competence and the psychosexual stages of development. In M. R. Jones (Ed.), *Nebraska Symposium on Motivation, 1960* (pp. 97–140). Lincoln: University of Nebraska Press.

Zablocki, B. D. (1980). *Alienation and charisma: A study of contemporary communes.* New York: Free Press.

CHAPTER ELEVEN

Narrative and Metanarrative in the Construction of Self

JEROME BRUNER
DAVID A. KALMAR

"Self" is an anomaly for philosophers and psychologists alike. Its seeming presence has a disconcerting way of evaporating when we examine it closely, as David Hume reminded us two and a half centuries ago (Hume, 1984, Bk. I, Sect. 4). We would all now agree that it is not something just "there" but, rather, something constructed out of sense and memory by acts of imagination. Yet if that were all there were to it, mere construction according to taste, then the classic problem of "other minds" would be virtually intractable.[1] So, even though Selves are constructed, *how* are they constructed in order that they be intersubjectively communicable while, at the same time, remaining privately and uniquely individual?

We must note at the start that, for all its day-to-day robustness, its resistance to the hiatuses of sleep, anesthesia, and being lost in our work, Self seems somewhat unstable over extended time—a fact that should not be overlooked. Autobiographies are typically full of turning points featuring presumably profound changes in Selfhood. As many as one-third of self-referent sentences in the corpus of "interview" autobiographies with which we have worked contain markers of doubt and uncertainty—subjunctives, modals of uncertainty, Hamlet-like musings.[2] Our fixed identity in the eyes of the law is not matched psychologically by the subjective twists and turns in our self-conceptions, perhaps the more so under conditions of rapid cultural change (Lifton, 1993). While Selves have, of course, always been taken to reflect their times (Misch, 1950), there may well be limits to how much and how severe a change any given Self can absorb without undergoing trauma, stress, or even splitting. Maintenance of Self should not be taken for granted, even given its seemingly perduring constancy.

Finally, there is one unique and puzzling aspect of Self that must concern us. To a degree, Self grows in an environment of its own making, so to speak. The events and circumstances that shape it are not aboriginal, ready-made, autonomous, "out there." They are themselves constructed, products of self-generated meaning making shaped to fit our growing conceptions of our Selves. The events we "encounter" are coded and filtered at the very entry port by our perception of the world (Bruner, 1973, 1992; Neisser, 1988; Niedenthal & Kitayama, 1994). So while the experienced world may produce Self, Self also produces the experienced world, all of which suggests that the Self is not only constructed, but also that its mode of construction is massively hermeneutic. Perhaps it is this interpretive feature of self-construction that imposes certain conceptual structures upon Self, a matter that will concern us in much of the remainder of this chapter.

A first pass at making sense of this difficult topic is to turn to philosophical works, for Self is an ancient topic among philosophers. We have no intention of reconceiving their approach; they pursue this subject in their own light, and it has been much pursued over the last decades (Hampshire, 1989; Nagel, 1986; Taylor, 1989; Parfit, 1984; with Williams, 1973, as initial inspiration). Some of their concerns, we believe, are eminently worth addressing at the start of our discussion, for they help avoid confusion later.

Most philosophers would certainly agree with Hume in denying Self any *ab initio* ontological primacy. For Hume (at least in one mood) was skeptical of granting it any reality in his terms, warning that it is "a mere fiction" (an expression that James, 1983, takes pleasure in citing ironically). Indeed, a new and highly acclaimed dictionary of philosophy reflects Hume's doubts when it defines "Self" as "the elusive 'I' that shows an alarming tendency to disappear when we try to introspect it" (Blackburn, 1994). This would make Self seem more like a product of our efforts than a given or a "primary datum"—and, therefore, a difficult topic for anybody to describe as if it were *eo ipso*. Yet, most people, even simple people, seem to have notably little difficulty producing self-descriptions. Strikingly, composition teachers can get their most "backward" pupils to write by assuring them, "I'm only asking you to writer about yourselves—*only* an autobiography."[3] And frustrated plaintiffs in small-claims court, who automatically win their case when their adversary fails to appear, can scarcely be restrained by overworked magistrates from offering highly elaborate autobiographical justifications of why they "deserve" their claim. If "self" is ontologically or epistemologically difficult for the philosopher, it seems rather a pushover for ordinary people.

Perhaps philosophers are overly preoccupied with identifying an "absolute Self," a pure Ego, some transcendant "core" of identity. William James chides his colleagues on just this score in his famous chapter on the subject—though he himself seems unable to resist its pursuit in that very chapter's pages (James, 1983)! Those resistant composition pupils seem willing and able enough when writing about their "material selves" (as James calls accounts of achievements, activities, and possessions) or about their "social selves" (what others make of them, in James's language).

We doubt very much whether this besetting Humean anomaly—the alleged ontological difficulty of Self-description, pitted against the evident ease of Self-accounting—can be resolved by exclusively philosophical discussion.[4] While, doubtless, there are worthwhile philosophical ends to be served in examining the *ontology* of Self, our aim in the present chapter is rather more modest. What we want to do is to examine what human beings take to be the *indicia* of Self when they *talk* or *write* about their own or other Selves in a "real-life" context. Obviously, we can never escape philosophical issues. But, we do much better addressing them once we have a better grip on what people take Self to be in their ordinary discourse.

To chart the ordinary landscape of Self, we must travel a somewhat difficult path (which at times, perforce, will lead us back to philosophical matters). We propose first to explore indicia of Self—the signs that people take as indicators that a "Self" is present. All told, we will identify nine such indicators. Having laid these out, thereby placing Self within the locus of social behavior, we turn to questions of function. What does the Self *do*? We suggest two functions: that Self places us intersubjectively into a social matrix, and that, at the same time, it individuates us from the Other. From function we turn to mechanism, exploring how Self promotes both intersubjectivity and individuation. Finally, we explore one model of Self: Self as a product of *narrative*, a product, indeed, of a form of metacognitive "omnibus" narrative. And at the close, we consider what it is that impels us (or deters us) from constructing an "omnibus" Self.

INDICATORS OF SELF

Everywhere we look, we find signs or cues of Selfhood that seem irresistible to us. They trigger a sense of our own Self or of Selfhood in others. Taken together, they suggest the domain of functioning within which Self seems to operate. Here is a tentative list of such compelling Self-indicators.

1. *Agency* indicators signal voluntary acts of free choice, initiatives freely undertaken in pursuit of a goal. They are legion, ranging from signs of mere hesitation to expressions of intention. They are signaled in language, gesture, forms of movement, indications of conflict. Heider's fa-

mous film testifies to our low threshold for seeing agency in others (Heider & Simmel, 1944). The sense of optativity, of the need to weigh alternatives, signals Selfhood in our own private consciousness.

2. *Commitment* indicators are about an agent's adherence to an intended or actual line of action, an adherence that transcends momentariness and impulsiveness. Commitment indicators tell about steadfastness, delay of gratification, sacrifice, or flightiness and inconstancy.

3. *Resource* indicators speak to the powers, privileges, and goods that an agent seems willing to bring or actually brings to bear on his or her commitments. They include not only such "external" resources as power, social legitimacy, and sources of information (Bourdieu, 1991), but also "inner" ones, such as patience, forgiveness, persuasiveness, and the like.

4. *Social reference* indicators signal that an agent is looking to others to share, legitimize, or evaluate experience, goals, commitments, and resource allocation. They signal that others are in mind, whether these are real people, cognitively constructed reference groups, or some sort of normative "law and order" (Merton, 1968).

5. *Evaluation* indicators provide signs of how we or others value the prospects, outcomes, or progress of intended, actual, or completed lines of endeavor. They may be specific (as with signs of being satisfied or dissatisfied with a particular act) or highly general (as with a sense of that some large enterprise as a whole is satisfactory or not—as with observing big terminal sighs in others, or with feeling a sense of well-being ourselves).

6. *Qualia* indicators signal the "feel" of a life—mood, pace, zest, weariness, or whatever. They are signs of the subjectivity of Selfhood and are, of course, different when experienced internally and when observed in others. When such signs are unsituated with respect to external events, as they usually are, they are notoriously subject to contextual interpretation.

7. *Reflexive* indicators speak to the more metacognitive side of Self, to the reflective activity invested in self- examination, self-construction, and self-evaluation. We say of others that they are "locked in thought" (Langer, 1989) and note in ourselves that "something turns on." In jurisprudence, this goes by the name "close scrutiny" (Black, 1979), but it needs no fancy labels to be recognized in everyday life.

8. *Coherence* indicators refer to the apparent integrity of one's acts, commitments, resource investments, self- evaluations, and so on. We say of some others that they seem to "have their act together," or of our own Self, that some particular line of endeavor is "very much a part of me." These indicators are taken to reveal the internal structure of a larger self-concept and are presumed to indicate how the particulars of various endeavors cohere into "life as a whole."

9. *Positional* indicators are presumed to reveal how an individual locates him- or herself in the coordinates of the social order. They become

particularly salient when there is a sensed discrepancy between our own or some other's sense of position and some canonical one, as when we exclaim, "Who does he think he is?" or "What in the world was I thinking?"

There is a certain kinship between what have just been described as "indicators" of Self and what the philosopher Peter Strawson (1959) has characterized as "P-predicates." P-predicates are terms we ascribe to those we take as "persons." In contrast, M-predicates are terms used in expressions about material things: rocks, buildings, objects of all sorts. He remarks (p. 102), "What I mean by the concept of a person is the concept of a type of entity such that both predicates ascribing states of consciousness [P-predicates] and predicates ascribing corporeal characteristics [M-Predicates] . . . are equally applicable to a single individual." Persons, in short, take predicates of both "having experience" and "thingness." M-predicates, on the other hand, characterize only physical objects or situations. What distinguishes personhood, then, is the belief that we and other corporeal beings show signs either of consciousness or of a capability for consciousness.[5]

One feature of Strawson's discussion, though implicit, is particularly relevant psychologically, and is worth elaborating. Note that he conceives of Self as a *concept*. Like most students of conceptualizing (e.g., Bruner, Goodnow, & Austin, 1956; Medin, 1989), he notes that any given concept requires a contrast class that permits any particular event or object encountered to be assigned *either* to it *or* to some contrast class. A particular figure is *either* a Man (an exemplar of that category) *or* a Woman, a Standing Bear, a Statue, or (at least) something else. A concept cannot exist when membership within it is obligatory—as in the famous aphorism about the fish being the last to "discover" water, because there is no contrastive medium that they could encounter without perishing.

And so it is with the conceptual category of "Self." A contrastive concept is needed. To distinguish one's *own* Self requires distinguishing it from *other* Selves. Strawson remarks accordingly (1959, p. 99), "It is a necessary condition of one's ascribing states of consciousness, experiences, to oneself, in the way one does, that one should also ascribe them, or be prepared to ascribe them to others who are not oneself." The condition for forming a category of *my* Self is the recognition of the existence of Self in *another*.

In response to the classic question, then, as to whether we become conscious of our *own* Selves as a condition for recognizing selfhood in others, or whether it is the other way round, the answer necessarily is that you cannot have one without the other, all of which need not imply, of course, that "*own* Self" and "*other* Self" need be equally well differentiated or well organized as concepts.

Note that Strawson's P-predicates represent conceptual distinctions—I *think*, he *feels embarrassed*, they are *enthusiastic*, and so on. These are *attributes* of Selfhood. Such conceptual attributes do not exist in isolation, *in vacuo*. As students of concept formation have always insisted, concepts are organized into systems with an internal congruence (e.g., Medin, 1989). A conceptual category such as "positively charged particles" is, as it were, extracted from a theory of physics, and its attributes only make sense in the light of that theory.

Obviously, our Self and another's Self are organized somewhat differently. The so-called "fundamental attribution error," whereby we see our own acts as determined by circumstances, while seeing the acts of others as stemming from traits of character, exemplifies one difference between them. But while we organize indicators of Self in ourselves and in others differently, we use a common set of indicators for both. That is to say, there is no P-predicate/Self indicator that we use for characterizing our own Self that cannot be used for characterizing somebody else's. And so it is with the self-concept: Its indicators reflect a theory of Selfhood. This is beautifully illustrated by the classic Asch (1946) study on trait names (which, of course, are prototypical P-predicates in Strawson's sense). Combine *intelligent* and *warm* in a trait list, and we are led to ascribe such other P-predicates as *nurturant*, *empathic*, and the like. Combine *intelligent* and *cold*, and we attribute such predicates as *crafty* and *manipulative*.

Just as any P-predicate applied to the Self must be applicable to the Other, so we cannot notice in our own Selves what cannot be contrasted with what characterizes another Self. We cannot predicate our own commitment or our own resources, say, without having somebody else's commitment or resources to compare to ours. Molière celebrated that truth in M. Jourdain's famous remark in *Le Bourgeois Gentilhomme* about not realizing he had been speaking prose all his life. Indeed, a once-wealthy friend told one of us that as a young child, he had never realized he was privileged until he discovered that other children did not have a chauffeur to drive them to school in a pinch! As another example, in adolescence or adulthood, one does not discover one is a "pessimist" until encountering an "optimist," or vice versa.

And so it is with self-indicators. We do not detect them save by contrasting ourselves with others—though we may organize them differently in the two cases, as noted.

FUNCTIONS OF SELF

The rather motley and partial list of Self indicators we have just considered—there are doubtless many others—surely is mediated by some deeper, simpler system for processing cues about Selfhood. What might such a system be like, and what functions might it serve? We want to speculate about this now.

Structurally, Self as achieved surely seems like the product of a preadapted processing system whose intake is both varied and modifiable, and whose output is what we call "Self." Its input in infancy (where the perception of others is concerned) initially includes such *external* sensory triggers as people's eyes (Eibl-Eibesfeldt, 1975), voice qualities (Fernald, 1989, 1991), and various forms of movement (Johannson, 1973). Such triggering seems innately determined at the outset. Let us dodge the intractable and indeterminate question of how the infant's initial subjective self-awareness is triggered. Recognizing one's own voice or having an innate appreciation of the difference between self-initiated and environmentally induced action must be part of what Neisser (1988) has called the primitive "Sensory Self" (see also Bruner, 1995). Certainly the findings on mirror self-recognition (as in the studies of Gallup, 1979, and Papousek & Papousek, 1994) suggest something of this order, though later studies suggest that such self-recognition is limited in scope among chimpanzees (Timothy, Gallup, & Povinelli, 1996). But recognition of one's own Self surely depends as well, at least in some measure, on the infant being treated "as if" he or she had or were a Self (see Bruner, 1996; Meltzoff & Gopnik, 1993; M. Tomasello, personal communication, April 7, 1994). Whatever the early "innate" indicators may be, they expand eventually into something like our Self-indicator list. Within a year at most, the range and variety of possible exemplars that "fit" into the categories specified by each of the "self-indicators" increases notably as the child masters the language's synonymy rules: Many more things, for example, are seen as signs of "agency" or "resource allocation." And eventually these categories come to cohere in a superordinate conceptual organization that, to look ahead, is probably more narrative in nature than logical. In a word, then, the Self-system, plainly, has a strong initial innate component giving it a "kick start," but it is nonetheless extraordinarily malleable or "docile"[6] in the face of experience.

The Self-system seems to fulfill two functions—a species-maintaining or *intersubjective* communicability function, and an *individuation* function. Such a system may very likely be a by-product of the form of cultural adaptation that characterizes the species *Homo sapiens*, a species that shares symbolic representations of the world and consensual ways of orienting to it (Cassirer, 1944), yet at the same time is capable of "private experience." Speciation requires that we recognize (as we do) conspecifics as possessing Self-systems—as having beliefs, desires, a sense of others as affordances, and so on. No other species seems to have such a highly elaborated system (Premack & Povinelli, 1996). It enables the young, virtually from the start, easily to share attention with others, to operate with mutuality in regard to what others are looking at and doing, and to elaborate a primitive grasp of intentionality or "standing for" (Astington, 1993; Moore & Dunham, 1995; Scaife & Bruner, 1975; Tomasello, Kruger, & Ratner, 1993).

As the system develops and individuates, the range of accessible common beliefs, expectations, and other intentional states increases accordingly. Such cultural "speciation" requires not only mutuality but also the establishment of a shared conception of *canonicity*: what particular acts, beliefs, and expectancies may be expected of others and what they may expect of us. And with this sense of canonicity comes a grasp of what endeavors may be legitimately pursued, and what "failure to perform" holds in store for deviants. This early sense of legitimacy creates a cultural community where there can be not only "shared" thoughts, but also a shared sense of *normativity*.

The achievement of intersubjective communicability seems often to be a counterpart to defining our Selves in relation to a social order: through group identity, family, gender, or whatever.[7] To know one's Self is to know far more than just one's inner feelings, or *qualia*. In this sense, Self is part of the social world, and within broad limits, it becomes publicly communicable. Yet, for all that, Self still remains private, subjective, our own. Nobody can "see" our Self directly, though they can (and do) make inferences about it from our acts, expressions, and so on—inferences that, nonetheless, we have a right to deny. *Deniability* is a crucial feature of Selfhood—not the denial of Self per se, but of particular features of Selfhood. Yet even the more private aspects of Self, it would seem, are a reflection of a culture's representations of experience and, accordingly, are capable of communication. There are, as it were, even "standard" secrets—some of them even specified in law, such as "fraud" and "intent to deceive."

A second function of Self-construction, *individuation*, is to create and maintain privacy, a protected and subjective enclave. Individuation gives a form of quasi autonomy or inviolacy to human beings, human beings who must still live in a community of mutual dependency. The manifesto of early behaviorism to ban "private experience" as part of the subject matter of psychology must surely be celebrated as one of the most self-defeating steps ever proposed for human psychology!

Individuation has two sides—one epistemic and the other deontic, the latter reflecting the canonicity feature mentioned earlier. The epistemic domain consists of what we each experience, know, and believe on our own: We alone "own" our own phenomenology as well as our own background knowledge and beliefs (Searle, 1992). The deontic domain comprises what we value, care about, fear, love, expect, and so on. The two never become fully autonomous of each other, and early on, they are very much intertwined—what Freud must have meant by the initial inseparability of the reality and pleasure principles.

Constructing an individuated Self of one's own surely serves to minimize the helter-skelter of immediacy and impulsiveness. As a conceptual structure that can be extrapolated to organize the present and estimate the future, it serves as a stabilizer. In the same sense, knowing others reduces their surprise, and thereby permits us to form predictable alliances, and so on. The deontics of individuation—how and what one values and expects as legitimate—has a self-evident value. While it is poorly understood, rapid progress is being made in the study of children's grasp of norms (Dunn, 1988; Kihlstrom & Cantor, 1984), of felicity conditions on normative utterances (Astington, 1988), and the like. One general feature of this deontic side of individuation is that it places value on consistency and predictability in one's own Self as in others.

Individuation, some argue, may facilitate cultural change by preserving individual susceptibility to noncanonical modes of representing the world—as with the greater openness to innovation among the young and the deviant (even, it seems, among Japanese macaques [Tsumori, 1967]). Cultures typically ritualize or institutionalize their normative prescriptions on selfhood to keep individuation within limits and to preserve requisite intersubjectivity. When individuation exceeds a certain limit, it is typically characterized as insanity, eccentricity, or some other deviance.

Thus, the Self serves both to bring people together, through the creation of intersubjectivity, and to separate them, through individuation. How are these opposing functions reconciled or "normalized"?

SELF AS A PRODUCT OF DISCOURSE

There is one matter that needs more attention before we can turn to the question, already mentioned, of how Self is organized as a system. We need to inquire about the circumstances in which Self serves the dual function of promoting both intersubjectivity and individuation. And that brings us to the intimate setting of dialogic discourse.

Our accounts of Self are, of course, highly constrained by reticence and taboos, whether in the small when offering self-revelatory excuses or justifications, or in the large when writing memoirs or more extended versions of our lives. Certain topics are off-limits with strangers, others with family, and, in general, we seem impelled to stay close to decorous norms about what Self *should* be. There are constraints, besides, about *how* one may talk about oneself. We must be neither too insistent nor too laid back, too naive nor too cynical, too forthcoming nor too guarded, and so on. Self-narrating seems shot through with "impression management," the nature of which seems to vary with different interpersonal settings, from the intimate *tête-à-tête* among close friends to casual encounters with strangers.

In time, we learn to script even our "to-ourselves-only" self-tellings to fit such socially imposed reticence and taboos. No doubt, these constraints reflect broader cultural norms, but postpone that for a moment. More relevant here are those reticences about self-telling, about which we have learned so much from the Freuds, both Sigmund and Anna (A. Freud, 1946; S. Freud, 1949, 1915/1950). For what are ego defenses if not reticences about how we speak to ourselves about ourselves? Denial, reaction formation, and the like are not only constraints on how we may talk to ourselves about ourselves, but they may even be guides controlling what memories we may recover. As one distinguished psychoanalyst has put it, Self is a product of our own intrapsychic and interpersonal acts. It is with respect to others that we cast our Selves as agent or victim, participant or witness (Schafer, 1992). Beliefs follow from such self-talking activity: When we depict ourselves as "agents," others are seen as less agentive, and when we see ourselves as "victims," others are the agents. We talk to ourselves and come to think and feel accordingly.[8]

Which returns us to Strawson's theme again. It is that the counterpoint between one's own Self-perception and our perception of others creates Self-elaboration. And that counterpoint is most evident in dialogue—both real and imagined. Such dialogue promotes both individuation and intersubjective conformity. It is a point that is prominent in the writings of Baldwin (1902). In a poet's terms (Eliot, 1972, p. 873), we "prepare a face to meet the faces that we meet," or in George Herbert Mead's more austere social science language, Self is shaped to meet Others, both *particular* Others and a Generalized Other (Mead, 1934). But it is perhaps the gifted Russian writer, Mikhail Bakhtin, whose posthumously published work (e.g., Morris, 1994) has made many Western scholars newly aware of the role of dialogue in self-construction. In his gnomic words, "The better a person understands the degree to which he is externally determined, the closer he comes to understanding and exercising his real freedom" (Bakhtin, 1986, p. 139). The task, in his terms, is to "transform [the world] into a semantic context for thinking, speaking, and acting" (p. 164) Cultures and subcultures, of course, characteristically provide us guides for such self- presentation in dialogue. A distinguished French anthropologist (Sperber, 1985) has even described a culture's characteristic modes of representing things to oneself as akin to what an epidemiology describes in the realm of susceptibility to illness; that is, a culture's forms of conceptual representations are, in his sense, "catching" for its members. So, for example, the required lexicon of "inside" and "outside" (*uchi* and *soto* as discussed by Bachnik & Quinn, 1994) in the Japanese language demands that one structure self-accounting in a way that emphasizes family and other affiliations, in contrast to American self-telling, couched in the lexicon of achievement (see Markus & Kitayama, 1991, for a direct comparison of Japanese and American self-description).[9] But as will later be clearer, Sperber's "epidemiological model" is probably much too passive. It is not simply that we "catch" the culture; we actively court it in the interest of furthering our enterprises.

Little question, then, that the demands of dialogic discourse provide the social microclimate in which Self is constructed. But it is equally plain that dialogic situations differ in the degree to which they promote individuation and intersubjective canonicity. Corporate cultures, for example, are believed to enforce intersubjective canonicity, and so are certain more demanding professions in which skillful and swift regulatory dialogue is required of practitioners. Indeed, French phrasing contains an expression, *deformation professionelle*, that specifically refers to the outcome of living with such demands. Recall too that the pioneering work of David Riesman (1969) on the "lonely crowd" suggests that too much pressure toward autonomy in self-definition often produces alienation from the forms of canonicity required for a socially attached Self.

What virtually all individuating, dialogic microclimates seem to have in common is some goad to reflection, or what today we call *metacognition*. The common feature of such dialogic interactions is that they lead their participants to attend not simply to the referent of what is being discussed, but to the processes of thinking or speaking involved. Such metacognitive activity is encouraged by exchanges between intimates, by the specialized world of the psychoanalytic couch, and so on. As Piaget (1995) put it in a too-often-neglected paper, "Cooperation is a source of reflection and of self-consciousness" (p. 239). Linguists often note that the language of such cooperative exchanges is heavily *metalinguistic* (Jakobson, 1960), "talk about talk." While the excesses of such individuating metacognitive and metalinguistic discourse are easily lampooned, it may be a necessary condition for achieving the Delphic enjoinder to "Know thyself."

We shall return to this matter later. Now, we take up the question of how Self and its indicators are organized.

THE SELF AS STORY

Typically, we tell ourselves about our own Self and about other Selves in the form of a story. These stories, however, seem to fall into narrative genres (Markus & Kitayama, 1991). Is this only a convention, or is it a necessary condition of Self-telling? A young poet one of us interviewed began his autobiographical story by telling about the attending obstetrician at his birth slapping him on his back to get him breathing—and thereby breaking two of his ribs, unaware that the poet-to-be was suffering prenatal osteoporosis. "It's the story of my life," he went on, "people breaking my bones to do me good, which is how it is when you're a homosexual." The genre of "suffering victim of society" served him well, carrying him all the way back to the moment of his birth! One even got the impression that he lived his life in this genre, or at least payed close heed to it. Although he did not fully bring it off in the messy details of life, it was in the background throughout his story.

Consider our list of Self-indicators presented earlier. It comprises what are usually taken to be features of well-formed narrative. To borrow the terminology of the great narratologist, Vladimir Propp (1968), they resemble his "functions" or "constituents" of narrative. A well-formed story does not need all of them, as Propp has noted, since they are typically redundant, the story-as-a-whole making it possible to draw reasonable inferences about missing constituents. *Beowulf*, for example, has little to say about the experiential *qualia* produced by the mayhem it portrays. These are, as we say, "left to the reader." Artful narratives, indeed, deliberately delete some Self-indicators better to recruit the imagination of the listener or reader. And imagination is easily recruited, for, as we know, narrative accounts are often condensed into images and metaphors that "stand for" features of a story—like the dying but victorious General Wolfe surrounded by fellow officers and a few sympathetic Indians in Benjamin West's famous imaginary portrait of the scene (Schama, 1991).

But come now to the *structure* of narrative. As classically defined by Burke (1945) for fiction, by Hayden White (1981) for history, and by Ricoeur (1984) in general, a narrative represents an interaction of the following constituents:

An *Actor* with some degrees of freedom;
An *Act* upon which he has embarked, with
A *Goal* to whose attainment he is committed;
Resources to be deployed in the above,
All occurring in a *Setting*
That presupposes the *Legitimacy* of some state of affairs
Whose violation has placed things in *Jeopardy*.

Could it be, then, that what we recognize as Self (in ourselves or in others) is what is convertible into some version of a narrative?

One of us has spelled out these constituents of narrative in some detail elsewhere (Bruner, 1990). All that need be noted here is that the italicized items in the foregoing description subsume all of the Self-indicators mentioned earlier—plus the notion of canonicity, as embodied in the legitimacy of some initial state of affairs that has been put in jeopardy. In Propp's classical description (Propp, 1968, 1984), the disruption of this initial legitimacy was taken to be a *lack,* that is, something missing. But this may reflect the fact that the corpus of narratives on which he worked consisted entirely of folktales (especially the so-called "wonder tale," the deciphering of whose morphology was the principal basis of Propp's prodigious reputation). But, in fact, as most literary theorists now agree (e.g., Riffaterre, 1990), any disruption (or *peripéteia,* to use Aristotle's term from his *Poetics*) will do—conflict, theft, whatever.

Without further ado, then, our general hypothesis can be stated as follows. *Those indicators that are amenable to being placed in some narrative structure that includes an Agent are taken as Self-indicators (or P-predicates).* To extend the point further, we would propose that the various genres of narrative are specialized to highlight the different types of Self- indicators referred to earlier. Emphasis on Agency signals an adventurous Self; a focus on Commitment signals a dedicated Self; specialization on Resources signals either a profligate or a miserly Self; too much social referencing reveals the in-grouper and/or the snob; preoccupation with *qualia* is the self-contained aesthete.

Question: Might various of these Self-indicators become organized into multiple "sub-Selfs" or "subnarratives" to form, so to say, a master scenario of Selfhood? To take a personal example, some version of an early adolescent "water-rat Self" continues under the senior author's professional surface even today, marked by such resource indicators as being able (still rather proudly) to tie a bowline in the dark while upside down, or high skill in celestial navigation by clock and sextant (a skill now outdated by satellite navigation), and so on. He was solemnly elected on his "sailing record" to two rather self- congratulatory cruising clubs and now updates his anachronistic sub-Self by paying rather stiff dues each year, even though he attends the meetings of neither. Why does he not resign? A nice question.

How did he take all that Self baggage from the "there" of adolescence to the "here" of his present life? Through what neat narrative tricks does he keep the adolescent water rat updated in the current picture? One trick of unification is through irony, achieved by the narrative juxtaposition of archaic maritime skills (like tying bowlines and the mastery of the anachronistic sextant) with later-found addictions to E-mail, postmodern novels, and the rest. The irony seems to keep the otherwise eccentric water-rat sub-Self viable. It even provides background stories and a lexicon to complement fuller "omnibus" versions of himself as currently conceived.

In an almost forgotten volume, Sigmund Freud proposes that a person might be conceived as a "cast of characters," much as in a novel or play. Just as a playwright or novelist (he claims) decomposes him- or herself into a constituent cast of characters from which he or she then constructs a novel or play that brings them all together, so we decompose our lives into constituent sub-Selves, with a "story" that more or less succeeds or fails in bringing them all together (Freud, 1956). This is a very tempting model. It may even account for why people encounter such difficulty constructing a "complete Self" in the form of a full-blown autobiography, for such a construction is like writing a novel, and novel writing is not easy.

The moment you begin putting together those sub-Selves into a "novel," you become prey to the library of stories and genres the culture has on offer. The senior author's water-rat Self can be put in the larger genre of "an adult (reluctantly) setting aside the playthings of youth," or into the "conversion–displacement" model of the dynamic psychologist, with present Self as a disguised version of an earlier one. Indeed, a little browsing in the library easily provides a stock of story/genre models that might serve to put most collections of sub-Selves into a more encompassing "overall" one.

But take a step back for a better look at the nature of narrative genre. Vladimir Propp (1968) argues that the protagonists and events in the wonder-tale genre (that he studied so brilliantly) are *functions* of the tale's overall structure: The wonder-tale genre requires *type* functions that can be filled by different *token* exemplars. The hero of Propp's wonder tale, for example, is required to be a culturally entitled figure, and for the story to begin appropriately, he must have been left to his own resources by some higher authority. He can be a prince, a young genius, a courageous believer, whatever, so long as he is culturally entitled and left to his own resources— all tokens of a type. In that genre, he must then go on a Quest—for the Grail, for hidden treasure, for an elixir. It is then required that he encounter a figure with extraordinary powers who offers him some form of supernatural aid in his quest: a tireless horse, an endless golden thread, the gift of tongues, the power of foresight. To work effectively, the tokens that fill each type function must create and preserve the narrative coherence of the whole. The chief protagonist must perform appropriate acts that get him toward appropriate goals, must deal properly with commitments and persist in overcoming them, must ally himself appropriately, and must deploy his resources fittingly.

Propp's wonder-tale genre is ancient and smoothed by usage. The inward turn of the novel (Auerbach, 1953; Bruner, 1990; Heller, 1976) in the last century produced new genres, and their novelist inventors—Joyce, Proust, Musil, and others—may even have transformed our notions of *possible* selves. Charles Taylor's magisterial *Sources of the Self* (1989) certainly suggests as much. Our conceptions of Selfhood, and even our ways of structuring our stock of sub-Selves, get modified to reflect both changing narrative conventions and changing ways of life.

Both the constituent cast of sub-Selves and the omnibus metanarrative by which we organize them into some coherence are typically supported by dialogic partners or "interpretive communities," to borrow Stanley Fish's (1980) phrase. This is a familiar enough phenomenon. Imagine an art historian/bird-watcher, for example, with different confidants in each of the two domains, "explaining" to himself the place of bird-watching in his overall "self-image" by noting that it brings him closer to nature itself than do painterly representations of it—and without professional rivalry. Usually (though not always) the "Self" that brings home the butter needs no explaining. So one might imagine this chap recruiting fellow art historians to join him in bird-watching in order to bolster his omnibus self-image, that is, the art-historian-as-bird-watcher rather than the other way round. Only his detractors might call him a bird-watcher-as-art-historian, which would doubtless cause him hurt, and might even lead to some rewriting of his omnibus self-narrative. How much like writing a novel all of this is! Yet there are also amusing real-life stories, like the one about the painter Ingres, who cultivated the violin as an avocation. He played it indifferently, yet he nonetheless expected his listeners to admire his playing much as the fabled Emperor expected his subjects to admire his new clothes. To this day, more than a century later, the French still refer to this kind of avocational eccentricity as *un violin d'Ingres*, even if it involves art historians as bird-watchers or senior psychologists as water rats! Cultures, fortunately, can make ritual jokes on the absurdities of metacognitive, omnibus self-narrating.

Self, then, is a narrative construction, and as such, operates under the same constraints as narrative construction in general. This should make us alert to the fact that the "writing" of an "overall" Self is never once and for all, that it is the product of a perspective, even of a period of life. Aristotle (in the *Rhetoric*) notes that young men tell their lives as "all future," whereas old men tell theirs as "all past." In a word, we seem well supplied not only with sub-Selves but also with many possibilities for creating omnibus vehicles for containing the lot of them (Markus & Nurius, 1986).

It is difficult to resist the view that the spate of new books on the "saturated" self, the "divided" self, and other such testimonies to our difficulty in constructing an omnibus version of Self, reflect cultural problems characteristic of our times (Benson, 1993; Gergen, 1991; Laing, 1969; Lasch, 1984)—whether too little time for metacognitive reflection, the fragmenting of those supportive interpretive communities on which we depend for dialogic support, or even, conceivably, the growing complexity of our mental lives produced by the information load of mobility (both social and physical), as well as by the new media. But it would take us too far afield to explore such matters here.

THE OMNIBUS SELF

Granting everything said thus far, why do we not all have well-worked-out omnibus autobiographies readily at hand? Why do so few people write or think out their lives in a completed narrative? Mostly, as already implied, we tell of ourselves fitfully and patchily: in excuses for some act, or in justifications for this belief or that desire. One might argue that these purpose-built, local accounts are derived from some more *implicit* self-narrative that we have "in memory," or "in imagination," or somewhere. But whoever has suffered or watched others suffer putting together an autobiography will doubt it. Even that broken-ribbed, gifted young poet, so enamored of his metaphor, lost his way in the nitty-gritty. If there is some "whole story" stashed away in memory, it eludes us. We fall back on orderly tricks for retrieving the past, like sheer chronology. But even chronology is not "ready made," as one discovers to one's dismay by checking one's memory of events in one's life against, say, the cumulative daily index of *the New York Times*.

In spite of that, we have no doubts at all when we proclaim that, for all the confusion, it was the same "me" that lived through it all. It is only the evidence of severe pathology that makes us aware of the special capacities that make this kind of "sense of self-continuity" possible, as in cases reported by Luria (1972) and by Hirst (1994).

We seem to impose omnibus structures on our remembered Selves only when we sense a demand for doing so. But most of these are imposed "from outside." When we go to a clinician for counseling, to a priest to confess, or to an attorney to prepare our story for litigation, all of them demand specialized, partial stories. Clinician, priest, and attorney each provides appropriate hints to help us. The clinician offers his theories; the priest, doctrines of repentance and redemption; the attorney, his hornbook of tried procedures. Even in ordinary, everyday autobiographical dialogues, the confidants provide implicit models for each other.

Why don't we store truly omnibus Self-narratives? We could conclude that we don't seem to need them, and when we do encounter appropriate demands, we rely on culturally prescribed formulae and even adhere to the Gricean maxims about being brief, perspicuous, relevant, and truthful (Grice, 1989). Still, wouldn't a frequently updated version of a larger-scale Self-narrative be functionally adaptive for the reasons mentioned earlier— as some sort of "personal gyroscope"? Indeed, we know from Luria's (1972) classic study of "the man with a shattered world" that normal human beings are quite able to construct such an omnibus version of Self— we know it from poor, brain-injured Zazetsky's inability to do so, despite his most conscientious efforts.

One hypothesis: We suspect there is something both culturally adaptive and psychologically comfortable about "keeping one's options open" where one's self-narrative is concerned. Fixing the story of one's life, with a limiting conception of Self as major protagonist, may have the effect of shutting down our possibilities and opportunities prematurely. A fixed-in-advance omnibus life story creates, to use Amélie Rorty's expression (1976), a tragic "figure" with no options. In any save a highly ritualized social world, the more firmly fixed one's self-concept, the more difficult it is to manage change, to deal with local contingencies. It is "staying loose" that makes negotiation possible. Not so surprising, then, that turning points are so characteristic of the autobiographies we finally write or tell: They are the points in a life where, faced with difficulties, one may be forced to fashion a new omnibus Self to cope with the jeopardy in which we have been put.

So what then leads people to move up to a more comprehensive, more temporally extended, more omnibus mode of self- accounting, to construct a more inclusive metaversion of Self? Surely not just a publisher's contract! This leads us to a second hypothesis: Many students of narrative have noted—notably, Kenneth Burke (1945), Hayden White (1981), and William Labov (Labov & Waletzky, 1967)—that the *engine* of narrative is Trouble (Bruner, 1986, 1990), or what we referred to earlier as *jeopardy*. Narrative begins, as we noted, with a canonical state of the world disrupted, with action directed to rectifying the *peripeteia*. The narrative then elaborates on the nature of the action and climaxes with an account of how the original canonical state was or was not redressed. Narrative, in a word, organizes the travails of jeopardy. Trouble, in this sense, may be not only the engine of narrative, but also the impetus for its elaboration.

But self-narratives in the face of trouble typically elaborate, as it were, "in a straight line," in the form of further justification for a line of action (where the chance of successful outcome seems possible) or in excuses (when it does not). Elaboration of a more omnibus narrative requires metacognition. Aside from the little we know about the promotion of metacognition in intimate dialogue, discussed earlier, we know much too little about what predisposes us to metacognition generally. One hint comes from the recently reviewed literature on "need for cognition" (Cacioppo, Petty, Feinstein, & Jarvis, 1996). The authors conclude that there are indeed reliable, stable, individual differences in people's "intrinsic motivation to exercise their mental faculties" (p. 247). Their most general conclusion (perhaps too general to suggest an answer to our question) is as follows: "The differences between individuals low and high in need for cognition appear to be derived in large part from past experience, to be buttressed by accessible memories and behavioral histories, to manifest in current experience, and to influence the *acquisition and processing of information relevant to dilemmas or problems*" (p. 247, emphasis added). In a word, some people are characteristically more thoughtful, particularly with regard to troubles and dilemmas. And not surprisingly, those people tend to be better educated, somewhat more introverted, and less impulsive—though none of the correlations reported suggest, by any means, that these are the only determinants.

Others (Langer, 1989; Piaget, 1976) have been concerned about less dispositional prods to metacognition. Langer, in effect, places especial emphasis upon dialogic settings that have the effect of highlighting the agentive role of the individual in a local cultural setting. Elderly people who have been placed in "old people's home" environments, where little initiative or accountability is expected of them, begin to lose a sense of themselves as capable of agency. They do increasingly less for themselves, suffer increased morbidity, and die earlier. When the discourse of their "microenvironments" is changed so that they must undertake a larger share of responsibilities and must account for their own acts in daily discourse, there appears to be a revival of "self-agency" not only in their talk but also in their behavior, and even in their health. They get sick less often, live longer, and appear to be more "zestful." In Piaget's account, "cognizance" (as he calls understanding awareness) increases as the person becomes more conscious of the "coordinations" required among various mental operations—ranging from the virtually nonconscious sort of "subception" required in near-automatic perception to the more elaborate forms of self-conscious reflection required in more abstract problem solving. Both these accounts emphasize, each in their own way, the importance of metacognition in the achievement of full awareness of one's own agency, one's intentions, and the like.

Metacognitive self-elaboration is obviously, then, no simple reflex reaction to "trouble outside." It seems to require a sensibility that converts such outside trouble into an occasion for reflection. Whether this sensibility comes by dint of intelligence, temperament, imagination, or cultural opportunity, it appears to drive those gifted (or cursed) with it to deepen or integrate Selfhood, to be less attached to a presently reigning version and its limits. The gifted writer, Eudora Welty, calls this sensibility "daring," and she ends her remarkable memoir with these words (Welty, 1984, p. 104): "A sheltered life can be a daring life as well. For all serious daring starts from within." We are told that the Chinese idiogram for "crisis" combines two component ideograms, one for "danger," the other for "opportunity." And then, of course, there are the Mr. Micawbers, who see neither danger nor opportunity, and cling proudly to "how they were before."

Finally, one's sense of oneself also serves as a political badge—whether as a sign of merit or as stigma. It is a commonplace that "models" of Self are often *imposed* by the more powerful in a community upon the less powerful to dominate or control—"rootless bohemian," "proverbial woman," "shiftless black," each image carrying with it and "idealized" sketch about Selfhood: lack of commitment, sacrifice, unsteady agency. And the victims of such marking often oblige by emulating or feigning emulation. Much of the social drama of Self is probably created by the counterpoint of what is hegemonically required of Selfhood from outside and what from within, and why, perhaps, "consciousness raising" was so important a feature of the Women's Movement of the 1960s, and why the redefinition of Selfhood is often so high on the agenda of cultural revolution. For self-narrative virtually always becomes caught up in issues of hegemony and *cultural power*. Pierre Bourdieu (1991) reminds us that we even trade our *constructed*, stylized Selves (our *habitus* as he calls them) on a "symbolic market" in return for "distinction."

To go back to the main thread of our argument, it is in facing or anticipating troubles in this metacognitive mode that we fashion Self to go beyond the here and now of immediate encounters—relying, of course, on the narrative resources of the culture and upon our own inherent capacities. The anomaly in all this, and perhaps it is the burden of our human species, is that to extend and elaborate our version of Selfhood seems to require some "distancing" from the immediacy of life, with different cultures offering different, preferred ways of achieving this end, each preferred way probably expressing something deep about the culture itself. And it is difficult to resist the view that what takes place during such "distancing" is akin to writing, or better, rewriting a narrative or a "life" in which one's seeming "sameness," for all that its identity holds constant, gets resituated in one's cultural world and is changed thereby. If we cannot resituate ourselves in a manner that fits what we see as the intractable demands of the world around us, we fall back on narrative-like, psychological maneuvers that protect our self-esteem: We become "embittered victims," split personalities, withdrawn hermits, "specialists" in our particular virtues, "fundamentalists" of one stripe or another, or, in those rare instances that pathology has never been able to encompass, artists in understanding (and sometimes even communicating) the inherent impossibility and endlessness of self-construction.

In conclusion, then, the Self is both outer and inner, public and private, innate and acquired, the product of evolution and the offspring of culturally shaped narrative. Our self-concepts are enormously resilient, but as we have learned tragically in our times, they are also vulnerable. Perhaps it is this combination of properties that makes Self such an appropriate, if sometimes uncomfortable, instrument in the dynamism of human culture. For without the malleability (or "rewritability") of Self, the human cultural adaptation that makes our species unique would probably not be possible.

NOTES

1. Philosophers are not alone in their concern about this problem. It has always gripped the poet, as in a notable letter that John Keats wrote to his friend J. H. Reynolds on February 19, 1818 (see Rollins, 1958, vol. 1, pp. 65–67).

2. These autobiographical interviews were collected as part of a project on "The Autobiographical Self" at the New School for Social Research with the assistance of Professor Susan Weisser.

3. We are indebted to Martha Bruner Burdett for bringing this interesting anomaly to our attention. She, like many others, uses autobiographical assignments to "get students writing." For a more technical linguistic view of the multiple means of self-expression contained in natural languages, see the proceedings of the London colloquium on "The Status of the Subject in Linguistic Theory" (Yaguello, 1994), and particularly the introductory essay in that volume by Sir John Lyons, "Subjecthood and Subjectivity."

4. This problem has been interestingly discussed by Derek Parfit (1971, 1984), John Campbell (1994), and Thomas Nagel (1986). See also John Searle's (1995) lively defense of "subjective objectivity" in attacking Dennett's (1991) stance on the subject.

5. Strawson, to be sure, is also flirting with the philosophical issue of mind–body dualism by having Persons be representable by both sets of predicates—a kind of double-aspect theory. But that is not a psychological issue and need not concern us.

6. We borrow the term "docile" from Tolman (1932). He uses it to describe adaptive openness to environmental contingencies.

7. See Tajfel (1981) on social identity and Self-definition.

8. See, for example, the discussion of this point in Hermans (1996), Freeman (1993), and Randall (1995).

9. It may well be that as one becomes more "expert" in the ways of a culture and its language, one becomes better able to "circumlocute" around the demands of a culture's canon by subtle transformations in how one acts, talks, and thinks. The "novice," generally, is more "struck" with the literal requirements of the codes he or she must work with. This is a point we shall come to later.

REFERENCES

Asch, S. E. (1946). Forming impressions of personality. *Journal of Abnormal and Social Psychology, 41*, 258–290.

Astington, J. (1988). Children's understanding of the speech act of promising. *Journal of Child Language, 15*, 157–173.

Astington, J. W. (1993). *The child's discovery of the mind.* Cambridge, MA: Harvard University Press.

Auerbach, E. (1953). *Mimesis: The representation of reality in Western literature.* Princeton, NJ: Princeton University Press.

Bachnik, J. M., & Quinn, C. J. (Eds.). (1994). *Situated meaning: Inside and outside in Japanese self, society, and language.* Princeton, NJ: Princeton University Press.

Bakhtin, M. (1986). *Speech genres and other late essays.* Austin: University of Texas Press.

Baldwin, J. M. (1902). *Social and ethical interpretations in mental development.* New York: Macmillan.

Benson, C. (1993). *The absorbed self.* Hemel Hempstead: Harvester Wheatsheaf.

Black, H. C. (1979). *Black's law dictionary: Definitions of the terms and phrases of American and English jurisprudence, ancient and modern* (5th ed.). St. Paul, MN: West Publishing.

Blackburn, S. (1994). *The Oxford dictionary of philosophy.* Oxford, England: Oxford University Press.

Bourdieu, P. (1991). *Language and symbolic power.* Cambridge, MA: Harvard University Press.

Bruner, J. (1973). *Beyond the information given: Selected pages of Jerome Bruner.* New York: Norton.

Bruner, J. (1986). *Actual minds, possible worlds.* Cambridge, MA: Harvard University Press.

Bruner, J. (1990). *Acts of meaning.* Cambridge, MA: Harvard University Press.

Bruner, J. (1992). Another look at New Look 1. *American Psychologist, 47,* 780–783.

Bruner, J. (1995, June). *"Self" reconsidered: Five conjectures.* Paper presented at the annual meeting of the Society for Philosophy and Psychology, State University of New York, Stony Brook.

Bruner, J. (1996). *The culture of education.* Cambridge, MA: Harvard University Press.

Bruner, J., Goodnow, J., & Austin, G. (1956). *A study of thinking.* New York: Wiley.

Burke, K. (1945). *A grammar of motives.* New York: Prentice-Hall.

Cacioppo, J. T., Petty, R. E., Feinstein, J. A., & Jarvis, W. B. G. (1996). Dispositional differences in cognitive motivation: The life and times of individuals varying in need for cognition. *Psychological Bulletin, 119*(2), 197–253.

Campbell, J. (1994). *Past, space, and self.* Cambridge, MA: MIT Press.

Cassirer, E. (1944). *An essay on man: An introduction to a philosophy of human culture.* New Haven, CT: Yale University Press.

Dennett, D. C. (1991). *Consciousness explained.* Boston: Little, Brown.

Dunn, J. (1988). *The beginnings of social understanding.* Cambridge, MA: Harvard.

Eibl-Eibesfeldt, I. (1975). *Ethology.* New York: Holt, Rinehart & Winston.

Eliot, T. S. (1972). The love song of J. Alfred Prufrock. In H. Gardner (Ed.), *The new Oxford book of English verse.* Oxford, England: Oxford University Press.

Fernald, A. (1989). Intonation and communicative intent in mothers' speech to infants: Is the melody the message? *Child Development, 60,* 1497–1510.

Fernald, A. (1991). Prosody in speech to children: Prelinguistic and linguistic functions. *Annals of Child Development, 8,* 43–80.

Fish, S. E. (1980). *Is there a text in this class? The authority of interpretive communities.* Cambridge, MA: Harvard University Press.

Freeman, M. (1993). *Rewriting the self: History, memory, narrative.* London and New York: Routledge.

Freud, A. (1946). *The ego and the mechanisms of defense.* New York: International Universities Press.

Freud, S. (1949). *An outline of psychoanalysis.* New York: Norton.

Freud, S. (1950). Instincts and their vicissitudes. In *Collected papers: Vol. IV* (Vol. 4, pp. 60–83). London: Hogarth Press. (Original work published 1915)

Freud, S. (1956). *Delusion and dream: An interpretation in the light of psychoanalysis of* Gravida, *a novel, by Wilhelm Jensen* (P. Rieff, Ed.). Boston: Beacon Press.

Gallup, G. G. (1979). Self-awareness in primates. *American Scientist, 67,* 417–421.

Gergen, K. J. (1991). *The saturated self: Dilemmas of identity in contemporary life.* New York: Basic Books.

Grice, P. (1989). Logic and conversation. *Studies in the way of words.* Cambridge, MA: Harvard University Press.

Hampshire, S. (1989). *Innocence and experience.* Cambridge, MA: Harvard University Press.

Heider, F., & Simmel, M. (1944). An experimental study of apparent behavior. *American Journal of Psychology, 57,* 243–259.

Heller, E. (1976). *The artist's journey into the interior, and other essays.* New York: Harcourt Brace Jovanovich.

Hermans, H. (1996). Voicing the self. *Psychological Bulletin, 119,* 31–50.

Hirst, W. (1994). The remembered self in amnesics. In U. Neisser & R. Fivush (Eds.), *The remembering self.* Cambridge, England: Cambridge University Press.

Hume, D. (1984). *A treatise of human nature* (book I, section 4). New York: Viking Penguin.

Jakobson, R. (1960). Linguistics and poetics. In T. A. Sebeok (Ed.), *Style in language.* Cambridge, MA: MIT Press.

James, W. (1983). The consciousness of self. *The principles of psychology* (Chap. 10). Cambridge, MA: Harvard University Press.

Johannson, G. (1973). Visual perception of biological motion and a model for its analysis. *Perception and Psychophysics, 14,* 201–211.

Kihlstrom, J. F., & Cantor, N. (1984). Mental representations of the self. In L. Berkowitz (Ed.), *Advances in experimental social psychology: Vol. 17. Theorizing in social psychology: Special topics.* Orlando, FL: Academic Press.

Labov, W., & Waletzky, J. (1967). Narrative analysis: Oral versions of personal experience. In J. Helm (Ed.), *Essays on the verbal and visual arts: Proceedings of the 1966 Annual Spring Meeting of the American Ethnological Society.* Seattle: University of Washington Press.

Laing, R. D. (1969). *The divided self.* New York: Pantheon Books.

Langer, E. (1989). *Mindfulness.* Reading, MA: Addison-Wesley.

Lasch, C. (1984). *The minimal self: Psychic survival in troubled times.* New York: Norton.

Lifton, R. (1993). *The protean self.* New York: Basic Books.

Luria, A. R. (1972). *The man with a shattered world.* Cambridge, MA: Harvard University Press.

Markus, H., & Kitayama, S. (1991). Culture and the self: Implications for cognition, emotion, and motivation. *Psychological Review, 98*, 224–253.

Markus, H., & Nurius, P. (1986). Possible selves. *American Psychologist, 41*, 954–969.

Mead, G. H. (1934). *Mind, self, and society.* Chicago: University of Chicago Press.

Medin, D. L. (1989). Concepts and conceptual structure. *American Psychologist, 44*(12), 1469–1481.

Meltzoff, A. N., & Gopnik, A. (1993). The role of imitation in understanding persons and developing a theory of mind. In S. Baron-Cohen, H. Tager-Flusberg, & D. J. Cohen (Eds.), *Understanding other minds: Perspectives from autism.* Oxford, England: Oxford University Press.

Merton, R. K. (1968). *Social theory and social structure* (enlarged ed.). New York: Free Press.

Misch, G. (1950). *A history of autobiography in antiquity* (2 vols.). London: Routledge & Paul.

Moore, C., & Dunham, P. J. (1995). *Joint attention: Its origins and role in development.* Hillsdale, NJ: Erlbaum.

Morris, P. (Ed.). (1994). *The Bakhtin Reader: Selected writings of Bakhtin, Medvedev, and Voloshinov.* London and New York: Arnold.

Nagel, T. (1986). *The view from nowhere.* New York: Oxford University Press.

Neisser, U. (1988). Five kinds of self-knowledge. *Philosophical Psychology, 1*, 35–59.

Niedenthal, P. M., & Kitayama, S. (Eds.). (1994). *The heart's eye: Emotional influences in perception and attention.* New York: Academic Press.

Papousek, H., & Papousek, M. (1974). Mirror image and self-recognition in young human infants: I. A new method of experimental analysis. *Developmental Psychobiology, 7*(2), 149–157.

Parfit, D. (1984). *Reasons and persons.* Oxford, England: Clarendon Press.

Piaget, J. (1995). *Sociological studies.* London: Routledge.

Piaget, J. (1976). *The grasp of consciousness.* Cambridge, MA: Harvard University Press.

Premack, D., & Woodruff, G. (1978). "Does the chimpanzee have a theory of mind?" *Brain and Behavioral Sciences, 1*, 515–526.

Premack, D., & Povinelli, D. J. (1996). *The evolution of Self.* Unpublished discussion at the Annual Meeting of the American Psychological Association.

Propp, V. (1968). *Morphology of the folktale* (2nd ed.). Austin: University of Texas Press.

Propp, V. (1984). *Theory and history of folklore.* Minneapolis: University of Minnesota Press.

Randall, W. L. (1995). *The stories we are: An essay on self-creation.* Toronto: University of Toronto Press.

Ricoeur, P. (1984). *Time and narrative* (Vol. 1). Chicago: University of Chicago Press.

Riesman, D. (1969). *The lonely crowd: A study of the changing American character* (abridged ed.). New Haven, CT: Yale University Press.

Riffaterre, M. (1990). *Fictional truth.* Baltimore: Johns Hopkins University Press.

Rollins, H. (Ed.). (1958). *The letters of John Keats*. Cambridge, MA: Harvard University Press.

Rorty, A. O. (1976). A literary postscript: Characters, persons, selves, individuals. In A. O. Rorty (Ed.), *The identities of persons*. Berkeley: University of California Press.

Scaife, M., & Bruner, J. S. (1975). The capacity for joint visual attention in the infant. *Nature, 253*(5489), 265–266.

Schafer, R. (1992). *Retelling a life: Narration and dialogue in psychoanalysis*. New York: Basic Books.

Schama, S. (1991). *Dead certainties*. New York: Knopf.

Searle, J. R. (1992). *The rediscovery of the mind*. Cambridge, MA: MIT Press.

Searle, J. R. (1995). The mystery of consciousness: Part II. *New York Review of Books, 42*(18), 54–61.

Sperber, D. (1985). Anthropology and psychology: Towards an epidemiology of representations (Malinowski Memorial Lecture, 1984). *Man* (N.S.), *20*(1), 73–89.

Strawson, P. F. (1959). *Individuals: An essay in descriptive metaphysics*. London: Methuen.

Tajfel, H. (1981). *Human groups and social categories: Studies in social psychology*. Cambridge, England: Cambridge University Press.

Taylor, C. (1989). *Sources of the self: The making of modern identity*. Cambridge, MA: Harvard University Press.

Timothy, J., Gallup, G. G., & Povinelli, D. J. Age differences in the ability of chimpanzees to distinguish mirror-images of self from video images of others. *Journal of Comparative Psychology, 110*, 38–44.

Tolman, E. C. (1932). *Purposive behavior in animals and men*. New York: Century.

Tomasello, M., Kruger, A. C., & Ratner, H. H. (1993). Cultural learning. *Behavioral and Brain Sciences, 16*(3), 495–552.

Tsumori, A. (1967). Newly acquired behavior and social interactions of Japanese monkeys. In S. A. Altmann (Ed.), *Social communication among primates*. Chicago: University of Chicago Press.

Welty, E. (1984). *One writer's beginnings: The William E. Massey, Sr. lectures in the history of American civilization, 1983*. Cambridge, MA: Harvard University Press.

White, H. (1981). The value of narrativity in the representation of reality. In W. J. T. Mitchell (Ed.), *On narrative*. Chicago: University of Chicago Press.

Williams, B. (1973). *Problems of the self: Philosophical papers 1956–1972*. Cambridge, England: Cambridge University Press.

Yaguello, M. (Ed.). (1994). *Subjecthood and subjectivity: The status of the subject in linguistic theory*. London: Ophrys-Institut français du Royaume Uni.

CHAPTER TWELVE

The Development of Self through the Coordination of Component Systems

✧

MICHAEL F. MASCOLO
KURT W. FISCHER

Me, me, . . . ME SELF!
—26-month-old girl resisting her father's
assistance when climbing into her car seat

In recent decades, researchers have offered divergent conceptions of the nature and development of self. Traditionally, Western scholars have depicted the self as a unified and bounded center of activity and experience. Social constructionist scholars have challenged traditional notions of self and have proposed that individuals construct multiple selves that are socially created in dialogues with diverse others. In this chapter, we propose a dynamic skills theory model of self (Fischer, 1980; Fischer, Shaver, & Carnochan, 1990). From this view, capacities to represent and regulate the self develop along multiple pathways, even within in the same individual. As such, the self is not a unified entity that emerges at any single point in development. Rather, the individual consists of a series of weakly connected control systems that represent and regulate different aspects of self. Although initially separate, self-relevant control systems become increasingly coordinated in development. In this chapter, we describe changes in several such self-systems and elaborate how a systems view can resolve tensions among competing views of self.

332

CARTESIAN AND DIALOGICAL MODELS OF SELF

Traditionally, many Western portrayals of the self have been informed by the Cartesian conception of mind. According to Fogel (1993), the Cartesian mind or self displays several properties. First, the self consists of a *unitary* entity with a single *cogito* responsible for reasoning. Descartes (1637/1994) argued that whereas the body consists of many parts, the mind presents itself as indivisible. Second, because the Cartesian self is immaterial, it is *disembodied*. As such, the mind can transcend material and corporeal boundaries and assume an abstract stance toward the world. Third, the self is *individual* rather than social. This is implied by Descartes's famous dictum, *cogito ergo sum* ("I think therefore I am"). Unable to doubt the existence of his mind, Descartes asserted that the very idea that one is able to think (or doubt) implies one's existence as a thinking substance. This view sparked an intellectual tradition that maintains that each person is in a privileged position to understand the contents of his individual mind. Fourth, the dualism between mind and body applies equally to the distinction between subject and object. Because the mind is an entity unto itself, it can function *free of context*. It can operate without any necessary reference to society or history. Thus, the Cartesian self functions as a *centralized* core that controls the activity of the body even as it exists outside of that body.

Much current thinking on the nature of self and mind is founded upon Cartesian assumptions (Chomsky, 1966; Freud, 1940; Piaget, 1957; Kohlberg, 1969). For example, within developmental psychology, Kohlberg's (1969) approach to self and socialization invokes several Cartesian assumptions. Although Kohlberg viewed development in terms of transformations in cognitive structure that occur as a result of social interaction, he nevertheless defined the self as a unified structure: "There is a fundamental unity of personality organization and development termed the ego, or the self. While there are various strands of social development (psychosexual development, moral development, etc.), these strands are united by their common reference to a *single concept of self* in a *single social world*. . . . There is a . . . unity of development" (p. 349, emphasis in original). Based on these assumptions, Kohlberg's (1976, 1981) theory of moral reasoning describes a universal sequence of general and unified stages of moral thought. These stages develop in the direction of abstract, decontextualized moral principles. Cartesian assumptions are reflected in Kohlberg's depiction of the self as a unified entity capable of adopting an abstract moral stance toward the world.

In the past decade, social constructionists and sociocultural psychologists have proposed alternatives to the Cartesian view of self (Davies & Harré, 1990; Fogel, 1993; Harré, 1983; Harré & Gillett, 1994; Hermans,

1996; Gergen, 1985a; Shweder, 1990; Shotter, 1997; Wertsch, 1991). Social constructionists draw on Wittgenstein (1953) and others to challenge the Cartesian view of the self as a private inner world to which individuals have privileged access. For example, Gergen (1985b, 1987, 1989) holds that psychological terms (e.g., "motive," "cognition," "emotion") cannot be regarded as simple reflections of an inner world. Like all experience, inner experience cannot be directly perceived, but rather must be classified in terms of meanings expressed by existing category systems. Gergen argues that category systems have their origins in the ways in which social partners use language within interpersonal discourse to solve various human problems. Communities construct different vocabularies and grammars for solving different problems of human interaction. As such, word meanings, including the meanings of psychological terms, arise from their use in interaction rather than from the nature of their presumed referents. "What is taken to exist on the level of mental functioning can be viewed in large degree as the objectification of linguistic practices born of pragmatic exigency" (Gergen, 1985b, p. 117).

The sociocultural approach offers the *dialogical* model of self as a radical alternative to the Cartesian formulation. The dialogical approach builds on the assertion that the self is created and constituted in dialogue with others within sociocultural contexts. It follows from an appreciation of the role of the other in constituting the self (Sampson, 1994). To illustrate, we draw upon the distinction between "I" and "me" as different aspects of self (James, 1890; Mead, 1934). The "I" refers to the sense of oneself as a subject or personal agent; the "me" refers to the sense of self as an object. According to Mead (1934), the "I" and "me" are two phases of the self's activity. The "I" is never directly knowable: "If you ask, then, where directly in your own experience the 'I' comes in, the answer is that it comes in as a historical figure. It is what you were a second ago that is the 'I' of the 'me'" (p. 243). Bakhtin further elaborates on the role of the other in making the "I" intelligible to the individual:

> As a unique becoming, my I-for-myself is always invisible. In order to perceive that self, it must find expression in categories that can fix it, and these I can only get from the other. So that when I complete the other, or when the other completes me, she and I are actually exchanging the gift of a perceptible self. . . . I get a self I can see, that I can understand and use, by clothing my otherwise invisible (incomprehensible, unutilizable) self in the completing categories I appropriate from the other's image of me. (cited in Clark & Holquist, 1984, p. 79)

Thus, the "I" is constructed in dialogue and only becomes "me" when made intelligible by social categories.

The dialogical self does not refer to a single, unified sense of self, but rather to a multiplicity of selves constructed in dialogue with multiple others. From this view, in different social interactions, the self assumes different "*I* positions." As this process becomes internalized, the self is experienced as a "multiplicity of relatively autonomous *I* positions in an imaginal landscape" (Hermans, Kempen, & van Loon, 1992, p. 28, emphasis in original). In elaborating this view, the authors hold that

> the I has the possibility to move, as in a space, from one position to the other in accordance with changes in situation and time. The I fluctuates among different and even opposed positions. The I has the capacity to imaginatively endow each position with a voice so that dialogical relations between positions can be established. The voices function like interacting characters in a story. Once a character is set in motion in a story, the character takes on a life of its own and thus assumes a certain narrative necessity. Each character has a story to tell about experiences from its own stance. As different voices these characters exchange information about their respective *Mes* and their worlds, resulting in a complex, narratively structured self.

Thus, consistent with Bakhtin's formulation (Clark & Holquist, 1984), the self refers to an entire ensemble of *I* positions in interaction with each other and with multiple others, real or imagined.

Thus, the dialogical self stands in opposition to the Cartesian self. Whereas the Cartesian self is an internal property of individuals, the dialogical self is a *product of social relations*. Whereas the Cartesian self is unitary and indivisible, the dialogical self is defined in terms of *multiple relatively autonomous I positions*. Thus, whereas the Cartesian self is a single, *centralized* controller, the dialogical self is *decentralized*. It is defined not by any single "*I* position," but rather by all of them and the dialogues that ensue among them. Whereas the Cartesian self is disembodied and decontextualized, the dialogical self is *embodied* and defined with respect to its *cultural and historical context* (Fogel, 1993).

PERSONS AND SELVES AS COACTIONAL, SELF-ORGANIZING SYSTEMS

The concept of self is an imprecise notion. People use the concept of *self* to refer to different aspects of person functioning. One the one hand, the concept of self can be used to refer to the agentive *organism* or *person*. In this sense, to speak of a self is to speak of one who has the power to regulate or direct one's self, regardless of whether self-*awareness* is involved. A second use of self is *self-awareness* or *self-representation*. This refers not just to the agentive organism itself, but rather to the person's awareness and representation of various aspects of its powers of control (agency) and of its physical, psychological, or social attributes (identity). To complicate matters further, these meanings of self—as agent and represented awareness—participate in each other. As one becomes increasingly reflexive about one's agency and identity, one can use such awareness to regulate one's own activity. Thus, with development, the *self-as-agent* is transformed by its own developing *self-awareness* and vice versa.

The traditional rational/objective and dialogical approaches to self represent two extremes along a wide continuum. In what follows, we present an alternative based on a systems perspective, specifically dynamic skills theory (Fischer, 1980; Fischer et al., 1990). Dynamic skills theory provides a framework for understanding the ways in which self-systems develop as a function of processes that occur both within and between persons. Within skills theory, persons consists of multiple control systems for regulating action and thinking in specific psychological domains and contexts. Thus, to speak of the self as a person or agentive organism is to refer of the operation of control systems within a given context. According to skills theory, control systems within any psychological domain undergo developmental change in a series of 13 different levels from birth through adulthood. As control systems undergo transformation, the agency of the organism—its power to regulate and coordinate its own action within any given psychological domain and context—undergoes transformation. Persons build control systems to represent and control diverse aspects of their phenomenal worlds, including their own processes and attributes. Self-awareness develops as persons develop skills to represent aspects of their own agencies and identities.

Dynamic skills theory provides a framework for addressing many contradictions that arise in the study of self. First, to say that persons build multiple control systems in diverse psychological domains and contexts suggests that self-systems are inherently fractionated rather than unified. However, self-relevant control systems develop through the coordination of component processes and systems. As such, although self-relevant control systems start out separate, they develop in the direction of increased integration. Thus, persons and selves are both fractionated yet capable of integration. Second, according to skill theory, persons operate as self-organizing systems that coact with other self-organizing systems. As such, self and other are inseparable as causal factors in the development of self-relevant control systems. However, despite the inseparability as causal processes, the person as self-organizing system nonetheless retains integrity as an active contributor to his or her own development. As such, persons and selves are simultaneously individual and social. Finally, self-relevant control systems develop in order to adapt to the particularities of local sociocultural contexts. As such, persons and selves are grounded in specific bodies, contexts, and cultures, and cannot be defined in terms of a decontextualized or disembodied *cogito*. Nevertheless, with development, individuals are capable of abstracting self-representations across local contexts, and domains into increasingly higher-order generalizations. Thus, dynamic skills theory provides a way to understand how persons and selves are both fractionated and capable of integration, individual and social, as well as both contextualized and yet capable of forming abstract representations of their own processes and identities. Before elaborating these ideas further, we examine the basic principles of systems thinking and skills theory.

Properties of Living Systems

Living processes function as hierarchical, integral, self-organizing systems. The organism–environment system is *hierarchical* in the sense that it consists of multiple levels. The genome functions as the lowest level in the human hierarchical system. Genes are located within chromosomes, which themselves are located within nuclei, cell matrices, cells, tissues, organs, organ systems, organisms, and a variety of embedded physical, dyadic, and sociocultural subsystems. Processes within each level function as *integral* systems in the sense that the elements that compose them cannot exist or function separate from one another or from the larger whole of which they are a part. Although each level is partially independent of other levels and operates according to its own principles, systems and subsystems *mutually regulate* each other. Each level in the organism–environment system is interpenetrated both vertically by other levels of the system (e.g., gene–cell; organ system–organism, organism–dyad), and horizontally by other subsystems at the same level (e.g., gene–gene, cell–cell, organism–organism; Gottlieb, 1991a, 1992; Weiss, 1970).

Gottlieb (1991a, 1992) provides compelling descriptions of both vertical and horizontal coactions. For example, Gottlieb (1992) describes a parasitic wasp that lays its eggs in either fly or butterfly hosts. Eggs nested in butterfly hosts develop wings and other attributes that are absent in ways that nest in fly hosts. Ho (1984) demonstrated that exposure to ether at a particular choice point in the embryonic development of fruit flies modifies cellular cytoplasm, which thereupon prompts the development of a second pair of wings. Remarkably, this effect is transgenerational to several generations. This example shows that even though different levels of the organism–environment system exist, they are not autonomous in their operation. Based on a systems viewpoint, humans function as multileveled self-organizing systems that coact with other such self-organizing systems. For our purposes, we differentiate among four levels of organism–environment systems. The *bio–genetic* level includes all "subpersonal" levels, from the operation of the genome through the nervous system; the *organismic–agentive* level refers to functioning at the level of the person as system. The *social–dyadic level* refers to coregulated transactions between at least two persons. The *cultural–linguistic* level refers to larger socially shared practices, institutions, and meanings, and the ways in which language and sign activity represent and communicate such meanings.

The Person as a System of Hierarchical, Coactional Control Systems

Although it is interpenetrated by both bio–genetic, social–dyadic, and cultural–linguistic systems, the *organismic–agentive* level functions as a distinct and integral self-organizing system. At this level, the person functions as a series of hierarchically organized *control systems*. One way to understand a control system is in terms of a series of hierarchically embedded feedback loops, each of which operates in terms of a series of behavioral "standards" (Carver & Scheier, 1981, 1990; Powers, 1973). A standard refers to any type of set point, reference value, or point of comparison. Each negative feedback loop would operate in terms of a TOTE unit (test, operate, test, exit); the loop would involve a test of whether input to the loop meets the reference standard. If not, the system takes action, the test resumes and exits when the standard is met. Within a hierarchically organized control system, the output from a superordinate loop specifies the reference standard for the next lower subordinate level of loops (Carver & Scheier, 1981).

For example, consider a host who is making coffee for some guests (Carver & Scheier, 1981). At the most superordinate level, his behavior is governed by the principled self-standard "Be a gracious person." The system attempts to meet this standard by invoking a series of lower-level standards and embedded feedback loops. At the next lower level of "scripted activity," one might invoke the standard "Provide guests with refreshments." To meet this standard, the system invokes a standard governing a "subscript" to "Make coffee." This subscript involves a series of discrete behaviors, many of which must be performed in a given sequence. At the level of "sequence," the system invokes the standard "Measure the coffee." This requires still lower-level standards governing relations among measuring acts (scooping the coffee), which requires still lower-level standards regulating the movement of the scoop, the pattern of the hand around the scoop, sensory feedback from grasping, and muscle intensity. Physical behavior is produced only at this lowest level where sensory feedback from action is compared against standards that regulate muscular intensity. Output from lower-level loops is sent to the next highest level of feedback loops, and so on, eventually informing superordinate units of one's status as a "gracious person."

Within dynamic skills theory (Fischer, 1980), human behavior is composed of sets of dynamic control systems called *skills*. A skill refers to an individual's capacity to control behavior and thinking within a given context and psychological domain. A skill is always an action performed in a context. A change in context generally prompts a change in the form and developmental level of the skill, often in profound ways. As such, a skill is a property not of a person, but rather of a person-in-a-context. Like Carver and Scheier's (1981) description of hierarchically embedded control systems, skills are hierarchically organized, dynamic structures in that they coordinate lower-level component actions into higher-order actions and representations. As such, they function as hierarchical control systems by which individuals regulate their behavior and thinking. Children's capacities to organize their behavior into skills within specified domains and contexts undergo a series of qualitative transformations in development. According to dynamic skills theory, skills develop through four broad tiers of development (reflexes, sensory–motor actions, representations, and abstractions) and a series of four levels within each tier (e.g., single sets, mappings, systems, and systems of systems), the last level of which specifies the first level of the next broad tier of development. Higher-level skills are the results of the coordination of multiple lower-level skills into a single control structure. The various tiers and levels are depicted in Table 10.1.

To illustrate the levels predicted by skills theory, we draw on studies that assess the development of children's capacity to represent nice and mean interactions in self and other during pretend play. A series of studies involving children's stories, actions, and explanations have produced maps of developmental pathways for nice and mean interactions from 2 to 20 years of age (Fischer, Hand, Watson, Van Parys, & Tucker, 1984; Hencke, 1991; Raya, 1993). Table 12.2 displays a developmental sequence of skills for representing nice and mean social interactions. Building on reflex and sensory–motor tiers of development, beginning around 24 months, children begin to construct *single representations*. At this level, children can construct concrete representations of persons as agents, that is, people or animate objects that perform concrete actions and have concrete characteristics. For example, children can control single categories of nice or mean behaviors in a doll by making it give another doll candy or by making it hit another doll. After a series of intermediate steps, beginning around 3½ to 4 years, children gain the capacity to coordinate two or more single representations into a *representational mapping*. Using mappings, children can control relations among representations and make one doll act mean or nice in response to another's mean or nice action. Around 6–7 years, children begin to coordinate two mappings into a *representational system*. At this

TABLE 12.1. Levels of Skill Development

Level	Reflex	Sensorimotor	Representational	Abstract	Age[a]
			Tier		
Rf1: Single reflexes	[A] or [B]				3–4 wk
Rf2: Reflex mappings	[A —— B]				7–8
Rf3: Reflex systems	$[A_F^E \leftrightarrow B_F^E]$				10–11
Rf4/Sm1: Systems of reflex systems, which are single sensory-motor actions	$\begin{bmatrix} A_F^E \leftrightarrow B_F^E \\ \Updownarrow \\ C_H^G \leftrightarrow D_H^G \end{bmatrix} \equiv [I]$				15–17
Sm2: Sensory-motor mappings		[I —— J]			7–8 mo
Sm3: Sensory-motor systems		$[I_N^M \leftrightarrow J_N^M]$			11–13
Sm4/Rp1: Systems of sensory-motor systems, which are single representations		$\begin{bmatrix} I_N^M \leftrightarrow J_N^M \\ \Updownarrow \\ K_P^O \leftrightarrow L_P^O \end{bmatrix} \equiv [Q]$			18–24
Rp2: Representational mappings			[Q —— R]		3.5–4.5 yr
Rp3: Representational systems			$[Q_V^U \leftrightarrow R_V^U]$		6–7
Rp4/Ab1: Systems of representational systems, which are single abstractions			$\begin{bmatrix} Q_V^U \leftrightarrow R_V^U \\ \Updownarrow \\ S_X^W \leftrightarrow T_X^W \end{bmatrix} \equiv [Y]$		10–12
Ab2: Abstract mappings				[Y —— Z]	14–16
Ab3: Abstract systems				$[Y_D^C \leftrightarrow Z_D^C]$	19–20
Ab4: Systems of abstract systems, which are principles				$\begin{bmatrix} Y_D^C \leftrightarrow Z_D^C \\ \Updownarrow \\ A_D^C \leftrightarrow B_D^C \end{bmatrix}$	24–25

Note. Plain letters designate reflex sets; bold letters, sensory-motor sets; italic letters, representational sets; and outline letters, abstract sets. Multiple sub- and superscripts designate differentiated subsets. Long straight lines and arrows designate a relation between sets or systems. Brackets mark a single skill. Note also that structures from lower tiers continue at higher levels (reflexes in sensorimotor actions, etc.), but the formulas are omitted because they become so complex.

[a]Ages are modal for the emergence of optimal levels based on research with middle-class American and European children. They may differ across cultures or social groups.

point, children relate nice and mean behaviors in one doll character to reciprocal nice and mean acts in another. For example, one doll acts both nice and mean by asking another doll to play and by playfully hitting her; the other doll is nice and mean by saying that she will play, but the hitting must stop.

Beginning around 10–13 years, children enter a new tier of development and begin to construct *single abstractions*. At this level, children can move beyond the particularities of concrete actions and meanings and begin to construct representations of less tangible aspects of positive or negative actions. For example, an adolescent can coordinate several representational systems that combine positive and negative social interactions into abstract concepts such as *intentionality* or *responsibility*. For example, a 12-year-old might produce an abstraction such as "intentions matter more than actions" by generalizing across concrete interactions with two people who acted in a harmful way but intended to be nice. Beginning around age 15, adolescents begin to represent relations among at least two abstractions in terms of *abstract mappings*. For example, a 15-year-old might relate the abstract notions of negative intentionality and positive responsibility into an abstract relationship: "Negative intentions can be compensated by taking positive responsibility for bad actions." Around age 20, young adults can begin to coordinate relationships between two or more abstract mappings into an *abstract system*. For example, an adult might relate intentionality and responsibility to each other in several ways, and indicate that one type of intentionality requires taking responsibility in one way, whereas another requires taking responsibility in a different way. The highest level of reflective action includes the coordination of multiple abstract systems into *single principles*. Using principles, some adults might use abstract principles of justice to explain how relations between different types of intentions and responsibility in one culture are different from such relations in another culture.

Unlike many models of the mind (Descartes, 1637/1968; Chomsky, 1966; Piaget, 1957, 1975; Kohlberg, 1969), skills theory does not assume that people have unified, fundamentally integrated minds. Rather, human beings consist of multiple control systems that are naturally separate. Thus, the mind is naturally *fractionated*, although the functioning of separate control systems can become increasingly coordinated with development. In studies of cognitive development, hundreds of studies have shown that tasks intended to measure the "same" ability produce different behaviors at different developmental levels. Unevenness in the emergence of cognitive skills is the rule rather than the exception in development (Biggs & Collis, 1982; Flavell, 1982; Fischer, 1980). For example, children who demonstrate an understanding of conservation of liquid at around 7 years of age often fail to generalize this knowledge to similar tasks in different

TABLE 12.2. Levels of Development for Interactions Involving Nice, Mean, Intention, and Responsibility

Level	Tier[a] Representational	Abstract	Examples of skills	Age[b]
Sm4/Rp1: Single representations[c]	$[YOU_{MEAN}]$ OR $[ME_{NICE}]$		Relations of sensory–motor action systems to produce concrete representations of objects, people (agents), or events: Child pretends that doll is hitting someone. Child says, "Doll mean."	18–24 mo
Rp2: Representational mappings	$[YOU_{MEAN} \longrightarrow ME_{MEAN}]$		Simple relations of representations: Child makes one doll's mean actions produce reciprocal mean actions in the other doll. Child makes two dolls act as Mom and Dad in parental roles. Child understands the self knows a secret and the teacher does not know it.	3.5–4.5 yr
Rp3: Representational systems	$\left[YOU \begin{smallmatrix} NICE \\ MEAN \end{smallmatrix} \leftrightarrow ME \begin{smallmatrix} NICE \\ MEAN \end{smallmatrix} \right]$		Complex relations of subsets of representations: Child makes two dolls interact in reciprocally nice and mean ways. Child makes two dolls act as Mom and Dad as well as doctor and teacher simultaneously (two reciprocal roles). Child understands that to prevent teacher from learning a secret, he or she must hide signs of it in own actions.	6–7

			Age
Rp4/Ab1: Single abstractions	$\begin{bmatrix} YOU\ 1 \overset{NICE}{\underset{MEAN}{\leftrightarrow}} ME \overset{NICE}{\underset{MEAN}{}} \\ \Updownarrow \\ YOU\ 2 \overset{NICE}{\underset{MEAN}{\leftrightarrow}} ME \overset{NICE}{\underset{MEAN}{}} \end{bmatrix} \equiv [INTE_{POS}]$	Relations of representational systems to produce abstractions (intangible concepts) about objects, events, or people (personalities): Person explains that intentions matter more than actions: In dealing with others, intention is more important than action, so that positive intention is not only better than negative intention but also overrides negative action. Person sees Dad as having general personality characteristics, such as conformity, emotionality, or secretiveness.	10–12
Ab2: Abstract mappings	$[INTE_{NEG} \longrightarrow RESP_{POS}]$	Simple relations of abstractions: Person relates intention to responsibility, indicating how a particular intention such as deceit behind a harmful act determines the responsibility for remedying the harm. Person sees a personality characteristic of Dad and a characteristic of Mom as complementary (or as incompatible). Similar for self and friend.	14–16
Ab3: Abstract systems	$\begin{bmatrix} INTE \overset{DECEIT}{\underset{HARM}{\leftrightarrow}} RESP \overset{FLAW}{\underset{HELP}{}} \end{bmatrix}$	Complex relations of subsets of abstractions: When harm is inflicted on someone, two types of intention (deceit and unintentional harm) are related to two types of responsibility (dealing with the flaw in one's character and helping the person harmed) so that one's specific intention determines the nature of one's responsibility. Person sees several personality characteristics of Mom and Dad as complementary (or as incompatible), defining their special relationship. Similar for Self with friend or lover.	18–20

(continued)

TABLE 12.2. *Continued*

Level	Tier[a]		Examples of skills	Age[b]
	Representational	Abstract		
Ab4: Systems of abstract systems: Principles		$\left[\begin{array}{c} \overset{\text{DECEIT}}{\text{INTE A}} \overset{\text{FLAW}}{\longleftrightarrow} \overset{}{\text{RESP A}} \overset{\text{HELP}}{} \\ \overset{\text{HARM}}{}\Updownarrow \\ \text{INTE B} \overset{R}{\underset{S}{\longleftrightarrow}} \text{RESP B} \overset{R}{\underset{S}{}} \end{array}\right]$	General principles for integrating systems of abstractions: Person understands moral principle of justice, integrating several types of relations between intention and responsibility. Person uses a general principle to integrate characteristics of Mom's and Dad's relationship as it changes across situations or over time. Similar for Self with friend or lover.	23–25

Note. Adapted from Fischer and Ayoub (1994). In skill structures, each work or letter denotes a skill component, with each large word or letter designating a main component (set) and each subscript or superscript designating a subset of the main component. Italics designate representations, and outline letters designate abstractions. Lines connecting sets designate relations forming a mapping, single-line arrows designate relations forming a system, and double-line arrows designate relations forming a system of systems.

[a]The specific formulas shown are for the first example at each level, which involves nice and mean interactions. Formulas for the other examples at each level follow the same general format.

[b]Ages given are modal ages at which a level first appears based on research with middle-class American or European children. They may differ across cultures and other social groups.

[c]Abstractions grow from coordinations of representational systems, but for simplicity that relation has been omitted from this table.

knowledge domains, such as conservation of area, let alone to problems in other logical domains, such as class inclusion or seriation (Gelman & Baillergeon, 1983). These findings suggest that neither conservation in particular nor logic in general is governed by a unified conceptual system. Instead, cognitive capacities involve specific skills or control systems tied to particular domains, tasks, and contexts.

Thus, any given individual does not function as a single control system, but rather as a set of weakly connected control structures. Each control system undergoes developmental transformation. However, because of their fractionation, control systems do not develop as a single entity in a linear fashion. As such, it is helpful to think of development using the metaphor of a "web" rather than a ladder (Bidell & Fischer, 1992, 1996). In Figure 12.1, the web metaphor represents development as a multidirectional process, even within the same person. Each strand represents the development of a different control system. The web metaphor suggests that different control systems can develop at different rates either toward or away from each other, as indicated by converging and diverging paths.

Even within particular control systems and pathways, individuals do

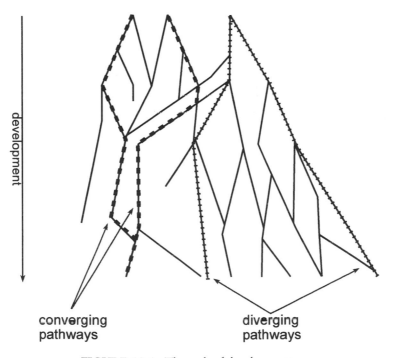

FIGURE 12.1. The web of development.

not function at a single point along a strand in the developmental web. Rather, one's behavior varies along a range of points within a given strand. To clarify this point, one might differentiate between *functional* and *optimal* levels of skill activity. One's optimal level consists of the upper limit of a person's capacity to organize his or her behavior in highly supportive conditions. Children achieve optimal level performance in contexts in which persons or features of the situation provide high levels of support or assistance. An individual's functional level refers to his or her level of performance in everyday contexts that do not provide such levels of support. The difference between the child's functional and optimal level reflects the *developmental range* for any given skill (Fischer, Bullock, Roenberg, & Raya, 1993). For example, children who tell a story about mean interactions after hearing a model story are likely to tell a story approaching her optimal level. In the absence of such contextual support, the child would likely perform at their functional level, which is often several steps below their optimal level.

A central mechanism in development of individual control systems is the *coordination of component systems into higher-order systems*. In development, persons construct coordinations among previously separate skills and control systems to form higher-order skills. As such, any given control system develops *from fractionation toward integration*. For example, in telling stories about nice and mean interactions, 2-year-old children are unable to make a doll act simultaneously nice and mean; these two subdomains are initially separate. With development, children gradually gain the capacity to coordinate nice and mean representations into a single skill, first by making a doll shift from a nice to a mean action without coordination, and later to coordinating nice and mean actions in a single doll simultaneously. Thus, although control systems start off separate, they can become more integrated. Although the mind is inherently fractionated, development occurs in the direction of increased (albeit not full) integration.

Organismic and Dyadic Processes in the Development of Skills

Control systems develop as a result of the coordination of component skills and subsystems. How do coactions among the various levels produce such developmental coordinations? To address this question, we focus first on coactions between organismic–agentive and social–dyadic levels of functioning.

As indicated earlier, a skill is property of the person-in-a-context. As such, context functions as an actual part of skilled activity. Behavior is not autonomously controlled by the individual, but jointly controlled by person and context. This is especially the case in social interaction, which transforms the functioning of the organismic–agentive system. Researchers have repeatedly demonstrated that children perform at higher developmental levels when interacting with others than when working alone (Fischer, Bullock, Rotenberg, & Raya, 1993; Rogoff, 1990, 1993; Vygotsky, 1978). This occurs in part because direct social interaction most often functions as a *continuous process communication system* (Fogel, 1993). Unlike discrete state communication (e.g., radio transmission, mail, fax), in continuous process communication, partners are simultaneously active in the process of communicating. Partners mutually modify the flow of information during the very process of communication. Because of the simultaneous adjustment of each partner's actions, it becomes meaningless to ask who is the initiator or receiver of a message. Indeed, there is no fixed "message," as the mutual adjustment of each partner's activity results in the modification of the message during the process of communication. As such, *the activity of the other becomes an actual part of the processes of the individual.* In this way, social interactions are most often *mutually regulated* as "individuals dynamically alter their actions with respect to the ongoing and anticipated actions of their partners" (Fogel, 1993, p. 34). Of course, exceptions to this rule exist, as when one blindly follows the orders of another, or when one partner is not sensitive to the other's actions.

Because behavior is often mutually regulated in activity between partners, social interaction often raises children's performance to levels that they cannot achieve alone, especially when children work with more accomplished others. As a result, social interaction acts as an important source of development. Skills develop as children reconstruct their activity in the direction of the higher-order structures originally produced in joint interaction. However, to reconstruct joint activity in terms of a new control system, *children must actively bring control system elements into coordination.* As such, the construction of a new control system requires an *act of integration* (Bidell & Fischer, 1996). To profit from social interaction, children must perform actions that coordinate for themselves what has been coordinated jointly in social interaction. It is as if social interaction says, "Put your actions together like this!" Children will be most likely to perform the necessary acts of coordination when aspects of joint activity fall within their developmental range for the domain in question. Thus, although skills development is an inextricably social process, the coordinating function of the organismic–agentive system plays a central role in development.

The construction of control systems is also interpenetrated by other levels in the organism–environment system. Fischer (1987; Fischer & Rose, 1994) has reviewed evidence that suggests ways in which developmental changes in neural systems in the brain are associated with spurts in children's optimal level of performance. Of course, despite changes in bio-genetic processes, the direction of influence is not unidirectional; self-directed activity at the organismic–agentive level also influences brain development (Bidell & Fischer, 1995). Sociocultural systems participate in the child's developing control systems virtually from the beginning of development. All skilled activity takes place within sociocultural contexts and is impacted by culturally shared goals and meanings. With development, the child's joint actions with others become increasingly mediated by socially shared meanings communicated in language. Cultural–linguistic systems contribute to children's development as children begin to use language and shared meanings to represent their worlds and regulate their own actions.

The Nature of Self within Dynamic Skills Theory

As indicated earlier, within dynamic skills theory, we note several uses of the notion of self. First, we use the term *self* to indicate *integrity of the organism as a distinct, self-organizing, dynamic system.* Thus, although the behavior of the organism is always an emergent function of coactions between and among all levels of the organism–environment system, at the organismic level, the person functions as a distinct and integral system. He or she does so by invoking multiple control systems to regulate his or her own behavior. To say that human beings function as sets of control system is not to imply that persons are autonomous. Because human beings coregulate each other in continuous process communication, human social behavior is under joint control. Furthermore, the control systems that individuals coordinate from their participation in social interaction are defined in relation to social rules, standards, and practices. However, although self and other are inseparable as causal factors in the construction of action, the organism nevertheless functions as a distinct system that coordinates its own component actions both in real time as well as in development. Dynamic skills theory provides a set of conceptual tools to describe the functioning of control systems in detail within contexts and domains, chart their development, and illuminate the processes by which they evolve.

Second, the concept of self refers not only to the integrity of the self-organizing agentive system, but also to the system's *capacity to become aware of its own processes, products, and identities* (Stern, 1985). The sense of self consists of many facets. In part, the sense of self is the awareness that one exists as an organism or a body, and that one experiences the world from the location of that body. It consists of the awareness that one is an agent who acts on the world from a particular location in space and in relation to other people—the awareness and representation of "I" (Harré & Gillett, 1994). It also consists of the constructed awareness of one's identity or self-as-object within one's social relationships—the awareness and representation of "me." From a dynamic skills theory perspective, representation and awareness of self are products of skilled activity and undergo development with transformations in one's control systems. Thus, like the control systems of which they are a part, representations of self are fractionated and develop toward integration. As instances of skilled activity, self-representations are jointly constructed both between and within individuals as active self-organizing systems.

From a systems perspective, behavior is a product of systemic and intersystemic coactions. If, at the organismic level of functioning, it is the self-organizing *system* that coordinates and regulates action, what is the role of self-awareness in that system? We have suggested that individuals operate as control systems governed by a hierarchy of standards and goals. With the capacity of the organism-as-system to construct an awareness of aspects of itself, *awareness and representation of self can function as a reference point in the hierarchy of standards that regulate the organism's activity* (Carver & Scheier, 1981, 1990). Thus, self-awareness is important *because the system uses awareness of its own agency and identity in its own regulation.* Within a given context, self-awareness and self-relevant goals (e.g., "to be a gracious host") function as superordinate standards that the system uses to regulate its behavior in social interaction. Thus, although the self-organizing *system* regulates its behavior at the organismic level, one's *sense* of agency and identity helps drive behavior as one's control systems become increasingly mediated by self-related goals and representations. As even young infants are capable of constructing a rudimentary sense of agency and identity, one might argue that self-awareness is an important aspect of virtually all behavior.[1]

Personal and Social Processes in the Construction of Self

As an instance of skilled activity, the awareness and representation of self develop as a result of coactions among all levels of the organism–environment system. Contributions at the intersection of organismic and dyadic processes are particularly salient. Organismic and social–dyadic processes coact in the production of an awareness of agency and identity; they make different contributions to these two aspects of self. In the construction of a sense of agency, the individual has access to self-relevant information that others do not. For example, infants, but not those with whom they interact, have access to sensory feedback from the volitional impulses that precede action as well as from proprioceptive feedback that accompanies action (Stern, 1985). As such, organismic activity provides an important source of the infant's sense of personal agency. However, this is only part of the picture. To develop a sense of agency, the child must be able to construct contingencies between volitional and proprioceptive feedback and reactions from other people and objects. As such, the construction of a sense of agency necessarily requires the participation of others. In the construction of a sense of identity or self-as-object, the situation is somewhat reversed. The other often has access to information about the self that the individual does not. For example, without a mirror, infants are not able to look at their own face or back, and persons generally do not have direct access to how they present themselves to others. Thus, the individual needs the other to complete his or her representation of self-as-object. This includes, of course, the social and cultural categories that individuals appropriate to represent and evaluate the self. Thus, although individual and other are inseparable as causal factors in self-development, they make different contributions to the development of the senses of agency and identity.

The Embodiment of Self

Selves are not decontextualized, abstract, or disembodied entities. Rather, selves are constructed in particular bodies within particular physical, social, cultural, and historical contexts. The production of affective and bodily experience consists of an especially important source of the self's embodiment. Elsewhere, we have elaborated a model of emotional development that draws upon the principles of dynamic skills theory (Mascolo & Harkins, 1998; Mascolo & Fischer, 1995; Mascolo & Griffin, 1998). Emotional experiences result from the patterning of multiple component systems (e.g., motive-relevant appraisals, bodily changes including central nervous system (CNS) and autonomic nervous system (ANS) arousal, action tendencies, expression, experience, etc.) as they mutually regulate each other in the context of notable changes in a perceived event. From this view, everyday categories of emotion, such as anger, joy, pride, or shame are characterized by prototypical configurations of the emotion-relevant component processes. For example, pride, a self-relevant emotional experience, involves appraisals that one is responsible for a valued outcome, the experience of one's body as bigger, taller or "on top of the world," and action tendencies to display one's worthy action or self (Mascolo & Fischer, 1995; Mascolo & Harkins, 1998).

There are several ways in which self is embodied by emotional experiences. First, the experience of emotion and other bodily experiences helps organize one's sense of self-consistency or stability throughout development. One reason why we experience ourselves as more or less stable beings in the context of physical and psychological flux is that basic feelings and bodily states can remain fairly stable throughout life (Emde, 1980; Izard & Malatesta, 1987). Although emotional experiences undergo change in development (Mascolo & Griffin, 1998), there is nevertheless some continuity to affective and bodily experience. Anger generally feels like anger, shame like shame, hunger like hunger, (Ackerman, Abe, &

Izard, 1998). Researchers have suggested that much consistency in personality is a product of temperament and other affective dispositions (Malatesta, 1990; Rothbart & Ahadi, 1994).

A second source of the affective embodiment of self concerns the functioning of self-conscious emotional experiences, such as pride, shame, guilt, and embarrassment (Lindsay-Hartz, de Rivera, & Mascolo, 1995; Mascolo & Fischer, 1995; Tangney & Fischer, 1995). Self-conscious emotions require a degree of self-awareness for their instantiation and serve important functions for an individual's sense of self in relation to his or her social world. An important function of affect is *the selection of perception and action* (Brown, 1994; Lewis, 1996; Mandler, 1984). For example, in any given situation, appraisal processes function to monitor all sources of event-related input. When changes in events occur that are relevant to one's important motives and concerns, emotional reactions are generated. Because appraisal and affect mutually regulate each other throughout the course of an emotional episode, affective experiences have the function of selecting, amplifying, and modulating the very self-evaluative appraisals that precipitated them (Mascolo & Harkins, 1998)! In this way, in a given context, the feelings that accompany self-conscious emotion *select* self-relevant concerns from among all concerns for conscious awareness. Thus, self-conscious emotions both organize and are organized by appraisals of the self vis-à-vis social standards and rules. In this way, the self-system is organized by and embodied in experiences of pride, shame, guilt, embarrassment, and related affective and bodily processes.

THE DEVELOPMENT OF SELF FROM INFANCY THROUGH ADULTHOOD

We turn now to an analysis of developmental transformations in self-systems from infancy through adulthood. In so doing, we examine when children first begin to exhibit agentive control over their behavior. We then examine how children's awareness of their own processes and attributes undergoes transformation as children's self-relevant control systems change from sensory–motor actions to increasingly sophisticated representations and abstraction within particular social and cultural contexts. In so doing, we will attempt to demonstrate that (1) children's awareness and representation of their own agency and identity emerge gradually and undergo a series of developmental transformations from infancy through adulthood; (2) persons and selves start off development as a series of fractionated control systems that develop in the direction of increased integration; (3) although children and socialization agents interact in the construction of any given level of self-awareness, children's own activity in the world contributes directly to the formation of self; (4) even though self-representations are constructed through embodied experiences in local sociocultural circumstances, with development, children are capable of generalizing over such experiences to form increasingly abstracted and principled representations of self.

The Development of Agentive Control and Self-Awareness in Infancy

To speak of the infant as agent implies the regulation of behavior in terms of control systems. As indicated in Table 12.1, infants begin to exert some control over their behavior soon after birth. Beginning around 3–4 weeks, infants begin to control *single reflexes*. Such reflexes are not involuntary, subcortical responses like eye blinks or knee jerks, but rather consist of controlled and coordinated action components, such as looking, grasping, listening, smiling, and other reflexes. For example, using single reflexes, infants can kick their legs or grasp a ball placed in their hands (Fischer & Hogan, 1989). Consistent with skill theory, infants collaborate with others to produce such actions; another person must place the ball in the child's hands in order for the child to invoke a grasping reflex. Nevertheless, single reflexes mark the early onset of infants' capacities to act as agents of their own actions; they are the first steps the emergence of an agentive self.

There is evidence that from a very early age, a rudimentary form of self-awareness participates in infants' controlled actions. With the onset of reflex mappings at 7–8 weeks of age (Fischer & Hogan, 1989; see Table 12.1), infants can detect contingencies between rudimentary acts and effects. Watson and Ramey (1972) provided 8-week-olds with mobiles that moved in response to pressure-sensitive pillows. Infants quickly detected the contingency between head and mobile movement, and responded with smiles and other positive reactions. Lewis, Alessandri, and Sullivan (1990) made an audiovisual display contingent upon arm pulling in 2-, 4-, 6- and 8-month-olds. After an initial learning period, each group of infants exhibited heightened levels of angry facial behavior upon withdrawal of the contingency. In an elegant follow-up study, Sullivan and Lewis (1993) demonstrated that anger reactions in 4- to 6-month-olds were not attributable to the withdrawal of stimulation per se, but rather to "the disruption of infant's perceived control of an outcome" (p. 3). After contingency learning, infants were exposed to one of three types of frustrating conditions: extinction (cessation of contingent stimulation), partial reinforcement (outcome on every third pull), and noncontingency (stimulation levels maintained, independent of the child's activity). Infants displayed the highest levels of angry facial behavior in response to noncontingency, followed by extinction and partial reinforcement. In addition, rate of arm pulling increased in the extinction and partial reinforcement conditions, but decreased in the noncontingency condition. This suggest that even very young infants can construct a *sense of control* over events within certain contexts.

Stern's (1985) observations of a set of 4-month-old conjoined twins provides additional evidence of early self-awareness. The twins shared no internal organs but were joined at the bottom of the sternum, facing each other. They often sucked on each other's fingers. When a twin was sucking on a finger, an experimenter pulled the arm of the finger being sucked away from the infant's mouth. When a twin was sucking her *own* finger, her *hand* resisted being pulled from the mouth; when sucking her *sister's* finger, her *head* followed the retracting hand. This suggests early awareness of both the sense of self-as-agent (differentiating self- from other-control) and self-as-object (differentiating "my finger" from "your finger").

The awareness and representation of self continues to develop gradually throughout infancy. A series of studies document developmental changes in children's awareness of featural and agentive aspects of self and others in infancy (Bertenthal & Fischer, 1978; Pipp, Fischer, & Jennings, 1987; Watson & Fischer, 1977). The tasks used and approximate ages of emergence of children's performance in each task are displayed in Table 12.3 for feature recognition and Table 12.4 for agency skills. For both domains, self- and other-knowledge emerge gradually and undergo a series of developmental transformations. For featural recognition (Table 12.3), soon after the onset of *single sensory–motor actions* at around 4 months, when placed in front of a mirror, 6 month-olds are able to coordinate reaching with the image they see in the mirror. Beginning around 7–8 months, with the emergence of *sensory–motor mappings*, infants can perform the hat task, which involves coordination of reaching with the image seen in the mirror and the movements of the child's own body. With the onset of *sensory–motor systems* around 11–13 months, infants can coordinate multiple sensory–motor mappings. They can perform the sticker tasks, which involve coordination of grasping and pulling with their awareness of feedback from various bodily parts. Later on at this level, infants can perform the toy task, which involves multiple coordinations of children's body movements with the mirror images of their own body and the independent movement of the toy. With further development within this level, beginning around 15–20 months of age, children can perform the traditional rouge task (Lewis & Brooks-Gunn, 1979). With the onset of *single representations* around 20–24 months, children become capable of performing a cluster of tasks, including identifying their spatial location, labeling and identifying the self, and indicating the self's possessions. With further development within this level, children are able to describe the self using terms indicating simple familial and gender categories.

Within the agency domain (Table 12.4), soon after the onset of senso-

TABLE 12.3. Infant and Mother Versions of Feature-Recognition Task

Step	Child version	Age	Mother version	Age
	Single sensory–motor actions			
1[a]	*Tactile exploration*. Infant is placed in front of mirror and touches image in mirror.	6	None.	
	Sensory–motor mappings			
2[a]	*Hat task*. Infant placed in front of mirror with vest that holds a hat above child's head. Infant looks in mirror and turns to look at or grab real hat.	8	None.	
	Sensory–motor systems			
3a	*Sticker–nose*. Infant pulls sticker off own nose.	12	Infant pulls sticker off mother's nose.	8
3b	*Sticker–hand*. Infant pulls sticker off own hand.	12	Infant pulls sticker off mother's hand.	13
4[a]	*Toy task*. Infant placed in front of mirror and a toy descends from ceiling behind child. Infant looks in mirror and then turns to look at object.	15	None.	
5[b]	*Rouge task*. Infant touches nose after detecting rouge in mirror.	18	Infant touches mother's nose after detecting rouge in mirror.	15
	Single representations			
6	*Spatial location*. Experimenter asks "Where's [child's name]?" Infant points to self or says "here," "there," etc.	24	Experimenter asks, "Where's Mommy?" Infant points to mother or says "here," "there," etc.	27
7[b]	*Name/verbal label*. Experimenter points to infant and asks, "Who's that?" Infant states name or says "me."	24	Experimenter points to mother and asks, "Who's that?" Infant says variant of "mother" or states mother's proper name.	27
8	*Identification of actor*. Child claps and experimenter asks, "Who did that?" Infant states proper name or says "me."	24	Mother claps and experimenter asks, "Who did that?" Infant says some variant of "mother" or state's mother's proper name.	27
9	*Featural possession*. Experimenter points to infant's shoe and asks, "Whose shoe is that?" Infant responds with name or says "mine."	26	Experimenter points to mother's shoe and asks, "Whose shoe is that?" Infant states variant of "mother" or says mother's name.	27

TABLE 12.3. *Continued*

Step	Child version	Age	Mother version	Age
	Compounded single representations			
10	*Family relation.* Experimenter asks, "Who do you belong to? Whose baby are you?" Infant indicates a family member.	32	Experimenter asks, "Who does your mommy belong to?" Infant states own proper name, says "me" or indicates other family member.	31
11	*Gender identification.* Child responds correctly to "Are you a boy?" and "Are you a girl?"	35	Child responds correctly to both "Is Mommy a boy?" and "Is Mommy a girl?"	36

Note. Age indicates approximate age of emergence in months based on empirical data.
[a]Task assessed in Bertenthal and Fischer (1978). No parallel mother-knowledge tasks were conducted.
[b]Task assessed in both Bertenthal and Fischer (1978) and Pipp, Fischer, and Jennings (1987).

ry–motor mappings around 8 months, infants can coordinate looking and reaching to perform simple self-relevant actions, such as eating a Cheerio. After the onset of sensory–motor systems around 11–13 months, infants can make the self, a doll, or a substitute object act as a passive agent and perform action sequences involved in sleeping or eating. Using single representations at around 20–24 months of age, children can make a doll act as an agent that actively performs goal-relevant actions (e.g., sleeping, eating). In addition, children can make a doll act a behavioral role, such as baby. With further development at this level, children represent more complex agentive acts involving interacting agents and complex social roles.

Pipp, Fischer, and Jennings (1987) compared the development of children's knowledge about their own features and agency to the development of children's knowledge about their mother's features and capacity to act as an agent. In so doing, the investigators assessed the issue of *horizontal decalage,* that is, whether knowledge of self developed before or after knowledge of the mother. Results indicated that for the agency tasks, prior to 18 months, self-knowledge tended to develop before mother-knowledge in infants who exhibited decalage, presumably because children can control their own actions more directly than they can control their mother's. For featural tasks, prior to 18 months, mother-knowledge tended to develop before self-knowledge, presumably because children can see their mother's features more often and more easily than their own. After 18 months, the picture becomes more complex. For feature tasks, there was no consistent decalage toward self- or mother-knowledge. We suggest that the emergence of representational skills eliminated the natural bias toward other's

TABLE 12.4. Infant and Mother Versions of Agency Tasks

Step	Child version	Age	Mother version	Age
	Sensory–motor mappings			
1	*Actor.* Infant acts on self by feeding self a Cheerio.	10	Infant acts on mother by feeding her mother a Cheerio.	10
	Sensory–motor systems			
2[a]	*Passive self-agent.* Infant acts as agent by passively pretending to feed self with spoon or pretending to drink with a cup.	14–17	Infant acts as agent by passively pretending to feed mother with spoon or give mother a drink with a cup.	18
3[b]	*Passive other-agent.* Infant puts a doll on a pillow to pretend that it goes to sleep.	18	None.	
	Passive substitute agent. Infant puts a block on a pillow to pretend that it goes to sleep.	20	None.	
	Single representations			
5[a]	*Active other agent.* Infant pretends that doll walks to a toy table and eats with a spoon or drinks with a cup, as if doll were actually performing act.	24	Infant asks (verbally or through action) mother to come to table and eat with a spoon or drink with a cup.	22
4	*Baby behavioral role.* Infant treats doll as a baby by calling it a baby, giving it a rattle or baby bottle, or having doll kiss a teddy bear.	26	Infant treats the mother as a baby by calling her a baby, giving her a rattle or baby bottle, or having her kiss a teddy bear.	29
	Compounded single representations			
5	*Two interacting agents.* Infant pretends that one doll feeds a second doll with a spoon or gives second doll a drink with a cup.	32	Infant pretends that the doll feeds the mother with a spoon or that doll gives the mother a drink with a cup.	33
6	*Mother behavioral role.* Infant pretends that a mother doll gives a baby doll a baby bottle, a rattle, or teddy bear to kiss, or calls mother doll "mother" or baby doll "baby."	36	Infant asks (gesturally or verbally) mother to give baby doll a bottle, a rattle to play with, a teddy bear to kiss, or the infant calls the mother a "mother" or baby doll a "baby."	38

Note. Age indicates approximate age of emergence in months in high-support contexts.
[a]Tasks assessed in both Watson and Fischer (1977) and Pipp, Fischer, and Jennings (1987).
[b]Tasks assessed in Watson and Fischer (1977) only; no parallel mother-knowledge task was conducted.

features. For agency tasks, self-knowledge continued to develop prior to mother-knowledge for several tasks, but some defensiveness also emerged. In the baby behavioral role task, when children were asked to pretend to be a baby, younger children (around age 2) were successful and enjoyed the task; older children (around age 3) failed the task, even though they passed the mother version of this task. Our interpretation is that 3-year-olds rejected seeing themselves babies and thus resisted pretending to be one. Others have also reported defensive distortion involving self-categories at 3 years of age. Edwards (1984) reported that boys, but not girls, categorized themselves with other big children even though they were in fact smaller than other children in the study. Similarly, in a study on self-identification in terms of gender, race, and age categories in preschoolers, Van Parys (1981, 1983) described a subgroup of 3- and 4-year-old African American boys who classified themselves as white, even though they appropriately classified the races of other persons in the study. African American girls and white children did not exhibit this pattern of behavior.

Overall, the findings on the development of self-awareness in infancy suggest that different aspects of self-awareness undergo developmental change with development in children's level of operative skill. Early in life, children can exert control over their actions and experience a rudimentary form of self-awareness. This self-awareness changes with transformations in sensory–motor and representational skills. Furthermore, the results suggest the ways in which self- and other-awareness develop within particular bodies and social contexts. Prior to 18 months of age, the relationship between the body and its ecological context determines whether self-knowledge develops ahead of other-knowledge or vice versa. The knowledge domain that is more accessible to the child will tend to develop first: For agency tasks, because children can directly sense their own powers of control but not their mother's, knowledge about self-agency develops before knowledge about other-agency. For feature tasks, because children can directly perceive their mother's attributes but not their own, knowledge about other's features develops prior to knowledge about self-attributes. With the onset of single representations around 18 months, children to overcome decalage between self- and other-knowledge in feature-recognition tasks. Soon after the onset of single representations, children begin to exhibit defensive reactions as they represent themselves in terms of social categories that reflect various degrees of value or power in society.

The Development of Self-Evaluative Emotion in Infancy and Early Childhood

Self-conscious emotions, like any psychological processes, do not emerge fully formed at any single point in development. Rather, emotional experiences such as pride, shame, and guilt emerge gradually and take on a series of increasingly complex forms throughout ontogenesis (Mascolo & Harkins, 1998; Mascolo & Fischer, 1995; Mascolo, Pollack, & Fischer, 1997). We illustrate such progressions by describing developmental changes in pride-relevant appraisals and action tendencies. Pride experiences involve appraisals that one is responsible for performing a socially valued act, and action tendencies, to present one's worthy act or self to others. As such, pride requires some degree of self-evaluation in terms of social standards or rules, however rudimentary they may be.

Mascolo and Fischer (1995; Mascolo & Harkins, 1995, 1998) used skills theory to propose a model of the development of appraisal patterns involved in pride from infancy through adulthood. We present the first several steps of this sequence to illustrate the nature of the changes described. Step 1 of the sequence, *joy over action–outcome contingencies*, builds on infants' early reflex and sensory–motor skills. At this step, infants can experience pleasure upon detecting contingencies between their own simple reflex or sensory–motor actions and salient results of those actions. For example, Watson and Ramey (1972) reported that 8-week-old infants would smile after detecting the contingency between self-produced head movements and contingent movement of a seen mobile. Such results suggest that the capacity to respond with positive affect to simple, self-caused effects begins early in life.

The next major step in the development of pride appraisals begins around 18–24 months, with the onset of single representations. At this level, children can begin to experience *joy/pride about a result attributed to the self*. This step is mediated by the child's capacity to form single representations of self as agent or of the production of a stable outcome. Thus, at this step, after having knocked over some bowling pins with a plastic bowling ball, children would be able to make an appraisal such as "I did it!", "I knocked it over!", or something similar. Several studies suggest that important changes in pride-relevant behavior occur around 18–24 months of age (Stipek, Recchia, & McClintic, 1992). H. Heckhausen (1984, 1987; J. Heckhausen, 1988) reported that beginning around 14–20 months, children stop and notice results of their acts. Bullock and Lutkenhaus (1988) suggest that this occurs when children can focus on outcomes per se rather than on the flow of activity. Kagan (1981) reported that smiles on completion of goal-directed acts increased between the ages of 20 and 24 months. Mascolo and Harkins (1995, 1998) reported that in the third year of life, children began to respond to success in simple achievement tasks (e.g., bowling, basketball toss, etc.) with simple evaluations, like clapping the hands, or saying "I did it!"

Although 18- to 24-month-olds are able to represent their role in producing a stable outcome, what is lacking is the ability to represent the self or outcome as explicitly valuable, good, or competent. At Step 3, with the development of compounded representations, children between the ages of 2 and 3 begin to experience *pride about a result caused by self performing well*. At this step, children can not only represent "I throw ball," but also "I throw good!" or "I throw far!" As such, at this step, children are increasingly able to represent social standards of value and worth in their pride-relevant appraisals. Studies suggest that between about 2½ and 3 years, children begin to evaluate themselves positively in the context of achievement. Lewis, Alessandri, and Sullivan (1992) and Mascolo and Harkins (1998) reported that 3-year-olds exhibited complex pride responses (i.e., three or more behaviors such as smile, social referencing, self-evaluation, postural changes) in basketball toss and drawing tasks. Only indirect evidence suggests that such reactions involve a sense of competence or concern about performing well. Geppert and Kuster (1983) reported that although 18- to 30-month-olds used their name, "I," or "me" to indicate that they wanted to complete interrupted tasks, only children older than 30 months made statements such as "I can do it alone" and tried to stop adults from completing the last step of tasks. This suggests that 30-month-olds wanted to demonstrate their competence, whereas younger children simply wanted to complete the interrupted act.

Step 4, *pride over social comparisons*, emerges with the ability to construct representational mappings. At this step, children can begin to compare the action-based performance of self and others. Pride-relevant social comparisons can be accomplished in several ways. In simple competitive tasks (e.g., competitive ring stacking), 3½- to 4-year-olds can compare the performance of self and other and experience pride upon winning. This assertion is supported by findings that children around the age of 3½ years smile and show other pride-relevant reactions upon winning in competitive ring stacking tasks (Heckhausen, 1984; Stipek et al., 1993). At this level, children can also compare the performance of self with that of a valued other and experience pride over performing as well as the other. For example, in a recent study, one precocious 3-year-old boy boomed out of the bathroom and exclaimed with a beaming smile, "I can pee standing up, just like papa!" Furthermore, around this time, children can use mappings to compare their own task performance over several situations and experience pride over possessing a valued trait. For example, a boy might compare the quality of his kicking on two different occasions and conclude, "I'm good at playing soccer!" Of course, these steps represent only the first steps in pride development; pride experiences continue to undergo development from childhood through adulthood.

Like all complex psychological processes, self-evaluative emotional experiences emerge and undergo transformation in ontogenesis. Self-evaluative emotional experiences undergo change with transformation in children's capacity to represent and evaluate self in terms of social standards of worth. We suggest that the affective components of self-evaluative emotions function to select and organize self-relevant events in consciousness. In this way, self-representations are embodied and organized by self-conscious emotions. Because self-evaluative emotions are founded in social relations, their development illustrates the intersection of cognition, affect, and social relations in the formation of self.

Development of Self-Representations from Early Childhood through Adulthood

Thus, with the onset of the representational tier of development around 18–24 months, children construct concrete representations of self (and other) as agents and as entities having concrete characteristics. From this step onward, massive development occurs in the development of self-representation. Most studies have used low-support, open-ended interviews to extract changes in self-understanding in children and adolescents, producing descriptions of general changes in children's functional level of self-understanding. For example, Damon and Hart (1982, 1986, 1988; Hart & Damon, 1986) have reported changes in self-understanding from simple categorical identifications in young children to comparative assessments of self in middle and late childhood, to interpersonal attributes in early adolescence, to abstract and systematic beliefs and plans in adolescence and beyond.

Dynamic skills theory provides a set of conceptual and methodological tools for assessing more precise changes in self-representation in high-support contexts that promote optimal-level performance. Such sequences are illustrated by research on development of evaluative categories for conceiving self and other, in particular nice–mean, good–bad, smart–dumb, and so on. The research described earlier on the development of nice and mean interactions involving self and other illustrates this point. To demonstrate understanding of each level in Table 12.2, children imitated stories involving representations of both agentive and categorical aspects of self and others. For example, when 2-year-olds use single representations to make a doll named after them act mean or nice to another doll, they exhibit skill in representing self as an *agent* who carries out concrete actions (the children make the doll herself carry out the action, indicating a representation of agency), and who have concrete categorical *traits* (the children can categorize the doll as "mean"). When 4-year-olds use representational mappings to enact simultaneous nice and mean actions in the self, they exhibit skill in conceiving the self in terms two or more same- or opposite-valenced categories. At this same level, children can also represent reciprocal relationships between the self and other, for example, in terms of the reciprocation of nice actions for nice actions, or mean actions for mean actions. Using representational systems at the next developmental level, 6- and 7-year-olds are able to construct more powerful representations of reciprocal relationships between self and other.

With the emergence of the abstraction tier of development, adolescents move from conceiving self and other as *agents* who have positive or negative concrete *traits* to conceiving them as active *personalities* who exhibit complex *identities* in relation to each other. For example, when adolescents coordinate multiple systems for combining positive and negative aspects of social interactions into the abstraction "Intentions matter more than actions," they move beyond conceiving their agency in terms of concrete actions toward a more abstract characterization of themselves as a complex *personality* that organizes meaning and action. Similarly, when adolescents begin to describe their characteristics in terms of single abstractions that coordinate multiple systems of concrete traits, they have taken a step toward constructing an abstract *identity* of themselves in relation to others. Similarly, with abstract mappings, adolescents can coordinate relations among multiple abstractions that they use to define their identities. They can also map abstract conceptions of their own identities with those of others. For example, through discussion, two adolescents may come to define a friendship in terms of a mapping of their abstract conceptions of their shared commitments to helping the poor. Of course, the capacity to conceive of self and other in terms of abstract personalities and identities continues to develop throughout adolescence and beyond.

Research on the development of self suggests that people show a bias toward representing the self as positive, emphasizing positive attributes and deemphasizing negative ones (Connell, 1991; Harter, 1983; Harter & Monsour, 1992; Ruble, 1983). Research on children's use of nice and mean categories in pretend play has documented two findings that suggest greater complexity. First, 2- and 3-year-old children typically enjoy negative pretend play and emphasize negative characteristics in their pretend play over positive characteristics. On average, young children show a more advanced understanding of the negative (e.g., mean interactions) in both interview- and pretend-play assessment contexts. This initial negativity bias seems to reverse itself to the adult-like positivity bias by about 4 to 6 years of age (Elmendorf, 1992; Hand & Fischer, 1989; Hencke, 1991). A second finding is that shy or behaviorally inhibited children Kagan, Reznick, and Snidman (1987) seem to exhibit a bias toward positive (nice) and an avoidance of negative (mean) portrayals of the self even at young ages (Hencke & Raya, 1993; Hencke, 1991). As a result, behaviorally inhibited children exhibited relatively slow development in understanding mean stories.

In a related study, Fischer, Knight, and Van Parys (1993; Van Parys,

1981, 1983) reported the results of studies on the representation of self and other using age, sex, and race categories in children between the ages of 2 and 6 years of age. Table 12.5 describes a developmental sequence for stories about age and sex roles in American children. Children's capacity to represent these levels was assessed using stories that children were asked to tell or act out for each step in the proposed sequence. For stories about age and sex roles separately, the most complex step passed by children increased with age, the growth curves for males and females were identical. The ages of children passing parallel steps in the age- and sex-role sequences were similar, although the age role developed slightly ahead of the sex role. In addition to the analysis of levels of understanding of age and sex categories, children were also assessed using four tasks in which they made choices based on age and sex categories. In self-identification tasks, children classified themselves using drawings of people who varied in age and sex characteristics. In the people-preference task, children chose categories of people (man, woman, boy, girl) that they would most like to be with. In the toy-preference task, children chose between toys that were gender stereotyped.

Results showed that boys and girls view age- and sex-role categories differently. For gender-identification tasks, as children grew older, girls identified themselves more often as children, whereas boys identified themselves more often as males. Similarly, for age-preference tasks, young boys tended to prefer to be with children, whereas older boys tended to prefer adults. Girls tended to prefer to be with children over adults. For toy-preference task, boys exhibited a preference for male-stereotyped toys, which increased with age. Girls showed only a slight preference for female-stereotyped toys, and this preference did not show an age trend. For the person-preference task, although boys were more consistent than girls, both boys and girls preferred to be with same-sex individuals, and this preference increased with age. Van Parys (1981) assessed these tasks in a sample of white and African American boys and girls. Although the findings for white children were similar to those discussed earlier, the gender preference was reversed for African American children. As girls grew older, they exhibited more gender-related identifications; with age, however, African American boys tended to avoid gender categorization. This suggests the importance of perceptions of power and status in children's preference for age and gender roles.

The research reviewed in this section indicates significant transformation in children's capacity to represent the self as both an agent and object throughout childhood. Children move from representing the self as concrete agents who possess concrete traits to increasingly complex personalities with abstract identities. Throughout development, children's self-representations structured in terms of affective meanings and socially valued

TABLE 12.5. Development of Understanding Sex and Age Roles

Step	Skill	Sex roles	Age roles
		Level Rpl: Single representations	
1	Active agent	Child makes doll perform at least one behavior, not necessarily fitting sex or age categories.	
2	Behavioral role	*Role for child's own sex:* Child makes boy/girl doll play with toys fitting own sex.	*Child role:* Child makes child doll play with a child's toys rather than an adult's tools.
3	Shifting behavioral roles	*Shift from opposite sex to same sex:* Child first makes boy doll play with boys' toys and then makes girl doll play with girls' toys.	*Shift from adult role to child role:* Child first makes adult doll cook and fix the stove and then makes child doll play with toys.
		Level Rp2: Representational mappings	
4	Effect of one-dimensional change on role	*Sex constancy during long time in bed:* Child makes doll of own sex spend a long time in bed and still remain in same sex.	*Change in size from child to adult role:* Child makes child doll play with toys and then has it grow up into an adult, who works with tools.
5	Shifting between effects of one-dimensional changes on role	*Shift from sex constancy despite one change to sex constancy despite another change:* Child makes doll of own sex remain the same sex despite one of the changes in Step 6 at a time, shifting between them.	*Shift from child–adult difference in one trait to child–adult difference in another trait:* Child makes child and adult dolls show one child–adult difference from Step 6 at a time, shifting between them.
		Level Rp3: Representational systems	
6	Effects of multi-dimensional changes on role	*Sex constancy despite both long time in bed and changes in different-sex clothes and name:* Child makes dolls of own sex remain the same sex even when he or she spends a long period in bed, changes clothes to those of the other sex, and changes name to one fitting the other sex.	*Unidirectionality of age growth for size, strength, and activity:* Child puts together several age–role characteristics that change with growth: Child doll puts on a necklace that will not fit adult, adult doll builds a tent that child doll cannot build, and child doll plays in tent, where adult will not fit.
7	Integration of effects of multi-dimensional changes on both roles	Integration of sex and age roles in children and adults: Child makes girl and boy dolls play with sex-appropriate toys and refer to their future roles as woman and man, and then child makes them grow up into woman and man dolls who show sex differences and refer to having been girl and boy.	

Note. Adapted from Fischer, Knight, and Van Parys (1993).

categories. In general, children exhibit a positivity bias, preferring to see themselves as positive rather than negative. With development, children become increasingly sensitive to valued social categories, including age, gender or race. Exceptions to these rules suggest the importance of considering ways in which biological (e.g., temperament, shyness), organismic (e.g., age, sex), and sociocultural variables (e.g., gender, race) coact in the formation of children's self-representations.

The Construction of Self-Portraits in Adolescence

A series of studies have been performed assessing the development of adolescent self-conceptions within interpersonal relationships. Harter and Monsour (1992) asked adolescents to describe what they were like with each of four role figures, including "parents," "friends," "in the classroom," and "in romantic relationships." After listing positive and negative aspects of their selves-in-relationships, experimenters recorded self-descriptions on gummed labels, and participants arranged them on a self-portrait of three concentric circles. The inner, middle, and outer circles represented the most, less, and least important aspects of the self, respectively. Experimenters also asked participants to note self-attributes that were opposites or that clashed. Results indicated that representations of self became more differentiated with age, as indicated by decreased attribute overlap among respective roles. In addition, results suggested a strong positivity bias, where participants placed positive self-descriptions in the center and negative self-descriptions in the periphery of their self-portraits. Finally, conflicts among self-descriptions were low early in adolescence, peaked in middle adolescence, and declined thereafter. To explain this pattern, skill theory suggests that young adolescents can construct but not coordinate single self-abstractions. Conflict emerged with the capacity to relate but not resolve clashing self-abstractions in midadolescence. The decline in conflict in later adolescence marks increased skill in coordinating conflicting self-abstractions.

Several studies have been performed based on adaptations of these procedures. Calverley, Fischer, and Ayoub (1994) reported a study assessing the self-portraits of a sample of maltreated and depressed adolescent girls. Participants included 16 depressed, inpatient adolescent girls, 7 of whom had been victims of prolonged sexual abuse, and 9 who were nonabused. Results indicated that the abused girls not only produced a large number of overall negative self-characteristics, but also defined their core selves using negative characteristics. Nonabused girls produced fewer negative self-descriptions, and generally regarded them as peripheral to their self-portrait. These findings are in direct opposition to the positivity bias that Harter and Monsour (1992) reported for normative samples of adolescents. Beyond these findings, the developmental level of participants' self-descriptions were assessed using a series of questions designed specifically to assess descriptions of self-attributes at each of the four developmental levels that occur most commonly during adolescence and early adulthood, including representational systems, single abstractions, abstract mappings, and abstract systems. Results indicated that the abused group of girls showed the same developmental level of self-description as the nonabused girls. This finding contradicts the traditional view that maltreatment and psychopathology produce developmental fixation or regression.

Fischer and Kennedy (1997) assessed the development of self-in-relationships with a group of Korean adolescents. Many scholars have suggested that members of many Asian cultures do not construct a concept of self that is comparable to that in the West (Lebra, 1983; Markus & Kitayama, 1991; Triandis, 1989). Western assessments of self-concept in Asian cultures often yield self-descriptions judged to be simple or undeveloped. However, most Western tests of self-understanding employ low-support, open-ended questionnaires that assess the functional level of an individual's self-understanding. Fischer and Kennedy (1997) assessed self-understanding in Korean adolescents between the ages of 14 and 20 in both high- and low-support contexts. Under low support, experimenters simply asked participants to describe themselves in an open-ended format. The high-support condition employed the adaptation of Harter and Monsour's (1992) self-portrait task (discussed earlier).

Results indicated that in the high-support condition, self-understanding in Korean adolescents was highly scalable and developed through a series of three levels comparable to those in Western samples. The optimal level of adolescents between the ages of 14 and 17 was single abstractions; for 17- to 18-year-olds, it was abstract mappings; for 19- to 20-year-olds, self-understanding approached the level of abstract systems. Self-descriptions were much less sophisticated in the low-support condition. Korean adolescents produced extremely simple self-descriptions that never rose above the level of single abstractions. In addition, like North American samples, the growth functions for the high- and low-support conditions were different. Whereas development in the low-support condition showed monotonic change, performance in the high-support condition exhibited discontinuous change with stage-like spurts in development. Despite these common patterns, the ages of the developmental spurts in the high-support condition were delayed by a year or two in comparison to American adolescents (Harter & Monsour, 1992; Kennedy, 1991, 1994). Thus, consistent with results of many Western studies, Korean adolescents produce extremely low levels of self-understanding when assessed under low-support interview conditions. However, contrary to such studies, under high-support conditions, the level of self-description produced by Korean adolescents approaches levels found in Western samples. Thus, although Korean adolescents may not prefer to speak of self in conversation, they are nonetheless capable of producing richly textured representations of self.

Overall, the results of the use of the self-portrait task among adolescents indicate the role of skill level, social relationships, affect, and cultural context in the development and organization of self-representations. With development, adolescent self-portraits become increasingly differentiated and hierarchically integrated. Although adolescents see themselves as different when they are with different people, they are increasingly able to coordinate conflicting representations of self with development. Social context exerts an important influence on self-understanding throughout adolescence. Interactions involving maltreatment lead to affectively based polarization in the ways in which adolescent girls see themselves in different relationships, even though abuse does not lead to retardation in the representation of self. Similarly, cultural context is an important organizer of the representation of self. Because of cultural values against a focus on the self, when asked to reflect upon themselves in low-support contexts, Korean adolescents exhibit less sophisticated self-descriptions that their North American counterparts. However, this does not mean that Korean adolescents cannot construct sophisticated self-representations, as such cultural differences are greatly attenuated in assessment conditions that prompt optimal-level performance.

The Development of Dissociation and Affective Splitting in Maltreated Children

An important principle of dynamic skills theory is that the mind is naturally fractionated in terms of weakly related control systems that develop in the direction of increased coordination. However, severe maltreatment and hidden family violence precipitate more severe forms of active dissociation, such as multiple personality disorder (MPD; Fischer & Ayoub, 1994, 1996). According to dynamic skills theory, active dissociation that occurs in MPD and hidden family violence does not simply consist of the population of the person with distinct and autonomous personalities. Although differences between agencies and personalities in MPD are dramatic, often involving different biological markers such physiological responsiveness and handedness, multiple personalities develop as a result of a history of active dissociation within trauma-inducing contexts. Although the personalities are often remarkably distinct, they are not completely separate. Different personalities serve adaptive functions for the survival and emotional well-being of the person as a whole. As such, their activity is coordinated through relations to the larger person-as-system. Thus, although there is no single executive or controller regulating the conduct of the organism's multiple selves, MPD is a product of a single person-as-self-organizing *system*, which constructs multiple agentive selves and personalities in order to cope with abuse-related trauma. In development, the dissociation of different agencies and personalities emerges and undergoes developmental change as a result of transformations in the person-as-system's capacity to dissociatively coordinate activity in terms of increasingly hierarchized control systems.

The natural fractionation of mind occurs as a result of a variety of factors. For example, in the development of children's representations of nice and mean interactions, fractionation occurs as a result of developmental level (children under age 4 are generally cannot coordinate multiple nice or mean representations into a single skill), domain (the development of numerical skills are not necessarily tied to social skills), context (nice and mean interactions often occur in different contexts), and emotional valence (nice and mean interactions are structured around differently valenced emotional scripts). Indeed, just as James (1890) spoke of the partial dissociation of selves in normal individuals, we suggest that a certain degree of fractionation or dissociation of self is typical in otherwise healthy persons. In the study of psychopathology, however, clinicians speak of more active and motivated forms of fractionation, including *dissociation* and *affective splitting*. In dissociation, elements are separated even though they should be coordinated according to some external standard or criterion. Affective splitting refers to separation along evaluative dimensions such as positive–negative, or more generally, separation between opposites. From the perspective of dynamic skills theory, *active* dissociation is a dynamic skill. When active dissociation occurs, *a person must create a coordination in order to actively keep elements apart;* that is, a child must coordinate several skills in order to actively keep two or more elements separate. Thus, as a skill requiring coordination, active dissociation undergoes development with transformations in children's control systems.

The emotional trauma that accompanies maltreatment is an important source of active and passive dissociation. For example, using the self-portrait task (described earlier) Calverley et al. (1994) described a form of complex dissociative coordination called "polarized affective splitting" in a group of adolescent girls who suffered prolonged sexual maltreatment. In their self-portraits, several maltreated girls produced clashing self-descriptions dichotomized in terms of a strongly evaluative positive–negative dimension. Despite their intense emotional content, the girls made no attempt to coordinate or reconcile their contrasting assessments. Several abused girls described their relationships with their mothers and fathers in polar-opposite terms, with the mother relegated to the negative pole and the father to the positive. This occurred even when the father was the perpetrator of abuse.

A much more extreme form of coordinative dissociation occurs in MPD, in which people develop coconscious agencies, called "personalities," which are organized around different affective themes. MPD often follows as the result of prolonged and intense maltreatment. The disorder is characterized by different personalities with distinct profiles organized around different emotional themes. However, despite their distinctness, the personalities are not entirely separate. They are related in at least three ways. First, distinct agencies are actively dissociated in the sense that they must be kept apart from one another. Second, the person can switch from one agency to another, which requires skills in coordinating distinct coconsciousnesses. Third, coconscious agencies can, within limits, act upon each other. As such, coconscious dissociation is a skill; it requires an active attempt to separate personalities and keep them apart. Like all skills, coconscious dissociation develops with changes in relevant control systems. We propose two tiers in the development of coconscious dissociation in MPD, with a series of levels within each tier. The first tier is made possible by the emergence of representations and is characterized by the development of *coconscious agencies* that control concrete actions and characteristics. The second tier develops with the capacity to construct abstractions during adolescence and is characterized by *coconscious personalities* that control intangible motives and personality traits. In terms of a skills theory, genuine personalities emerge around adolescence, although earlier forms of coconscious agencies develop earlier. The limited evidence that exists suggests that early forms of MPD begin to emerge around 4 years (Bliss, 1980; Fagan & McMahon, 1984; Kluft, 1985).

To illustrate developmental changes in dissociative skill in MPD, consider the case of Carrie Smith, who was the victim of neglect and severe physical and sexual abuse from her mother and the two "fathers" who lived with her (Fischer & Ayoub, 1994). Beginning around age 4, Carrie began to switch her behavior and mood in ways that exhibited consistent patterns of emotion, food preference, and activity, seemingly indicating separate agencies. She began to call herself different names, including Carla (her mother's name) and Carrie. Carla was angry and seductive, and exhibited verbally abusive as well as sexual behavior. Carrie was good and obedient, and cared about babies and children living in foster homes. We will refer to the host child as "Carrie Smith" and to her coconscious agencies as "Carla" and "Carrie."

Table 12.6 displays suggested developmental changes in dissociative skills based on a skills theory analysis of changes in Carrie Smith's dissociative behavior over time. Each step specifies a different level of skill in the child's capacity to coordinate agencies or personalities, as well as to coor-

TABLE 12.6. Levels of Development of Dissociative Coordination in Multiple Personality

Level	Dissociative skills[a]		Examples of dissociative and nondissociative skills	Age[b]
	Representational tier	Abstract tier		
Sm4/Rp1: Single representations[c]	$\begin{bmatrix} ME\text{-}CARLA \\ {}_{MEAN} \end{bmatrix}$ ✦ $\begin{bmatrix} ME\text{-}CARRIE \\ {}_{NICE} \end{bmatrix}$		Single concrete representations of people as agents: When child is expected to be nice and she thinks of being mean, she shifts abruptly to a nice representation to avoid being mean [transition to active dissociation]. Child pretends that she is her mother Carla hitting someone and saying mean things, and then she shifts to pretending that she is Carrie hugging a baby and saying nice things [not dissociation].	18–24 mo
Rp2: Representational mappings	$\begin{bmatrix} ME\text{-}CARLA \\ {}_{MEAN} \end{bmatrix} \blacksquare\!\!\longrightarrow \begin{bmatrix} ME\text{-}CARRIE \\ {}_{NICE} \end{bmatrix}$		Simple relations of representations: When she feels threatened, child switches from agent nice Carrie to agent mean Carla, and she switches back when she feels comfortable again [active dissociation]. As Carla, child acts mean in response to actions by someone else [not dissociation].	3.5–4.5 yr
Rp3: Representational systems	$\begin{bmatrix} ME\text{-}CARLA \overset{SEDUCTIVE}{\underset{MEAN}{\longleftrightarrow}} \overset{CAREGIVING}{ME\text{-}CARRIE} \\ {}_{NICE} \end{bmatrix}$		Complex relations of subsets of representations Child switches from agent Carrie to agent Carla, with each agent having several key concrete characteristics that determine when switching will occur [active dissociation]. As Carla, child acts both tough (mean) and seductive with a man, whom she expects to be tough and sexual in return [not dissociation].	6–7

(continues)

369

TABLE 12.6. *Continued*

Level	Dissociative skills[a]		Examples of dissociative and nondissociative skills	Age[b]
	Representational tier	Abstract tier		
Rp4/Ab1: Single abstractions	``` SEDUCTIVE CAREGIVING ⎡ ME-CARLA ←■→ ME-CARRIE MEAN NICE ⇕ SEDUCTIVE CHILDREN ME-CARLA ←■→ ME-MOM ⎤ MEAN ADVICE ```	≡ [CARLA CONTROL]	Relations of representational systems to produce abstractions (intangible concepts) about people (personalities): For one agent (Carla), person coordinates several specific relations with other agents (Carrie and Mom) in terms of an intangible characteristic or motive, thus forming a coconscious personality (Carla in Control) with that characteristic [active dissociation]. There is not yet coordination of personalities, but only of concrete agents. Within each agent, person develops abstract personality concepts and motives, thus beginning to form separate personalities instead of mere agents [not dissociation].	10–12
Ab2: Abstract mappings		``` parenting ⎡ CARRIE ■ MOM ⎤ NEEDY NURTURANT ```	Simple relations of abstractions: Person relates two personalities, not just concrete agents, so that one personality, such as Carrie, can influence the activity or experience of another personality, such as Mom [active dissociation].	14–16

Within each personality, person relates several abstract characteristics, such as Mom's nurturance with children and her providing wise advice about children [not dissociation].

Complex relations of subsets of abstractions: Person relates two personalities so that one can influence the activity or experience of a second along several dimensions simultaneously [active dissociation].

18–20

Ab3: Abstract systems

Within each personality, person relates set of personal characteristics along multiple dimensions, such as several types of abstinence (from both sex and drugs) with several types of personal moral or religious concepts (such as honoring emotional commitment and bodily integrity) [not dissociation].

General principles for integrating systems of abstractions:

Dissociative strategies can be remarkably sophisticated, with switches and influences between personalities being determined by sublte relations reflective of some system-wide organization [active dissociation], such as relations between subordinate and dominant personalities (Carla and Mom in contrast to Carrie).

23–25

Ab4: Systems of abstract systems: Principles

(continues)

371

TABLE 12.6. *Continued*

| Level | Dissociative skills[a] | | Examples of dissociative and nondissociative skills | Age[b] |
	Representational tier	Abstract tier		
Ab4: Systems of abstract systems: Principles (*cont.*)			Within each personality, person can understand general principles, such as a moral principle of treating others with compassion and justice, integrating relations among, for example, commitment, integrity, and dominance [not dissociation].	

Note. In skill structures, each large upper-case word designates a main component (set) and each small upper-case word (subscript or superscript) designates a subset of the main component. Each lower-case word names the relationship in a skill. Italics designate representations, and outline letters designate abstractions. Lines connecting sets designate relations forming a mapping, single-line arrows designate relations forming a system, and double-line arrows designate relations forming a system of systems. An arrowhead without a line (>) designates a shift of focus between two skills. A dark rectangle in the middle of a line or arrow indicates that the relation is dissociative. Adapted from Fischer and Ayoub (1996). Copyright 1996 by Lawrence Erlbaum. Adapted by permission.

[a]The skill structures in the diagrams show relatively simple skills for a given level and are described in the first example, which in each case involves active dissociation. The second example is a more ordinary skill at this level and therefore fits a typical skill structure for that level, *not* the dissociative structure in the diagram.

[b]Numbers given provide estimates of modal ages at which a level first appears based on cognitive developmental research (Fischer et al., 1993). They may differ across cultures and other social groups.

[c]Abstractions grow from coordinations of representational systems, but for simplicity that relation has been omitted from this table.

dinate dissociative relationships between at least two such agentive elements. At the level of single representations, because a child is only capable of controlling a single representation at time, the child is limited in her ability to coordinate relations. At best, the child can manage to shift focus from one representation to another as a defense when she is frightened in some way. This is represented in Level Rp1 in Table 12.6, in which Carrie Smith shifts from performing concrete mean actions as Carla to performing concrete nice actions as Carrie. In Table 12.6, shifts are represented using a shift sign, mappings and systems by connecting lines, and coordinative dissociations by the black bar. It is not until representational mappings beginning around age 4 (Rp 2) that the child constructs the first genuine dissociative coordinations. At this step, the child can control whether she is Carrie or Carla, switching to whichever agency is most affectively needed at the time. As such, she can use her mapping skill to keep the two agencies dissociated. Using representational systems at around age 6 or 7 (Rp 3), the separation and invocation of separate agencies is more elaborated. A simultaneously seductive and mean Carla can be actively dissociated from a pleasant and nurturant Carrie.

The second tier of active dissociation begins with the emergence of single abstractions around 10–11 years of age. At this point, the individual can coordinate two or more representational systems to produce a new agency, a coconscious "personality." Carrie Smith constructs such an abstract skill for the coconscious personality Carla, whose abstract-motive structure was defined in terms of controlling others (Level Rp4/A1 in Table 12.6). Specifically, Carla operated by controlling two other coconscious agencies, Carrie and "Mom." Mom was a newly constructed coconscious agency who was able to deal well with children and provide advice about them. The abstract personality, Carla, emerged through the coordination of representational systems for controlling and switching between Carla and Carrie, with a second system for controlling and switching between Carla and Mom. The resulting coordination produced a skill that enabled Carrie Smith to either control or switch from Carla to the nurturant Carrie and/or to the parent Mom. When Carla needed a nice, caregiving person, she could call upon Carrie; when she needed someone to interact with children, she called upon Mom. As such, Carla became transformed from a coconscious agent to a coconscious personality with a specific abstract-motive structure involving control of others.

With the use of single abstractions, although the adolescent is capable of constructing single personalities, she cannot yet control relations among personalities except by reducing them to the level of concrete agents and shifting from one complex agency to another. With the emergence of abstract mappings around age 15 (A2 in Table 12.6), the adolescent can finally begin to control relations among personalities. She can thus begin to dissociatively coordinate one coconscious personality with another. As such, one dissociative personality can evaluate the activity or experience of another and even intervene in the other's activity or thought. An example of such interplay is described in the case of *The Three Faces of Eve,* when Eve Black intervened in Eve White's activity while Eve White had control of the body. Thus, despite a long history of earlier forms of switching among coconscious agencies, it is not until the level of abstract mappings that the adolescent can command sufficient skill in dissociatively coordinating multiple personalities. For this reason, one might regard abstract mappings as the first full step in the onset of dissociative coordination in MPD. From this point onward, skills to produce active dissociations continue to develop, eventually producing abstract systems and principles such as those described in Table 12.6. (See Fischer & Ayoub, 1994, 1995, for a discussion of similar developmental trajectories in *isolating dissociation,* which is typically found in instances of hidden family violence.)

MPD is an extreme examples of the fractionation of the person into a series of agentive control systems. Although the coconscious personalities that form in MPD and hidden family violence are often quite distinct, they are not completely separate. The different personalities are actively dissociated as a result of abuse-related trauma and serve different functions for the emotional well-being of the overall person-as-system. Each personality can be seen as a different, constructed self-as-agent that possesses not only self-awareness but also limited awareness of other coconscious agencies and personalities. As such, MPD and isolating dissociation are not products of the peopling of the mind with autonomous selves. Rather, to the extent that they are actively kept apart, dissociated selves are products of the mutual regulation of multiple control systems that make up the overall person-as-system.

SYSTEMS THEORY AS A FRAMEWORK FOR RESOLVING COMPETING CLAIMS ABOUT THE NATURE OF SELF

Dynamic skills theory provides a framework for understanding the many uses of the concept of self. From the standpoint of skills theory, the assertion that persons act as agents implies the operation of a series of weakly connected control systems within various psychological domains and sociocultural contexts. Self-awareness refers to the capacity of control systems to focus conscious attention on their own processes and attributes. Self-awareness undergoes developmental transformation as control systems move from sensory–motor actions to representations through higher-order abstractions. In this way, self-awareness moves from an initially implicit and sensory–motor consciousness of one's own processes to increasingly reflective and coordinated representations and abstractions about one's agencies and identities.

Skills theory provides a framework for resolving tensions that exist between Cartesian and dialogical approaches to self and mind. First, traditional approaches depict the self as unitary, whereas dialogical approaches hold that the self consists of a multiplicity of semiautonomous "I" positions within a sociomoral space. Existing evidence suggests that there is both fractionation and coordination in the operation and representation of self. Such evidence takes the form of unevenness and domain specificity in cognitive development, decalage in the development of self- and other-knowledge in infancy, the increasing differentiation and coordination of self-representations within domains and across relationships, as well as the use of active coordinative dissociation in the development of MPD and other forms of psychopathology. Using dynamic skills theory, we conclude that the person functions as a *system of fractionated control systems that operate within specified contexts and domains but develop in the direction of integration. Because they are part of a larger organismic system, despite their fractionation, control systems nevertheless function in the service of the organism as a whole as it operates within sociocultural contexts.* Thus, despite the natural fractionation of control systems as they function in different contexts and domains, some degree of unity is possible as subsystems come together in development and function in the service of adaptive goals of the organism as a whole.

Second, traditional Western approaches view the mind and self-awareness as a property of the bounded individual. In rejecting this view, proponents of dialogical views generally endorse the idea that "the mind extends beyond the skin" (Bateson, 1971; Resnick, 1991) to include the activity of others as well as the objects of thought and action. A systems view maintains that the organism–environment system is multileveled, and that activity and self-representations are emergent products of simultaneous coactions within and between all levels of functioning, including bio–genetic, organismic–agentive, social–dyadic and cultural–linguistic levels of functioning. Thus, *although self and other (and other levels of the organism–environment system) are inseparable as causal factors in producing self-relevant activity and experience, the individual functions as a distinct organismic system that contributes to its own development and self-representation.* For example, although social interaction has the effect of raising the performance of partners to levels that they could not achieve alone, children develop from such interactions as they perform actions that coordinate for themselves what has been coordinated between themselves and others in social interaction. Thus, although self and other are inseparable as causal factors in the development of self, the functioning of the organismic system performs an important organizing role in development.

Third, the traditional Cartesian approach depicts the mind as disembodied and thus capable of transcending the limitations of body, time, space, and context. The self is capable of adopting an abstract stance toward the world, viewing the world from nowhere in particular. From a systems view, *a direct implication of the principle that development occurs as a result of mutual regulation among all levels of the organism–environment system is that the developing self is both embodied and contextualized*; that is, the sense of self reflects the awareness that one experiences the world from the location of one's body. Furthermore, the functioning of the body and its bio–genetic subsystems directly influences the developing sense of self. This notion finds support in evidence reviewed earlier that bodily processes, such as the temperament of behaviorally inhibited children or the experience of strong affect in abusive relationships, can organize the development of self-representations. Furthermore, the sense of self is contextualized in that self-awareness always emerges within a particular physical, social, cultural, and historical context. This assertion gains support by the results of studies, reviewed earlier, that indicate the importance of social context, interpersonal relationship, gender, race, and culture in the creating of self-representations.

Even though representations of self are grounded within specific bodies, relationships, and contexts, with development, adolescents are capable of coordinating self-representations across interactions, relationships, contexts, domains, and in some cases, even cultures. As such, through coordination, individuals can use abstractions to generalize their experience across multiple concrete particularities. Nevertheless, despite this capacity, self-related abstractions emerge as a result of a history of coordinations of concrete activities that occur within a particular body that functions within particular contexts, social relationships, and behavioral domains. As such, despite their ability to take an abstract stance through coordination, adolescents and adults never adopt a decontextualized "view from nowhere" concerning self or any other domain of experience. Abstract self-representations are constructed and grounded within particular bodies, social contexts, and personal histories.

Dynamic skills theory can help resolve tensions among traditional and dialogical versions of self by depicting the individual as a hierarchical, self-organized system that coacts with other self-organizing systems within larger sociocultural systems and contexts. By treating individuals and selves as integral systems with permeable boundaries, a systems approach offers a synthesis that can explain how the self is fractionated and capable of integration, individual and social, contextualized yet capable of forming grounded abstractions of its own functioning and identity. Such a self is neither an autonomous center of control nor a decentralized product of social forces. The self consists of a system of control structures that function as but one level in a richly interactive organism–environment system.

ACKNOWLEDGMENT

Work on this chapter was supported by a grant from Merrimack College to Michael F. Mascolo, and from Mr. and Mrs. Frederick Rose and Harvard University to Kurt W. Fischer.

NOTE

1. The concept of negative feedback is meant to provide a metaphor to describing the goal-directed nature of human activity and the processes by which humans attempt to reduce the distance between events as perceived events and goal states. Negative feedback, of course, consists of but one process by which control systems operate. An important aspect of persons as self-organizing systems is that they are themselves composed of multiple lower-order systems. Such systems mutually regulate each other within any given context. Component systems mutually regulate each other through positive as well as negative feedback. Whereas negative feedback functions to reduce discrepancies between sensory events and higher-order goals, positive feedback acts to increase activation of interacting systems. For example, the evolution of emotional processes within a given context often involves mutual regulation through positive feedback (Lewis, 1995, 1996, 1998; Mascolo & Harkins, 1998). For example, in hurrying home after a hard day, one might appraise a traffic jam at the tolls as a negative, motive-inconsistent event. This appraisal might then evoke physiological changes and concomitant negative affect. Through a positive feedback loop between appraisal and affect, negative affect could then modulate the ongoing event appraisal, perhaps prompting the individual to blame the toll takers for working too slowly. Such an anger-relevant appraisal could then amplify one's negative affect, and the process would continue. Thus, both positive and negative feedback are operative in the mutual regulation of self-organizing component systems.

REFERENCES

Ackerman, B. P., Abe, J. A., & Izard, C. E. (1998). Differential emotions theory and emotional development: Mindful of modularity. In M. F. Mascolo & S. Griffin (Eds.), *What develops in emotional development?* (pp. 85–106). New York: Plenum.

Bateson, G. (1971). *Steps to an ecology of mind.* New York: Ballantine.

Bertenthal, B. L., & Fischer, K. W. (1978). Development of self-recognition in the infant. *Developmental Psychology, 14,* 44–50.

Bidell, T., & Fischer, K. W. (1992). Beyond the stage debate: Action, structure, and variability in Piagetian theory and research. In R. Sternberg & C. Berg (Eds.), *Intellectual development* (pp. 100–140). New York: Cambridge University Press.

Bidell, T., & Fischer, K. W. (1996). Between nature and nurture: The role of human agency in the epigenesis of intelligence. In R. Sternberg & E. Grigorenko (Eds.), *Intelligence: Heredity and environment* (pp. 193–242). New York: Cambridge University Press.

Biggs, J., & Collis, K. (1982). *Evaluating the quality of learning: The SOLO taxonomy structure of the observed learning outcome.* New York: Academic Press.

Bliss, E. L. (1980). Multiple personalities. *Archives of General Psychiatry, 37,* 1388–1397.

Brown, T. (1994). Affective dimensions of meaning. In W. F. Overton & D. S. Palermo (Eds.), *The nature and ontogenesis of meaning* (pp. 167–190). Hillsdale, NJ: Erlbaum.

Bullock, M., & Lutkenhaus, P. (1988). The development of volitional behavior in the toddler years. *Child Development, 59,* 664–674.

Calverley, R. M., Fischer, K. W., & Ayoub, C. (1994). Complex splitting of self-representations in sexually abused adolescent girls. *Development and Psychopathology, 6,* 195–213.

Carver, C. S., & Scheier, M. F. (1981). *Attention and self-regulation: A control-theory approach to human behavior.* New York: Spinger-Verlag.

Carver, C. S., & Scheier, M. F. (1990). Origins and functions of positive and negative affect: A control process view. *Psychological Review, 97,* 19–35.

Chomsky, N. A. (1966). *Cartesian linguistics.* New York: Harper & Row.

Clark, K., & Holquist, M. (1984). *Mikhail Bakhtin.* Cambridge, MA: Harvard University Press.

Connell, J. P. (1991). Context, self, and action: A motivational analysis of self-system process across the life span. In D. Cicchetti & M. Beeghly (Eds.), *The self in transition: Infancy to childhood* (pp. 61–97). Chicago: University of Chicago Press.

Corrigan, R. (1981). The effects of task and practice on search for invisibly displaced objects. *Developmental Review, 1,* 1–17.

Damon, W., & Hart, D. (1982). The development of self-understanding from infancy through adolescence. *Child Development, 53,* 841–864.

Damon, W., & Hart, D. (1986). Stability and change in children's self-understanding. *Social Cognition, 4,* 102–118.

Damon, W., & Hart, D. (1988). *Self-understanding in childhood and adolescence.* New York: Cambridge University Press.

Davies, B., & Harré, R. (1990). Positioning: The discursive production of selves. *Journal for the Theory of Social Behavior, 20,* 43–63.

de Rivera, J. (1981). The structure of anger. In J. H. de Rivera (Ed.), *Conceptual encounter* (pp. 35–82). Washington, DC: University Press of America.

Descartes, R. (1994). *Discours de la method/Discourse on the method* (G. Hefferman, Trans.). Notre Dame, IN: University of Notre Dame Press. (Original work published 1637)

Edwards, C. P. (1984). The age group labels and categories of preschool children. *Child Development, 55*, 440–452.

Elmendorf, D. (1992). *Preschool children's interpretations of the intentions behind physically harmful acts: Developmental changes in understanding, misunderstanding, and distortion.* Unpublished doctoral dissertation, University of Denver, Denver, CO.

Emde, R. (1980). The prerepresentational self and its affective core. *Psychoanalytic Study of the Child, 38*, 165–192.

Fagan, J., & McMahon, P. P. (1984). Incipient multiple personality in children: Four cases. *Journal of Nervous and Mental Disease, 172*, 26–36.

Fischer, K. W. (1980). A theory of cognitive development: The control and construction of hierarchies of skills. *Psychological Review, 87*, 447–531.

Fischer, K. W. (1987). Relations between brain and cognitive development. *Child Development, 57*, 623–632.

Fischer, K. W., & Ayoub, C. (1994). Affective splitting and dissociation in normal and maltreated children: Developmental pathways for self in relationships. In D. Cicchetti & S. L. Toth (Eds.), *Rochester Symposium on Development and Psychopathology: Vol. 5. Disorders and dysfunctions of the self* (pp. 149–222). Rochester, NY: University of Rochester Press.

Fischer, K. W., & Ayoub, C. (1996). Analyzing development of working models of close relationships: Illustration with a case of vulnerability and violence. In G. G. Noam & K. W. Fischer (Eds.), *Development and vulnerability in close relationships* (pp. 173–199). Hillsdale, NJ: Erlbaum.

Fischer, K. W., Bullock, D. H., Rotenberg, E. J., & Raya, P. (1993). The dynamics of competence: How context contributes directly to skill. In R. Wozniak & K. W. Fischer (Eds.), *Development in context: Acting and thinking in specific environments* (pp. 93–117). Hillsdale, NJ: Erlbaum.

Fischer, K. W., Hand, H. H., Watson, M. W., Van Parys, M., & Tucker, J. (1984). Putting the child into socialization: The development of social categories in preschool children. In L. Katz (Ed.), *Current topics in early childhood education* (Vol 5, pp. 27–72). Norwood, NJ: Ablex.

Fischer, K. W., & Hogan, A. E. (1989). The big picture in infant development: Levels and variations. In J. Lockman & N. Hazan (Eds.), *Action in social context: Perspectives on early development* (pp. 275–305). New York: Plenum.

Fischer, K. W., & Jennings, S. (1981). The emergence of representation in search: Understanding the hider as an independent agent. *Developmental Review, 1*, 18–30.

Fischer, K. W., & Kennedy, B. (1997). Tools for analyzing the many shapes of development: The case of self-in-relationships in Korea. In F. Amsel & K. A. Renninger (Eds.), *Change and development: Issues of theory, method, and application* (pp. 117–152). Mahwah, NJ: Erlbaum.

Fischer, K. W., Knight, C. C., & Van Parys, M. (1993). Analyzing diversity in developmental pathways: Methods and concepts. In W. Edelstein & R. Case (Eds.), *Contributions to human development: Vol. 23. Constructivist approaches to development* (pp. 33–56). Basel, Switzerland: S. Karger.

Fischer, K. W., & Rose, S. P. (1994). Dynamic development of coordination of components in brain and behavior: A framework for theory and research. In G. Dawson & K. W. Fischer (Eds.), *Human behavior and the developing brain* (pp. 3–66). New York: Guilford Press.

Fischer, K. W., Shaver, P. R, & Carnochan, P. (1990). How emotions develop and how they organize development. *Cognition and Emotion, 4,* 81–128.

Flavell, J. H. (1982). On cognitive development. *Child Development, 53,* 1–10.

Fogel, A. (1993). *Development through relationships: Origins of communication, self and culture.* Chicago, IL: University of Chicago Press.

Freud, S. (1940). *An outline of psychoanalysis* (J. Strachey, trans.). New York: Norton.

Gelman, R., & Baillergeon, R. (1983). A review of some Piagetian concepts. In P. H. Mussen (Ed.), *Handbook of child psychology* (4th ed.): *Cognitive development* (Vol. III, pp. 167–230). New York: Wiley.

Geppert, U., & Kuster, U. (1983). The emergence of "Wanting to do it oneself": A precursor to achievement motivation. *International Journal of Behavioral Development, 3,* 355–369.

Gergen, K. J. (1985a). Social pragmatics and the origins of psychological discourse. In K. J. Gergen & K. E. Davis (Eds.), *The social construction of the person* (pp. 111–127). New York: Springer-Verlag.

Gergen, K. J. (1985b). The social constructionist movement in modern psychology. *American Psychologist, 40,* 266–275.

Gergen, K. J. (1987). Toward self as relationship. In K. Yardley & T. Honess (Eds.), *Self and identity: Psychosocial perspectives* (pp. 53–63). New York: Wiley.

Gergen, K. J. (1989). Social psychology and the wrong revolution. *European Journal of Social Psychology, 19,* 463–484.

Gottlieb, G. (1991a). Experiential canalization of behavioral development: Theory. *Developmental Psychology, 27,* 4–13.

Gottlieb, G. (1991a). Experiential canalization of behavioral development: Results. *Developmental Psychology, 27,* 35–39.

Gottlieb, G. (1992). *Individual development and evolution: The genesis of novel behavior.* New York: Oxford University Press.

Hand, H. H., & Fischer, K. W. (1989). *Development of understanding of mean and nice and other opposites in social interaction.* Cognitive Developmental Laboratory Report. Cambridge, MA: Harvard University Press.

Harré, R. (1983). *Personal being: A theory for individual psychology.* Oxford: Blackwell.

Harré, R., & Gillett, G. (1994). *The discursive mind.* Thousand Oaks, CA: Sage.

Hart, D., & Damon, W. (1986). Developmental trends in self-understanding. *Social Cognition, 4,* 388–407.

Harter, S. (1983). Developmental perspectives on the self-system. In P. Mussen (Series Ed.), & E. M. Hetherington (Vol. Ed.), *Handbook of child psychology: Vol. IV. Socialization, personality, and social development* (pp. 276–385). New York: Wiley.

Harter, S., & Monsour, A. (1992). Developmental analysis of conflict caused by opposing attributes in the adolescent self-portrait. *Developmental Psychology, 28,* 251–260.

Heckhausen, H. (1984). Emergent achievement behavior: Some early developments. In J. Nicholls (Ed.), *Advances in motivation and achievement: Vol. 3. The development of achievement motivation* (pp. 1–32). Greenwich, CT: JAI Press.

Heckhausen, H. (1987). Emotional components of action: Their ontogeny as reflected in achievement behavior. In D. Gorlitz & J. F. Wohlwill (Eds.), *Curiosity, imagination and play: On the development of spontaneous cognitive and motivational processes* (pp. 326–348). Hillsdale, NJ: Erlbaum.

Heckhausen, J. (1988). Becoming aware of one's competence in the second year: Developmental progression within the mother–child dyad. *International Journal of Behavioral Development, 3*, 305–326.

Hencke, R. (1991, April). *Who me? Individual differences in preschool children's understandings of nice and mean attributes in the self.* Paper presented at the meeting of the Society for Research on Child Development, Seattle, WA.

Hencke, R., & Raya, P. (1993, March). *You're mean! Individual differences between three-year-old boys and girls in narratives about nice and mean social interactions.* Paper presented at the meeting of the Society for Research in Child Development, New Orleans, LA.

Hermans, H. J. M. (1996). Voicing the self: From information processing to dialogical interchange. *Psychological Bulletin, 119*, 31–50.

Hermans, H. J. M., Kempen, H. J. G., & van Loon, R. J. P. (1992). The dialogical self: Beyond individualism and rationalism. *American Psychologist, 47,* 23–33.

Ho, M.-W., & Saunders, P. T. (Eds.), *Beyond neo-Darwinism: An introduction to the new evolutionary paradigm* (pp. 267–289). San Diego, CA: Academic Press.

Holquist, M. (Ed.). (1981). *The dialogic imagination. Four essays by M. M. Bakhtin* (C. Emerson & M. Holquist, Trans.). Austin: University of Texas Press.

Izard, C. E., & Malatesta, C. Z. (1987). Perspectives on emotional development I: Differential emotions theory of early emotional development. In J. Osofsky (Ed.), *Handbook of infant development* (2nd ed., pp. 495–554). New York: Wiley.

James, W. (1890). *The principles of psychology* (Vol. 1). New York: Dover.

Kagan, J. (1981). *The second year: The emergence of self-awareness.* Cambridge, MA : Harvard University Press.

Kagan, J., Reznick, J. S., & Snidman, N. (1987). The physiology and psychology of behavioral inhibition in children. *Child development, 58*, 1459–1473.

Kennedy, B. (1991). *The development of self-understanding in adolescence.* Unpublished qualifying paper, Harvard University, Cambridge, MA.

Kennedy, B. (1994). *The development of self-understanding in adolescents in Korea.* Unpublished doctoral dissertation, Harvard University, Cambridge, MA.

Kluft, R. P. (Ed.). (1985). *Childhood antecedents of multiple personality.* Washington, DC: American Psychiatric Press.

Kohlberg, L. (1969). Stage and sequence: A cognitive developmental approach to socialization. In D. A. Goslin (Ed.), *Handbook of socialization theory and research* (pp. 347–480). Chicago: Rand McNally.

Kohlberg, L. (1976). Moral stages and moralization: The cognitive-developmental approach. In T. Lickona (Ed.), *Moral development and behavior* (pp. 31–53). New York: Holt, Rinehart & Winston.

Kohlberg, L. (1981). *The philosophy of moral development* (Vol. 1). San Francisco: Harper & Row.

Lebra, T. S. (1983). Shame and guilt: A psychocultural view of Japanese self. *Ethos, 11,* 192–209.

Lewis, M. D. (1995). Cognition–emotion feedback and the self-organization of developmental paths. *Human Development, 38,* 71–102.

Lewis, M. D. (1996). Self-organizing cognitive appraisals. *Cognition and Emotion, 10,* 1–25.

Lewis, M. D. (in press). Personality self-organization: Cascading constraints on cognitive-emotion interaction. In A. Fogel, M. C. Lyra, & J. Valsiner (Eds.), *Dynamics and indeterminism in developmental and social processes* (pp. 193–215). Mahwah, NJ: Erlbaum.

Lewis, M. D., & Douglas, L. (1998). A dynamic systems approach to cognitive-emotion interactions in development. In M. F. Mascolo & S. Griffin (Eds.), *What develops in emotional development?* (pp. 159–188). New York: Plenum.

Lewis, M., Alessandri, S., & Sullivan, M. W. (1990). Expectancy, loss of control and anger in young infants. *Developmental Psychology, 26,* 745–751.

Lewis, M., Alessandri, S. M., & Sullivan, M. W. (1992). Differences in shame and pride as a function of children's gender and task difficulty. *Child Development, 63,* 630–638.

Lewis, M., & Brooks-Gunn, J. (1979). *Social cognition and the acquisition of self.* New York: Plenum.

Lindsay-Hartz, J., de Rivera, J., & Mascolo, M. F. (1995). Guilt and shame and their effects on motivation. In J. P. Tangney & K. W. Fischer (Eds.), *Self-conscious emotions: The psychology of shame, guilt, embarrassment, and pride* (pp. 274–300). New York: Guilford Press.

Malatesta, C. Z. (1990). The role of emotions in the development and organization of personality. In R. A. Thompson (Ed.), *Nebraska Symposium on Motivation: Socioemotional development* (Vol. 36, pp. 1–56). Lincoln: University of Nebraska Press.

Mandler, G. (1984). *Mind and body.* New York: Norton.

Markus, H., & Kitayama, S. (1991). Culture and the self: Implications for cognition, emotion, and motivation. *Psychological Review, 98,* 225–253.

Mascolo, M. F., & Fischer, K. W. (1995). Developmental transformations in appraisals for pride, shame, and guilt. In J. P. Tangney & K. W. Fischer (Eds.), *Self-conscious emotions: The psychology of shame, guilt, embarrassment, and pride* (pp. 64–113). New York: Guilford Press.

Mascolo, M. F., & Griffin, S. (1998). Alternative trajectories in the development of appraisals in anger. In M. F. Mascolo & S. Griffin (Eds.), *What develops in emotional development?* New York: Plenum.

Mascolo, M. F., & Harkins, D. (1998). Toward a component systems model of emotional development. In M. F. Mascolo & S. Griffin (Eds.), *What develops in emotional development?* New York: Plenum.

Mascolo, M. F., Pollack, R., & Fischer, K. W. (1997). Keeping the constructor in development: An epigenetic systems approach. *Journal of Constructivist Psychology, 10,* 25–49.

Mead, G. H. (1934). *Mind, self and society.* Chicago: University of Chicago Press.

Piaget, J. (1954). *The construction of reality in the child* (M. Cook, trans.). New York: Basic Books.

Piaget, J. (1957). Logique et équilibre dans les comportements du sujet. *Etudes d'Epistemologie Génétique, 2,* 27–118.

Piaget, J. (1975). L'équilibration des structures cognitives: Problème central du développment. *Etudes d'Epistemologie Génétique, 33.*

Pipp, S. K., Fischer, K. W., & Jennings, S. L. (1987). The acquisition of self and mother knowledge in infancy. *Developmental Psychology, 22,* 86–96.

Powers, W. T. (1973). *Behavior: The control of perception.* Chicago: Aldine.

Raya, P. (1993). *The relationship between empathic responses in a young neglected child and maternal values, goals, and child-rearing practices: A case study.* Unpublished qualifying paper, Harvard University, Cambridge, MA.

Resnick, L. B. (1991). *Perspectives on socially shared cognition.* Washington, DC: American Psychological Association.

Rogoff, B. (1990). *Apprenticeship in thinking.* New York: Oxford University Press.

Rogoff, B. (1993). Children's guided participation and participatory appropriation in sociocultural activity. In R. Wozniak & K. W. Fischer (Eds.), *Development in context: Acting and thinking in specific learning environments* (pp. 121–154). Hillsdale, NJ: Erlbaum.

Rothbart, M. K., & Ahadi, S. A. (1994). Temperament and the development of personality. *Journal of Abnormal Psychology, 103,* 55–66.

Ruble, D. N. (1983). The development of social comparison processes and their role in achievement-related self-socialization. In E. T. Higgins, D. N. Ruble, & W. W. Hartup (Eds.), *Social cognition and social development: A sociocultural perspective* (pp. 134–182). New York: Cambridge University Press.

Sampson, E. E. (1994). *Celebrating the other: A dialogical account of human nature.* Boulder, CO: Westview Press.

Shotter, J. (1997). The social construction of our inner selves. *Journal of Constructivist Psychology, 10,* 7–24.

Shotter, J., & Gergen, K. J. (Eds.). (1989). *Texts of identity.* London: Sage.

Shweder, R. (1990). Cultural psychology: What is it? In J. W. Stigler, R. A. Shweder, & G. Herdt (Eds.), *Cultural psychology: Essays on comparative human development.* New York: Cambridge University Press.

Stern, D. N. (1985). *The interpersonal world of the infant: A view from psychoanalysis and developmental psychology.* New York: Basic Books.

Stipek, D. J., Recchia, S., & McClintic, S. (1992). Self-evaluation in young children. *Monographs of the Society for Research on Child Development, 57*(1, Serial No. 226).

Sullivan, M. W., & Lewis, M. (1993, March). *Determinants of anger in young infants: The effect of loss of control.* Poster presented at the 30th meeting of the Society for Research in Child Development, New Orleans, LA.

Tangney, J. P., & Fischer, K. W. (Eds.). (1995). *Self-conscious emotions: The psychology of shame, guilt, embarrassment, and pride.* New York: Guilford Press.

Triandis, H. C. (1989). The self and social behavior in differing cultural contexts. *Psychological Review, 96,* 506–520.

Van Parys, M. M. (1981). *Preschoolers in society: Use of the social roles of sex, age, and race for self and others by black and white children.* Unpublished master's thesis. University of Denver, Denver, CO.

Van Parys, M. M. (1983). *The relation of use and understanding of sex and age categories in preschool children.* Unpublished doctoral dissertation, University of Denver, *Dissertation Abstracts International, 45(2),* 700B. (University Microfilms No. DEQ84–11938)

Vygotsky, L. S. (1978). *Mind in society.* Cambridge, MA: Harvard University Press.

Watson, J. S., & Ramey, C. T. (1972). Reactions to response-contingent stimulation in early infancy. *Merrill–Palmer Quarterly, 18,* 219–227.

Watson, M., & Fischer, K. W. (1977). A developmental sequence of agent use in late infancy. *Child Development, 48,* 828–835.

Werner, H., & Kaplan, B. (1984/1963). *Symbol formation.* Hillsdale, NJ: Erlbaum.

Wertsch, J. V. (1991). *Voices of the mind.* Cambridge, MA: Harvard University Press.

Wittgenstein, L. (1953). *Philosophical investigations.* New York: Macmillan.

PART FOUR

INTEGRATION AND CONCLUSION

✧

CHAPTER THIRTEEN

Being and Becoming Self-Aware

✧

MICHEL FERRARI

Certainly, more so than any of the volume's preceding chapters, this concluding chapter illustrates Schrödinger's (1944) observation, cited by Pribram and Bradley,[1] that in our need to make sense of many different areas of knowledge, "some of us should venture to embark on a synthesis of facts and theories, albeit with secondhand and incomplete knowledge of some of them—and at the risk of making fools of ourselves." Thus, I hope to recall some of the major themes addressed throughout the volume and to suggest how they might relate to each other, albeit in an idiosyncratic way that reflects my own biases, at the risk of making a fool of myself, and certainly with no pretension to do justice to the deep thought and wealth of information that has gone into each of them.

LEVELS OF SELF-DEFINITION

Self-awareness is a complex concept, one that cannot be discussed without referring to many different levels of organization at which individuals live and act. A key insight in many of the chapters in this volume is that self-awareness functions within different hierarchical, self-organizing systems. Mascolo and Fisher propose four broad levels of organization: (1) the bio–genetic level, which includes all subpersonal physical levels (both genetic and ontogenetic); (2) the personal level of the individual as organismic–agentative psychological system; (3) the immediate social level, which is negotiated, coconstructed, and coregulated through joint individual transactions; and finally, (4) the cultural–linguistic level, which involves historical, socially shared practices, institutions, and meanings that are represented and communicated through symbol systems (especially language, but also including art, math, and other forms of cultural expression).

I entirely agree with Mascolo and Fischer's point that the elements that constitute each of these four levels cannot function independently of each other, although each level of the hierarchy is semi-independent of the others. Self-organizing systems and subsystems at each level mutually regulate each other, both horizontally (e.g., person to person) and vertically (e.g., person to culture) (Gottlieb, 1992; Mascolo & Fischer). In order to explain *how* different levels of these systems interact, one must identify the transfer functions that allow operations at one level of scale to influence those at adjacent levels.

According to Pribram and Bradley, general systems theory suggests that collectives of different scales often operate according to similar organizational principles. They suggest that operations at the neurobiological, neuropsychological, and social-psychological levels all show correspondences in the organization of behavior that produce a competent, stable self. This stability depends on the structural organization of constituent elements at all levels—as much at the collective and interpersonal levels as at the level of individual neurobiological organisms (Pribram & Bradley; see also Lewis, 1995). Acknowledging multiple levels of organization is not an argument for reducing one level to another; that is, each level must not be thought of as a reducible entity, but "in terms of processes continuously influenced by the dialectical relations between levels" (Hinde, 1992, p. 1019, quoted in Pribram & Bradley).

An initial hurdle is the development of a uniform vocabulary for analogous processes at different levels of scale. Relying on the structural method, which is based on the structural premise of transposable invariance of structural relations among parts of any system (Nadel, 1957; Piaget, 1970), Pribram and Bradley identify distinguishable brain systems supporting three structural dimensions of self. These levels span biological, personal, and social levels of organization to include (1) arousal–familiarization (passion/flux); (2) activation–selective readiness (commitment/control); and (3) effort–comfort (intimacy/collaboration). Each level shifts focus from more physical and individual awareness to broader and socially integrated meanings of life and self. Loss of any key elements supporting these structures can lead to a loss of one's sense of self.

While I find the structural approach outlined by Pribram and Bradley very appealing, as a closet existentialist, I think it is important to note that the experiential dimension of each of their levels of organization is very different—that is, physical comfort, interpersonal intimacy, and collaboration, for example, are very different personal experiences—despite their common organizing principles. Of course, the details of how individual experience relates to its biological underpinnings seem to remain somewhat mysterious for all of the contributors to the volume, myself included.

Let us consider how these different levels of organization impact on self and awareness.

PHYSICAL SELF-AWARENESS

Few would argue with Mascolo and Fischer's claim that the self is necessarily embodied. Like individual members of many other species, we each have biological markers that distinguish self from nonself. For example, our immune system seems designed to distinguish "self" from "nonself" when identifying a foreign virus to be isolated and destroyed. Yet while our immune system may be essential to preserving our physical identity, all of its activity proceeds outside of awareness in humans and in other animals.

On a larger scale, however, as human beings we are aware of our own physical existence. We feel the wind at our back and the sun on our skin. Scientists and philosophers as far back as Aristotle (if not before) have wondered whether other animals share our subjective awareness of the world. Yet, as Povinelli and Prince point out, while many organisms are sensitive to information from others (e.g., eye movements) this does not always imply an awareness that mental states exist. To extend Natsoulas's term, they may be "mindblind," a term developed by analogy to the experiences of individuals suffering from blindsight (Weizkrantz, 1993) and from comparable losses of inner awareness, such as the tragic case of Alice, described by Pribram and Bradley.

Povinelli and Prince suggest that only higher primates are self-aware in even the most basic way. For example, in his classic studies, Gallup (1970) and subsequent researchers have found that humans, chimps, and orangutans (but not gorillas[2]) can recognize themselves in a mirror, although there is wide interindividual variability. Great apes seem to display the same quality of self-awareness as do human children of somewhere between 1½ and 3 years of age. For example, sophisticated experimental techniques show that by age 5 to 8, many chimps show the level of mirror self-recognition seen in human infants around 18 months of age (Povinelli, Rulf, Biershwale, & Landau, 1993; Eddy, Gallup, & Povinelli, 1996). This finding suggests that these apes must have at least a primitive self-concept (Gallup, 1994; Gallup et al., 1995). Other research with nonprimates (e.g., dolphins) has produced no compelling evidence in favor of self-recognition (Marino, Reiss, & Gallup, 1994; Parker, Mitchell, & Boccia, 1994).

The fact that some great apes show self-awareness—but not Old World monkeys—suggests an evolutionary sequence such as outlined by Parker (1996; this volume) and by Povinelli (1987). But it is difficult to interpret what this self-awareness consists of. Povinelli and Prince present startling data to suggest that although chimps may have a sense of themselves as agents, they may have very little insight into the minds of other chimps, or even into their own continued existence over time in the form of a "narrative I."

Evolution of the Fusion between Concepts of Self and Other

Povinelli and Prince present two hypotheses about the evolution of the concepts of self and others in the great apes.

The synchrony hypothesis suggests that self-concept in chimps implies a correlative understanding of others. If natural selection favors general representational structures, then the consequences might be systemwide awareness. In this view, the common ancestor of chimpanzees and humans evolved the level of self-awareness of an 18-month-old human infant. This sort of empathetic simulation of the other has a long history in psychology and philosophy, and seems to be favored by Bruner and Kalmar in their discussion of triggers of infant self-awareness.

The asynchrony hypothesis argues that despite the intimate connections of self–other knowledge in our species, the evolutionary history of the two knowledge systems may be different. Indeed, if one agrees with the idea of dedicated brain modules (cf. Fodor, 1983), there is no reason to assume that concepts of self and other evolved concurrently. Orthograde clambering may have required an explicit representation of the self's actions (i.e., agency) that might allow for self-recognition in mirrors. If so, recognition of other may have arisen during the rapid evolution of brain size during the last few million years since our evolution diverged from that of the apes. Other primates may have "clever brains, but blank minds" to use Humphrey's (1982) phrase.

According to Povinelli and Prince, Romanes' (1882) error was to assume that behavioral similarities reflect underlying psychological similarities. Their argument is that only one of the descendants of our common ancestor, genus *Homo*, "evolved an additional cognitive specialization that produced a uniform, mentalistic understanding of self and others" and this was woven into then existing developmental pathways controlling the expression of ancient yet sophisticated social behaviors—an idea not unlike that put forward long ago by Vygotsky (1930–1931/1966).

Indeed, Povinelli and Prince draw support for their model by turning to what Karmiloff-Smith (1992) calls children's need for "representational description" that makes information from specialized cognitive processors (evolved to process information about core cognitive domains such as physical and social realities) available to other parts of the cognitive system by redescribing it in another representational format. In her terms, information *in* the mind becomes information *to* the mind. If so, then, as for Piaget (1974a, 1974b), self-knowledge is the slow developmental consequence of reflective abstraction of knowledge that children (or adults) may already implicitly possess (Chandler & Carpendale; Montanegro, 1996; Pinard, 1986).

Povinelli and Prince argue that human beings have evolved mechanisms that allow for precisely this sort of redescription. This view, if correct, dramatically changes how one views similarities among higher primate behavior. It may be that many primate behavior patterns existed for millions of years before anyone ever became aware of them. Their claim is not meant to suggest that such experiences are epiphenomenal, rather, such awareness may be causally connected to the behavior of organisms that possess them, but merely to support, optimize, or otherwise reorganize existing behavioral patterns (e.g., by affecting the complexity, speed, or information density that a particular cognitive system can attain).[3]

The interweaving of old and new cognitive architecture seems to have generated a functional interdependency between them; that is, even if folk narratives are about ancient behavioral patterns, they may provide human beings with a powerful adaptive device that allows them to reorganize fundamental behavioral units into novel cultural configurations that are well suited to current and historical ecological challenges to human cultures. Such narrative may also open up the possibility of "mental time travel" which binds past, present, and future together in time, while leaving the fundamental building blocks of events largely unaltered (Povinelli & Prince). A sense of autobiographical continuity over time may be the foundation of the narrative self, an idea that will be further explored when discussing self-concept and autobiography.

Self-Conscious Emotions

Parker, along with Ford and Maher, notes that emotions shaped by natural selection function as primary motivators of behavior, and thus serve internal, behavioral, and social regulatory functions. However, Parker—building on the work of Lewis (1992), Mascolo and Fischer, and others—argues that an important distinction is to be made between self-conscious emotions (i.e., shame, guilt, pride, envy, and perhaps jealousy and empathy) and the non-self-conscious emotions of happiness, anger, surprise, fear, disgust, sadness, and interest. With the exception of fear and surprise, non-self-conscious emotions are social. These emotions have specific facial expressions shared with other primates, and have been phylogenetically ritualized for their signal value. By contrast, self-conscious emotions require self-evaluation against a social standard and are involved in the personal enforcement of moral codes of right and wrong. Because affect and appraisal mutually regulate each other in an emotional episode, Mascolo and Harkins (1998) suggest that self-conscious emotions serve to select, amplify, and modulate the self-evaluative appraisals that led to those episodes. Following M. Mead (1963), Parker suggests that although all emotions are involved in socialization, only self-conscious emotions are involved in enculturation, because only enculturation involves self-appraisal and interiorization of culturally mediated standards.

There are several good reasons why such emotions might have evolved in human beings. For one thing, they allow parents, kin, peers, and others to socially manipulate others to improve their social fitness, an ability that is especially adaptive for parents seeking to influence their children's fitness. Parents and other socialization agents influence emotional development, and Parker suggests that this influence represents an energy-saving innovation analogous to communication over direct physical action (cf. Dawkins & Krebs, 1978).

The power of all emotions has long been known, and the Stoics would probably have had little trouble agreeing with recent theoreticians such as Goleman (1995) and Salovey (Salovey & Mayer, 1990) that the emotional intelligence needed to manage one's passions is an important key to personal and social success and well-being. And this is especially true of self-conscious emotions. As Alain (1911/1956) said so well, "Emotions don't make us suffer all that much; if a real fear spurs you to run, then you hardly stop to think of yourself. But the shame of having been afraid, if you are shamed, leads to anger or justification. Especially, shame in your own eyes, when you are alone, . . . that's when you really taste it . . . ; you release arrows and each one comes back to pierce you; you are your own enemy" (p. 108, my translation).

Finally, Povinelli and Prince, and, less directly, Parker, remind us that no behavior, not even tool-making or language, is unambiguously unique to human beings. But Povinelli and Prince suggest that what may be unique is our interpretation of these things; that is, reorganization and redeployment of ancient behavioral patterns, at a proximate level, may have produced the incredible flexibility and diversity of human culture and the human selves that populate them.

In my opinion, though, even if our awareness of other minds did evolve separately from our awareness of ourselves as agents, our human self-concept and self-conscious emotions must necessarily be intimately bound up with our concept and appraisal of others. As Baldwin (1906) pointed out long ago, the only way to recognize others as selves is to accord them a subjective existence that takes one's own experience of self as a reference. If so, then what it means to be another self is tied to what it means to be one's own self. Even if our awareness of self and other originate in independently evolved "brain modules," which I suspect are not so easily identified (cf. Gardner, 1993), our understanding and experience of self and other may nevertheless be dialectically united. Let us now explore some of the peculiarities of human self-awareness in more detail.

THE HALLMARKS OF HUMAN AWARENESS

In contrast to Descartes, Searle views mental states as conscious or unconscious and suggests that the essence of the mental is consciousness. While unconscious states are at least potentially accessible to consciousness, access to consciousness may be blocked due to any number of things, such as a brain lesion, tip-of-the-tongue phenomenon, or Freudian repression. Some of these impediments are described by various authors throughout the volume.

Searle suggests that consciousness is easy to define in a commonsense (nonanalytic) way: Consciousness consists in subjective states of awareness (sentience) that begin upon waking from a dreamless sleep and continue until one sleeps again, or dies, or otherwise loses consciousness. However, Natsoulas argues that a close consideration of the ordinary definition of consciousness, such as that given in the *Oxford English Dictionary* (*OED*), provides key insights into several related meanings of the word "consciousness."

Searle (see also Sperry, 1980, 1995) further argues that consciousness is a higher-level feature of the brain system caused by lower-level neuronal brain processes. Thus, consciousness is a biological phenomenon like digestion and mitosis, and includes all mental states (thoughts, pains, itches, loves, etc.). This relationship is no different than what is true of other relationships in nature such as the wetness of water or the hardness of stone. No one claims that liquidity and solidity are separate entities; rather, they are states of the system certain molecules are in, and Searle argues that the same thing is true of consciousness.

Although how the brain, in fact, generates the potential for self-awareness remains one of our greatest mysteries, there is no mystery in how mind relates to brain in principle. Pribram and Bradley present some fascinating thoughts on the mystery of how mind is related to biological functioning in their discussion of a possible role of bosons in the generation of thought (see also Hameroff, Kaszniak, & Scott, 1996).

The most important functional feature of the mind as biological phenomenon is intentionality. Intentionality means that the mind is able to represent objects and states of affairs in the world other than itself. The everyday meaning of intention (as in, "I intend to leave the house at 8:00 A.M. tomorrow") is just one kind of intention, in the technical sense, along with believing, perceiving, hoping, wanting, and so on (e.g., unlike pains and undirected anxieties, which are not "about" anything). Intentional states and processes function to relate us to our environment and are in turn acted upon by our environment. Notice that there is no necessary connection between consciousness and intentionality (I believe that Salman Rushdie is still in hiding, although I may rarely be conscious of that belief); still, a fact is only understood as intentional through its accessibility to consciousness.

Finally, Searle points out that all intentional states only function against a background of culturally and biologically developed abilities, capacities, dispositions, and general know-how. All mental life is conditioned by a set of presuppositions that is not part of conscious awareness. This background is what allows us to understand different uses of the word "back" in phrases such as "My back hurts" and "Oh no! He is back."

Consciousness and Self-Awareness

An impatient reader might ask what all of this has to do with self-awareness. As a prelude to answering this question, consider Ford and Maher's discussion of behavior episode schemata. They note that life is a stream of behavior episodes of varying length, significance, and personal importance. Behavior episode schemata are considered integrated representations of a certain set of behavioral episodes that guide action more or less effectively. These behavior episode schemata can only be altered and improved when "in mind," and even then, people have only a limited attentional capacity. Behavior episode schemata make recurrent situations predictable but are bound to the immediate context. Only by combining several behavior episode schemata together through cognitive representations can one achieve generative flexibility: But how is such a combination accomplished?

Carver and Scheier (1981; Carver, Lawrence, & Scheier, 1996) suggest that internally focused attention (i.e., self-awareness) may increase the accessibility of one's general self-schema or self-concept, which in turn can influence collecting and processing self-relevant information. Yet my reading of Natsoulas suggests that this explanation, invoking what he calls "inner awareness" (i.e., consciousness$_4$) is not sufficient; one must also use this awareness to be a witness to one's self through what he calls consciousness$_2$ (or what I might call "personal awareness," although Natsoulas himself fears that such labels are too often misleading [Natsoulas, personal communication, 1997]).

Natsoulas's very subtle discussion of how consciousness relates specifically to self-awareness pays special attention to the second *OED* definition of consciousness (consciousness$_2$) that refers to "internal knowledge or conviction; knowledge as to which one has the testimony within oneself; especially of one's own innocence, guilt, deficiencies, etc." He suggests that this sort of consciousness is intimately tied to Lewis's notion of consciring.

Lewis (1967) uses the Latin word "consciring" to refer to knowing together, or sharing secret knowledge of something with someone (or a few people), including reflexive or interpersonal knowing. As he states so beautifully, a person can sometimes feels him- or herself to be "two people, one of whom can act upon and observe the other. Thus he pities, loves, admires, hates, despises, rebukes, comforts, examines, masters or is mastered by, 'himself.' . . . He is privy to his own acts, is his own *conscius* or accomplice" (p. 187, quoted by Natsoulas).

Extrapolating on the significance of the examples given in the OED, Natsoulas suggests that everything of which one can have consciousness$_2$ has direct relevance to what James (1890/1983) called the "spiritual self." According to James, the spiritual self refers to "a man's inner or subjective being, his psychic faculties or dispositions," that is, to what is most intimate and enduring in one's self; that which, when altered (e.g., through brain damage) make a man *alienatus a se*. The spiritual self thus corresponds to our every impulse toward psychic advancement, be it intellectual, moral, or religious.

Thus, Natsoulas suggests that, considered abstractly, the spiritual self can be equated to all of the psychological powers, abilities, or dispositions that constitute the intellectual, moral, and religious dimensions of one's personality. Note that an individual or others infer these features about the self; they are not part of the stream of consciousness and, like all conceptions, they can be misconceived. Similarly, Chandler and Carpendale seem to suggest that children's developing theory of mind may represent a progressively deeper understanding of their own and other's spiritual selves.

Any instance of consciousness$_2$, whether witnessed on the spot or based on evidence from the past, requires several kinds of self-awareness: (1) witnessing of one's self, firsthand, regarding some piece of behavior that can be used as evidence about one's intellectual, moral, or religious powers, traits, abilities, dispositions, or tendencies; (2) personal appropriation to one's self of what was witnessed; (3) retrowareness of one's self at some later point in time than the originally perceived event that serves as evidence of what kind of person one is; (4) inner awareness (consciousness$_4$) of one's retrowareness; (5) inner awareness extended backwards (recalling both the evidence witnessed and one's thoughts about that evidence); and (6) self-thoughts and self-judgments, in which one uses what one observes as evidence about the spiritual self, to reaffirm or modify one's conception of that self as valid.

I must admit that I find Natsoulas's discussion of different definitions of consciousness to be very abstract, but I also believe that he provides an extremely insightful view of how consciousness and self-awareness are conceptually related. Now, give these important characteristics of self-awareness, how might they interact with the self-system?

What Is the Self-System?

The self-system is a complex construct that refers both to what the self is, and to how it functions. The self-system may be a by-product of evolution coupled with the sort of cultural adaptation that lets human beings share consensual symbol systems for orienting toward the world while still acknowledging private experiences (Baldwin, 1906; Bruner & Kalmar; Cassirer, 1944).

Mascolo and Fischer point out that at least two competing views of the self-system are discussed in the literature. The first view focuses on the individual person and is associated with rationalists such as Descartes. In this view, self is considered a unified, bounded center of experience and activity—a private, inner world to which individuals have privileged access. According to Fogel (1993), the Cartesian self (mind) is a unitary entity with a single *cogito*, distinct from the body and responsible for reasoning. The self is considered individual rather than social, implying that the mind can function outside of society, culture, and its own history.

This view still has many advocates, although as both Chandler and Carpendale, and Pribram and Bradley point out, Descartes himself may not have unconditionally supported it. Consider Descartes's comment in a letter to his friend Chanut (Descartes, 1647/1953, p. 2777), in which Descartes recounts his childhood love "for a girl who was a little cross-eyed" such that "for a long time afterwards, on meeting cross-eyed people, [he] felt more inclined to be attracted to them than to be attracted to others" yet without knowing "in the least what it all meant" (cited in Pinard, 1989, my translation). This letter and other examples cited by Pinard (1989) certainly suggests that Descartes never believed that the self operates without reference to its own history.

In reaction to this first view, social constructionists have recently argued that self is socially constructed in dialogue. For example, Gergen (1991, 1994) holds that psychological terms, like all experience, must be classified in terms of the meanings expressed by existing category systems. And these systems originate in how individuals use language with others in a given culture, and not from any absolute external referent. Of course, to be useful, these category systems must capture some feature of the phenomena or they will be rejected (Agnew, Ford, & Hayes, 1994). Furthermore, even social constructivists do not contest that individuals experience the present moment from a particular and unique point in space and time. If so, then there may be less difference between these two views than is sometimes supposed.

Nevertheless, the sociocultural approach suggests a *dialogical* model of the self in which the self is constituted in dialogue with others, in specific sociocultural contexts, or what Bruner and Kalmar call *microclimates*. This view further suggests that even internal dialogue resembles a social interaction between various semiautonomous "I-positions" (cf. Hermans, 1996) in an imagined landscape. Thus, individuals endow different "I"s with different voices that may be situated at different points in time (e.g., between one's self as a child and as an adult) and locations in space (e.g., one's self at home or at work).

Clearly, we also more or less concretely interiorize the voices of important others and set up dialogues between these voices as though they were characters in a story, who conscire with us as witnesses to our actions and our self-stories. Harry Stack Sullivan (1954) once remarked that, although only he and his client were physically present in the room during therapy, the number of imaginary people who came through could be almost frightening.

Freud (1956) also proposed that people may be conceived as a cast of characters that, like any novelist, they more or less successfully organize into a play or novel that brings them all together. Bruner and Kalmar suggest that this would explain why we have a hard time constructing a complete account of ourselves: Novel writing is hard work and necessarily relies on the stock of stories and narrative genres offered by one's culture.

Now, because individuals experience many contexts, Mascolo and Fischer claim that self-control systems are initially fractionated and separate but become increasingly coordinated during development. Thus, self is sometimes viewed as a set of weakly connected, contextualized control systems that regulate and represent different aspects of self (each perhaps with its own subset of Ford and Maher's behavior episode schemata) that develop simultaneously along multiple pathways within the individual. This view of self suggests the image of a web of development, rather than a ladder (Fischer & Bidell, 1998). Thus, each strand of the web represents a particular control system (and eventually awareness and representations of those systems). During development, each strand may develop at a different rate and can converge or diverge with the development of other strands. Furthermore, part of the difficulty of coordinating subselves may be due to the fact that one's actions are often in response to the situation as much as they are to any internal disposition.

I will explore the issue of development of the self-system later in the chapter; but first, let us consider some of the key structures and functions of the self-system.

I and Me: Two Key Aspects of the Self-System

Over and above these two broad differences in perspectives, the self-system is often considered to include two key structural aspects of the self. The "I" refers to the self as subject, with its own experience and sense of agency; the "me" refers to the self as object, as captured through self-concept and autobiographical stories (James, 1890/1950; Mead, 1934). Thus, the "I" stands for the author; the "me" for the actor or character of the narrative drama (Hermans & van Gilst, 1991).

Many of the authors in this volume suggest that the "I" is never directly knowable but, rather, is constructed in dialogue with others, becoming "me" when made intelligible in social categories acquired from others. Of course, once I have appropriated general cultural tools such as language, I may generate an image of myself that differs from others' image of me. "Me" thus consists of personal memories, stories, concepts, and theories that we accumulate about ourselves that guide our actions in the world. Thus, a large number of "me's" can emerge through different dialogues in different contexts (Cantor & Kihlstrom, 1987; Mascolo & Fischer).

Autobiographical Stories

People have little trouble giving self-descriptions and autobiographical statements. According to Bruner and Kalmar, autobiography is made up of memories filtered through conceptual and narrative lenses. Typically, narratives contain the following constituents: an *actor* with some freedoms; an *act*; a *goal* to which the actor is committed; *resources* to be deployed toward that goal; and a particular *setting*, one that presupposes the *legitimacy* (canonicity) of some state of affairs whose violation or disruption has placed things in *jeopardy* (Bruner, 1990).

Bruner and Kalmar's general hypothesis is that those narrative indicators that can be placed in a narrative structure that includes an agent are considered self-indicators. In this view, different sorts of narratives highlight different types of self-indicators; for example, an adventurous self-story highlights agency, whereas generous or miserly self-stories highlight the importance of resources. Not all the constituents of any self-story need be explicitly mentioned; in fact, missing elements engage the imagination of the listener in the form of images and metaphors that stand for story features.

Now, since we are important witnesses to our own self-stories (especially those we tell to ourselves alone), an intriguing possibility is that consciring may be what gives our own autobiographical stories their power and their internal coherence. In consciring, I am a witness to myself in a way that is very different from how I witness others' stories about themselves, even when I participate in telling them. For example, self-stories are often accompanied by (or even centered around) vivid memories that seem to take one back to the situation described and can sometimes be a defining moment in one's life (Singer & Salovey, 1996). The poet Stanley Kunitz (1995) wrote of how, as a child, he accidentally found a picture of his father, dead by his own hand before Kunitz was born, and showed it to his mother, asking about the stranger: "she ripped it into shreds without a single word / and slapped me hard. / In my sixty-fourth year / I can feel my cheek / still burning" (p. 22).

One might ask whether these various "subselves" or "subnarratives" are ever consolidated into a master scenario of selfhood (Bruner & Kalmar; McAdams, 1993). Although many people would deny that a person's life is ever fully integrated, many might allow that our lives are organized into chapters (or scenes, if one prefers a film metaphor.) Junction points between chapters typically occur when life's difficulties (i.e., *peripeteia*) force one to fashion a new omnibus self. Narrative organizes the efforts to overcome trouble by saying what actions returned things to a desired or better state. Yet despite this fragmentation, most people would agree with James (1890/1983), who insisted that everyone has a very basic sense that "my thought belongs with *my* other thoughts, and your thoughts with *your* other thoughts" (pp. 220–221, my emphasis). The tragic example of brain-injured Zazetsky's inability to feel this, despite his most conscientious efforts (Luria, 1972), shows that this experience is not an inevitable result of storytelling.

According to Bruner and Kalmar, generating omnibus self-narratives requires metacognition (awareness)—whether as a result of one's own disposition (Cacioppo, Petty, Feinstein, & Jarvis, 1996), reflective abstraction (Piaget, 1974a, 1974b), or dialogic settings provided by institutions (Langer, 1989). By metacognitively anticipating trouble, we fashion a self to go beyond our immediate encounters—relying all the while on our inherent resources and on the narrative resources of our culture.

Bruner and Kalmar caution that such reflection can lead to distancing from the immediacy of life. However, Baldwin (1915), also troubled by the personal distancing associated with conceptual understanding, suggests that individuals regain their sense of immediacy in aesthetic contemplation, which provides a synthesis of truth and value.

Difficulties in creating an omnibus self may reflect cultural issues endemic to our times. For Bruner and Kalmar, some of these cultural issues include too little time for reflection, fragmentation of supportive interpretive communities, and the growing complexity of our mental lives due to increased personal mobility and to information technologies that connect us with communities in far-flung corners of the earth. Nevertheless, I tend to agree with Mascolo and Fischer, and McAdams (1993), who hold out hope that one's life may become more integrated as one's intellectual powers deepen. However, I also believe that Bruner and Kalmar's discussion of self-as-narrative clarifies some of the details of how Mascolo and Fischer's reconciliation of dialogic and rationalist views of the self might take place.

At a more abstract level, individuals also develop theories about themselves and others. These conceptions are typically called "theory of mind" when individuals make immediate contextualized judgments about the mind, and "self-concepts" when they address one's general knowledge about self and others.

Self-Concept

One's self-concept is organized into a system, just like all concepts (Bouffard & Vezeau; Bruner & Kalmar). Elements of the self-concept include beliefs about past, present, and future selves, as well as desires and emotional appraisals of each of these selves. All of these attributes about self are organized into a theory about the self, as Chandler and Carpendale show is true (in a very basic way, at first) even for very young children.

Chandler and Carpendale describe a large body of research showing that children under age 4, as compared to older children, fail to appreciate that others will act on the basis of false beliefs if those others are missing key information. Other studies, many by Chandler and his associates, have shown that under optimizing conditions children at least as young as age 3 may also appreciate false beliefs.

Research on the self suggests that people show a positive bias in representing the self; that is, they typically emphasize positive attributes about themselves and deemphasize negative ones (Harter & Monsour, 1992). Mascolo and Fischer point out that children under age 4 nevertheless often enjoy portraying themselves as mean, with the exception of shy or inhibited children who avoid negative portrayals of self even at an early age (Kagan, 1989).

Other-Concept

Since, by definition, all concepts have a contrasting class, Bruner and Kalmar maintain that people develop concepts of others by contrasting their attributes with those of others: Self implies not-self, or other-selves, that differ along these stated self-dimensions. However, as Povinelli shows, the contrasting class for self may not necessarily be other selves, or even the self at other times, but perhaps simply the self at other points in space (i.e., my hand is above me, and not off to my side).

Human adults do commonly apply self-attributes to others, and these abstractions do coordinate the attributes given to others with those given to ourselves. Even though concepts of self and other are organized differently (cf. fundamental attribution error), we use a common set of descriptive dimensions in both cases, and this is how we discover what is specific to ourselves (Baldwin, 1906; Bruner & Kalmar). Chandler shows that our theory of self may not really emerge until about age 7, when children come to understand that individuals with different histories can have different personal interpretations of the exact same information.

We all experience the counterpoint between I and me. Borges (1967, p. 200) captured this relationship beautifully in his story "Borges and I." He writes, "It would be an exaggeration to say that our relation is a hostile one; I live, I go on living, so that Borges may contrive his literature; and that literature justifies me. . . . But these pages cannot save me . . . perhaps because what is good no longer belongs to anyone, . . . but to the language or tradition."

FUNCTIONS OF THE SELF

According to Bruner and Kalmar, the self has two interdependent functions: Self both individuates us from others, and connects us to others in the larger community. These functions are generated through dialogue, which promotes both individuation and intersubjective conformity. It is worth recalling Povinelli and Prince's argument that that this dual nature of the self seems unique to mankind.

Individuation

Individuation is aimed at creating and guarding not only exclusivity, but also *privacy*. This definition is consistent with the original Latin source for the words "consciousness" and "conscience," and is the hallmark of Natsoulas's "consciousness to one's self." Bruner and Kalmar suggest that individuality has two aspects. The *epistemic* aspect of individuality refers to what we each know and experience based on our own unique phenomenology, as well as our background knowledge and beliefs (see also Natsoulas; Searle). The *deontic* aspect refers to what we value, expect, and consider just.

Although these two aspects are often viewed as firmly divided, Chandler and Carpendale point out that much human knowledge defies this basic division.

They recall Habermas's (1981/1984) theory, which argues that many beliefs are defined, not as true or false, but as valid or invalid; that is, if a statement is not about a matter of personal taste, then the speaker is expected to be able to justify his or her claim by pointing to evidence or to a chain of reasoning that assumes its validity. In the case of instrumental rationality involving the subject and some material object, the claim is based on evidence about its truthfulness. But for claims about the shared social world, we refer to the validity or legitimacy of norms.

In the film *Dead Man Walking*, Susan Sarandon plays a nun caught between the belief that a person is worth more than his worst acts, and the belief of the parents of a teenage girl raped and murdered by this man, that death is not punishment enough for his crimes. Fact and value fuse in trying to reach one's own verdict on how best to act in this tragic situation. Furthermore, some facts and values only exist because we agree to them socially and institutionally, like the fact that in every fourth year, February has 29 days (Searle, 1995); perhaps self-knowledge is based on this sort of normative validity that combines both facts and values that can be defended and challenged both by one's self and others. If so, understanding the need to establish such validity may take several years to develop, and may not be appreciated before adolescence.

The individuating function of self may stabilize personal experience by coordinating the present with the past, and in helping to prepare for the future. Individuation may also help promote cultural change by preserving individual openness to unusual ways of representing the world; but when individuation exceeds a certain limit, it is characterized as insanity, eccentricity, criminality, or some other deviance (Baldwin, 1906; Bruner & Kalmar).

Intersubjective Communicability

Individuals develop not only a sense of self, but also a sense of canonicity (what acts and beliefs to expect of others) and of shared normativity. The achievement of intersubjective communicability often seems to be a counterpart to defining ourselves in terms of the social order (cultural group, family, gender), such that knowing one's self is knowing how one fits into the social world (Bruner & Kalmar; Mascolo & Fischer).

Furthermore, others often have access to information that the self does not; thus, the individual needs others to complete his or her representations of self (Baldwin, 1906; Mascolo & Fischer). And the better a person comprehends to what extent he or she is externally determined, the more he or she can comprehend and exercise his or her real freedom (Bakhtin, 1986). The counterpoint between our self-concept and our concept of others is most evident in dialogue.

Bruner and Kalmar argue that any individuating dialogic microclimates may serve as a goad to reflection (i.e., metacognition). This is particularly true of microclimates that encourage free exchange and cooperation among equals, and not slavish conformity to tradition. Such microclimates may be as wide ranging as scientific debate, discussion between intimates, or a therapy session. This is true of a variety of different microclimates, such as school, family, or professional contacts. We are often called upon to justify different subselves to different audiences, although, typically, the subself that earns one a living needs less explaining. Some subselves will be more highly valued and thus more central to our personality, and so we will work harder to defend them and will be hurt if we find ourselves lacking in them (Leont'ev, 1978; Smirnov, 1994).

The cast of subselves and the overarching metanarrative that provides them some overall coherence are influenced and supported by "interpretative communities" (Fish, 1980). Even the existence and meaning of memories sometimes depends on the negotiated support of others, and on how one adapts the tale for particular audiences (Crawford, 1996). In all social settings, our self-stories are highly constrained by reticence and taboos; our stories stay close to how we "should be." Indeed, self is a product of our intrapsychic and interpersonal acts: We cast ourselves as agent or victim, participant or witness (Crawford, 1996; Bruner & Kalmar; Shafer, 1992). This formulation suggests to me that we are influenced by others, not only cognitively, but also through consciring that is often accompanied by self-conscious emotions.

As mentioned earlier, often our tales about ourselves and about others are organized in the form of a story, and these stories fall into narrative genres. Cultures and subcultures (stock of stories) provide guides for such self-presentation in dialogue (Bruner & Kalmar; Hermans & van Gilst, 1991; Shore, 1996). Some stories (especially deeply moving stories and dreams) may even be inherent to the human experience, as they seem to appear in most if not all cultures (Jung, 1964). The inward turn of the novel in the last century has produced new genres that may have transformed contemporary notions of possible selves, as our conceptions of self change to reflect the narrative conventions and changing lifestyles of our times (Taylor, 1989; Bruner, 1990). In fact, the very lexicon used to construct these stories requires a specific structure of self-description that reflects both one's culture and one's times. For example, contemporary Japanese frame self-descriptions in terms of "inner (*uchi*)" and "outer (*soto*)," whereas the contemporary American lexicon is framed in terms of achievement (Markus & Kitayama, 1991).

Because it is intimately tied to culture, self-narrative is almost always caught up in issues of hegemony and cultural power (Bourdieu, 1991, 1997; Bruner & Kalmar). Sociocultural construal of race, gender, family or work, for example, frame the stories we tell about ourselves. Indeed, self-descriptions are often imposed by more dominant others in the community through stereotypes such as "dumb blond" or "rebellious teenager." Individuals marked by those descriptions may consciously or unconsciously court them or rebel against them (Tarde, 1897/1907). A lot of the social drama of self may be created by the dialectic between what is required of one from without and what is required from within (Bruner & Kalmar; Shore, 1996). Given their importance, it is perhaps not surprising that Fischer and his colleagues (Fischer, Knight, & Van Parys, 1993) found differences in understanding of self-attributes such as race, age, and gender for children as young as 2 years old.

FRAGILITY OF SELF

The self-system is both resilient and vulnerable. Each new day seems to begin with the same self as the day before, and that self retains this identity even after radical changes in self-definition. Yet just as one's physical, social, and cultural identity help create and maintain one's sense of self, under certain conditions, they can contribute to an experienced loss of self. The self can be shattered and split by traumatic events that affect these supports to self-identity.

One's physical sense of self can be lost or distorted through brain damage, brought on by a stroke or some other physical injury. Depending on the part of the brain affected, individuals can lose conscious awareness of entire parts of their visual or kinesthetic field, as poignantly described in Pribram and Bradley's case study of Alice; equally startling are cases of phantom limbs, where individuals feels as though an amputated limb still exists (Melzak, 1989).

As mentioned earlier, one's feeling of an autobiographical continuity can be lost (Luria, 1972). Pribram and Bradley also describe the case of a boy unable to achieve a feeling of autobiographical continuity, who could only quantify the passage of time with difficulty, and even then, often based on abstract understanding (e.g., that crawling comes before walking). Other well-known illnesses such as schizophrenia, or traumas such as those resulting in amnesia, can also destroy one's sense of self as existing in time (see Hameroff et al., 1996).

Equally tragic and far more common are cases described by Mascolo and Fischer of individuals subjected to radical and severe abuse, whose selves are fragmented through dissociation and affective splitting that in extreme cases generate multiple personalities. (These cases may be extreme examples of the sort of isolation of subselves described earlier.)

People who very negatively conscire life events may become depressed, lonely, or guilty, leading them to try to escape this feeling by what Ford and Maher call "navigating down" to a less globally meaningful level of activity. If unchecked, this downward spiral may lead to suicide, the ultimate escape from one's sense of self (Baumeister, 1990).

Loss of self is sometimes the result of social causes. The death or loss of an important other, such as a parent, sibling, or spouse, can be experienced as a loss of self, as one's way of being with that person is lost along with them. Ostracism or separation from a social or cultural group, including loss of employment, can have a similar effect. The despair of life in America by the Hmong refugees (routed from their native Laos after helping the United States during the Vietnam war) is heard in the song of Chua Doua Xiong, preserved by ethnomusicologist Amy Catlin: "I look forward but the way is overgrown; I look backwards but there are no relatives coming" (quoted in Mydnas, 1997). More broadly, experienced loss of culture can lead to a feeling of loss of self. When one's way of life vanishes, sometimes the meaning of one's personal life can also vanish, leaving suicide the only apparent option, as Achebe (1996) so poignantly expressed in his book, *Things Fall Apart*.

SELF-DEVELOPMENT

Before addressing the issue of self-development, we must first look at when and how this self emerges during infancy.

Before 18 Months: Identity without Consciring

Kagan suggests that post-Freudian psychoanalysts had a strong theoretical motivation for positing an infant self, and some of this volume's contributors (e.g., Mascolo & Fischer) also argue for a connection between subjective experiences at the earliest stage of infancy and later self-development. All of the contributors who address this question seem to agree that there is little evidence of reflexive self-awareness (i.e., of consciring) in infants before about 18 months of age. In general, younger infants appear to fail even simple tests of self-awareness, such as mirror recognition, and Kagan argues that there is no reason to for us to attribute them with a sense of self; claiming infant self-awareness based on fear of strangers (observed around 8–12 months), smiles to familiar caregivers (observed around 3 months), or frustration at loss of some object (observed around 1 month) is too bold. Much of these data might be the result of biologically prepared responses that proceed without reflexive self awareness. For Kagan, saying self emerges in the second year is no more radical an idea than saying speech emerges during the second year, or that reproductive maturity emerges during adolescence.

However, although self-awareness in the strong sense may not be present at birth, self-awareness in a weak sense of agentative actions may be.

Agency and Control Systems

At the level of organism as agent, Mascolo and Fischer argue that even infants possess control systems that become organized into ever more complex hierarchies. One common description involves embedded feedback loops operating according to a set of standards, with output from a superordinate loop setting the reference standard for the next lower level of loop (Carver & Scheier, 1981; Carver et al., 1996). Within such systems, consequential behavior is self-organizing (Lewis, 1995; Pribram & Bradley). Although I find that the rhetoric of these descriptions is a little too tied to cybernetics for my taste, there is no denying that even infants set simple goals and execute plans to accomplish those goals, with all phases of the process regulated by information about present conditions (feedback) and possible cognitive and affective consequences (feed-forward) of actions (Carver et al., 1996; Duval & Wickland, 1972). For example, infants do begin to exert some control over single reflexes soon after birth (3–4 weeks) through looking, grasping, and so on, although, at first, the infant must cooperate with others to produce this effect (i.e., a ball must be placed in their hand) (Mascolo & Fischer).

It is important to distinguish between the organism as agent, and the organism's constructed awareness of its agency and identity. Awareness is important because it allows the cognitive system to use self-awareness to improve self-regulation (Mascolo & Fischer; Bouffard & Vezeau). In this view, individuals increasingly become active contributors to their own development. The early expressions of control by infants suggest that they have some sense of agency, but probably not any representation of themselves as agents.

Pribram and Bradley argue that in a very basic sense, the "objective awareness," or "me," dimension of self locates us in the world with respect to our body's figural integrity and its relationship to other objects. Awareness of our spatial existence allows attention to be directed to the inside and outside of our own body. Ford and Maher argue that attention (or awareness) is constantly shifting between different components of the person-in-context system, depending primarily on current concerns and context. If the information brought to light by this attention is not sufficient, optimal, or relevant, self-awareness energizes further cognitive processing and information-construction activities. The key to what Ford and Maher call attentional–navigational control is timely and selective targeting of useful informational content. Individuals can even alter their mood by selectively redirecting their attention, whether physically to lessen pain, or psychologically, by changing their goals or their interpretation of events (Bandura, 1986; Carver et al., 1996; Csikszentmihalyi, 1990).

Nevertheless, answers to the difficult question of what leads to the emergence of an infant's sense of self remain controversial. Bruner and Kalmar cite research suggesting that external sensory triggers to infant self-awareness may include people's eyes, voices, and movement. These findings suggest that, partly due to recognizing their own voice, and partly from appreciating the difference between self-initiated and environmentally induced movement, infants generate a primitive "sensory" self (Bruner & Kalmar; Neisser, 1988). Mascolo and Fischer suggest that such self-representation may emerge and develop through transformations in one's control systems (Mascolo & Fischer; Pinard, 1986).

In any case, many of Bruner and Kalmar's self-indicators (i.e., agency, commitment, internal and external resources, social reference, evaluation, *qualia,* reflexivity or metacognition, coherence among individual acts, location or position in a social order) are visible or at least attributed to infants in the first year of life (Bruner, 1996).

Emergence of Self-Representation

While several of the contributors to this volume propose slightly different milestones in self-representation, all seem to agree with the commonsense observation that a major change occurs around the middle of the second year, one that coincides with infant brain development. Kagan reminds us that while Americans and Europeans may call this change the emergence of self-awareness, it has been called by other names in different cultures, often with subtle differences in connotation, much as is true of the word "growth." Pribram and Bradley point to evidence that such basic biological development is not independent of caregiver's treatment of infants, which can have profound effects on later self-development. For example, when the primary caregiver withholds nurturing love during critical periods in infancy, the infant frontal cortex may be affected, with long-term pathological effects (Pribram & Bradley; Schore, 1994).

As mentioned earlier, one of the most basic changes in infants' behavior is that they become able to recognize themselves in a mirror. So, for example, if they observe a mark on their forehead when looking in a mirror, they will now try to rub it off, something they would not have done a few months earlier. What are the implications of mirror recognition for self-awareness?

At least two views of self-recognition are possible: a general representation view and a theory-of-mind view. Povinelli (1995) proposes a model in which mirror self-recognition is presumed to have nothing to do with understanding the reflective properties of mirrors; rather, 18- to 24-month-

old children (and older primates) have developed a general representational system that allows them to relate objects and events in the world to mental schemes. Once this occurs, individuals can grasp equivalence relations
between the representation of self's actions (and perhaps desires) and those
of the image in the mirror. Povinelli and Prince equate this experience to
what James calls the "present self."

The theory-of-mind view is the major alternative to the general representation view. According to this view, we form a theory (albeit, at first, a
naive one) with which to explain and predict our own and other's behavior. Chandler and Carpendale suggest that this approach often ignores development by arguing that the emergence of a theory of mind is a singular
achievement that involves triggering some innate "mind module" (Gopnik
& Meltzoff, 1996; Karmiloff-Smith, 1992). In their cutting, but delightful,
term, Chandler and Carpendale call this the "one miracle" view of child
psychology and suggest that it has parallels to maturationist views of cognitive change popular during the first half of this century.

A more developmental take on this theory-of-mind view, one that I
personally prefer, suggests that developing knowledge about the mind occurs through a series of revisions of these theories based on accumulated
evidence through life experience. And that this is as true for our theories
about ourselves as for our theories about others (Chandler & Carpendale;
Mascolo & Fischer). Chandler and Carpendale present evidence that, by
age 4, children have come to understand that the mind actively processes
information. However, only at age 7 or 8 are children able to generate
many plausible interpretations of the same ambiguous stimuli, even when
already competent at understanding standard false belief (Chandler &
Lalonde, 1996).

Several behavioral indicators besides mirror recognition suggest that
infants begin to acquire and develop a sense of self-awareness around the
middle of the second year. For example, late in the second year, a child
working alone will smile after mastering a task. Even more compelling evidence is that, around the same time, children begin to give orders to adults,
suggesting that they expect to influence other's behavior. Other features of
the speech of 2-year-olds also suggests that they have an awareness of self.
Two-year-olds engage in "self-descriptive utterances" (e.g., "sit," as they
are sitting). They are also able to describe self in terms of simple family and
gender categories (Kagan, 1981). With the onset of what Fischer calls single representations (also around 2 years of age), children can represent simple hiding strategies of an experimenter, suggesting that they can now represent the experimenter as an independent agent. This fact suggests that 2-
year-olds are impressed by their ability to accomplish actions and feel
pressed to comment on these behaviors. Although this view goes against

most current theorizing about infant theory of mind, some theory-of-mind research adds support to this claim (Hala, Chandler, & Fritz, 1991; O'Neill, 1996).

According to Lewis (1992) and Parker, self-conscious emotions such as shame and pride emerge around the third year of life and require coordinating self-evaluation with that of a social standard. Furthermore, only around 30 to 36 months do most children begin to use personal pronouns. For example, they add "I," "my," or "mine" to self-descriptive utterances (e.g., "I sit"). Mascolo and Fischer suggest that self-conscious or self-evaluative emotions, each with its own developmental pathway, are stimulated by a child's increasingly complex appraisal of his or her behavior and its effect on others.

Note that none of the self-awareness described so far necessitates having any sense of a autobiographical identity that persists over time. The emergence of this autobiographical identity, and concomitant changes in children's theory of mind, suggest another major milestone in children's developing sense of self.

BECOMING ONE'S AUTOBIOGRAPHICAL SELF OVER TIME

Another central dimension of an adult's sense of self is that we monitor our experience over time. Pribram and Bradley suggest that this temporal dimension is essential for experiencing what Dennett (1991) calls the "Cartesian theater." While such experience is central to adult experience, it is striking to realize that this autobiographical sense seems not to be true of children.

Parker (citing Snow, 1990) argues that the autobiographical self emerges around age 4. In support of this claim, Povinelli (1995) cites evidence that 2- to 3-year-olds viewing a videotape that shows an experimenter placing a sticker on their forehead 3 minutes earlier will not reach to remove the sticker—although they will if they see the sticker in a mirror. Even though they call the image on the video "me," they clearly fail to see how that image relates to their current self (i.e., their on-line image of present representations of actions and desires) (Povinelli, Landau, & Perilloux, 1996).

Damon and Hart (1988) show that children's appraisals of self-as-object (me) and self-as-subject (I) develop in complexity at least through to adolescence, as seen in children's increasingly complex verbal judgments about their physical, social, and psychological selves. Mascolo and Fischer (see also Fischer & Bidell, 1998) have traced a comparable development of evaluative categories for conceiving of self (e.g., nice, mean; smart, dumb). First, children think agents act due to their personal traits (e.g., a bad person does bad things). Later, children can conceive of self in terms of several opposite valence categories (e.g., I am good to good people, and bad to bad people). Up until early adolescence, this all stays very concrete (although progressively more complex and coordinated).

Once the abstract tier emerges in adolescence, individuals begin to think of themselves as personalities who exhibit complex identities in relation to one another, in which intentions count more than actions. Adolescents soon learn to coordinate relations among the many abstractions used to define themselves, as well as to map their identity onto the ones that others accord them. This developmental course may continue to extend into early adulthood (Chandler, 1987; Kuhn, Amsel, & O'Loughlin, 1988).

Chandler and Carpendale propose that only in late adolescence do individuals come to understand that all beliefs, including beliefs about self, are subject to doubt and rest on the strength of the arguments and evidence brought in support of them. To quote Chandler and Carpendale, if an essential part of mental development is "a growing appreciation of the constructive ways in which minds routinely and characteristically shape the subjective nature of their own experience, then this process of epistemic development clearly prepares the way for a whole new class of insight concerning one's self as a knower."

PERSONAL SELF-DEVELOPMENT: INVENTING THE SELF AS ONE NAVIGATES THROUGH LIFE

Personal Navigation

As we have seen, early on, children begin to see themselves as agents, with increasingly abstract and semantically rich desires and beliefs about how to influence themselves, others, and the world. Individuals also increasingly see themselves as belonging to families, communities, and ethnic groups. According to Bruner and Kalmar and to McAdams (1993), these two tendencies generate a dialectic tension that frames our attempts to navigate through life.

For Sternberg and Spear-Swerling, "personal navigation" refers to a person's control of his or her voyage through life, including plans, beliefs, and ability to overcome obstacles. As such, it is a moderator variable influencing success in whatever one values (profession, personal endeavors, etc.). According to Sternberg and Spear-Swerling, effective self-navigation requires resilience, restlessness, necessary risk taking, and relentlessness in pursuit of one's goals.

By taking the long view on self-development, emphasizing personal navigation suggests that direction is more important than mental ability. Speed of performance is in the service of direction. But no matter how clear the destination, obstacles and barriers must necessarily be overcome—including both prejudicial barriers and legitimate social barriers. Not only that, certain paths are hard to follow, or hard to see; sometimes one's efforts are sabotaged, or one lacks the necessary social or intellectual resources. Furthermore, individuals may fail because of hidden implications of a particular task or career, or because of conflict between different spheres of interest, such as the common conflict between job and family. In these cases, flexibility is needed to change one's life course if some change suddenly makes the destination uninteresting. Of course, in many ways, this description is highly idealistic; people juggle several paths at once and are emotionally implicated in many, if not all, of them.

Resources

Furthermore, personal navigation is never completely independent of one's social and biological resources. As Binswager (1963) and other existentialists have argued, we are "thrown" into a particular sociocultural setting, in a body that has certain inherited potential ranges of abilities and temperament. While this point remains controversial for some (Ericsson, 1996; Scarr, 1992), there is every reason to believe, and some evidence suggests, that personal navigation results from (and is shaped by) interaction between these resources (Elder, 1995).

Development of "I" is influenced by hereditary factors, some of which are species general (e.g., the very ability to appreciate that others have a theory of mind) (Povinelli & Prince), and some of which reflect hereditary differences in potential abilities (e.g., in verbal ability, or musical ability) or in temperament (Fodor, 1983; Gardner, 1993; Plomin & McClearn, 1993; Sternberg & Spear-Swerling.)[4] Even beyond these genetic influences, individual differences in ontogenetic development, including access to basic nutrition and protection from exposure to disease, might reasonably be expected to shape basic cognitive abilities. Level of cognitive development and individual differences in actual ability will also influence self-development by affecting how easily one profits from learning opportunities that may emerge at different points in one's life (Case, 1992; Mascolo & Fischer; Vygotsky, 1934/1986).

Importantly, individuals do not function at a single point along a developmental strand for any given skill; rather, they range from functional (typical, everyday) to optimal performance depending upon whether others or features of the situation provide high levels of support or assistance for behavior. This difference represents the *developmental range,* or zone of proximal development, for any given skill (Mascolo & Fischer; Vygotsky, 1934/1986).

Similarly, many observed individual differences between children reflect how the opinions of important others shape children's background self-beliefs, which in turn influence how effectively children engage in self-regulation during learning, as Bouffard and Vezeau demonstrate in their chapter (see also Bandura, in press; Schunk & Zimmerman, 1997). The opinions of important others may influence not only an individual's short-term task goals, but also their long-term life goals (Hackett, 1995; Markus & Norris, 1986). Without a clear life project (Cantor & Harlow, 1994) and the efficacy beliefs to press on in the face of difficulty, one may succeed at day-to-day demands but end up going nowhere. Of course, broader cultural climates (macroclimates) such as laws, mass media, religion, art, and technology will also influence self-development. For example, they will influence the type of schools one attends and the values promoted in one's community, and a host of other beliefs that will form part of what Searle calls the background of consciousness and of self.

Personal navigation necessarily involves forging a path in the contexts into which one is initially thrown. Over time, participation in personally and culturally valued activities increasingly shapes one into who one eventually becomes (Ferrari & Mahalignam, 1988; Leontiev, 1978). It is no exaggeration to suggest that one acquires much of one's personal and social identity, and sense of self, through collaborating with particular others, either through direct imitation or indirect knowledge-appropriation during traditional apprenticeship (Rogoff, Mistry, Göncü, & Mosier, 1993), or through cognitive apprenticeship (Collins, Brown, & Newman, 1989) in which one may even participate in inventing new knowledge about still unsolved problems. Over time, individuals appropriate the skills required to accomplish particular tasks, and this appropriation helps shape their personal and professional identity as they become responsible for guiding others in various social and cultural activities.

As Gruber (1981) points out, each person makes a different set of decisions about the use of his or her personal (and one might add, social) resources, thus setting the scene for chance occurrences and thoughts, and choosing among them when they happen. While we cannot choose our parents or where we are born, increasingly, as we enter adulthood, we can take charge of our own personal and cognitive development (Pinard, 1992) and navigate toward valued future goals. To do so, we often seek out social settings in which we wish to learn and contribute, and we seek out teachers and institutions that will give us the knowledge, confidence, and credentials to accomplish our goals (Scarr, 1992).

Personal Stories

Sternberg and Spear-Swerling suggest that life must often be navigated by "dead reckoning" through estimation and guesswork. But people also tend to spend considerable personal and social effort trying to map out where they have been and where they want to go. The personal stories we construct to give our lives meaning are a great aid to navigation (McAdams, 1993). Of course, much of this personal navigation occurs after the fact, when we invent or reinvent some long-term objective to justify our frantic, daily racing from one thing to the next (Bruner, personal communication, 1997).

Like any good historian, however, we are not free to tell just any story: indeed, vivid memories often anchor and frame our accounts to and about ourselves (Singer & Salovey, 1996). McAdams (1993) argues that our stories can be evaluated by ourselves and others along several dimensions. Ideally, a good life story (what McAdams calls a personal myth) develops toward increasing coherence, openness, credibility, differentiation, reconciliation, and generative integration.

Naturally, our life story (or stories) is constantly under revision as new events unfold. In fact, as Bruner and Kalmar note, dramatic life events may serve as junction points in one's autobiography, during which one abandons one story and begins anew, navigating toward a new destination. The different stories we tell may nevertheless share a narrative tone of optimism or pessimism, comedy or tragedy. They may also share imagery of growth, acceleration, battles, or other metaphors often drawn from or shaped by culture (McAdams, 1993; Shore, 1996). Bruner and Kalmar also suggest that culture provides guidelines for constructing stories and examples of past methods and solutions for dealing with life experiences.

Clashes among subselves can impede or paralyze self-navigation, whereas the synthesis (or simple confederacy) of one's many stories about self can produce an integrated and generative personal identity. Some personal stories begin inauspiciously, with few apparent resources, but end in greatness; others begin fabulously and end in tragedy. Thus, I suggest that personal storytelling helps frame the accounts given by Sternberg and Spear-Swerling of why many child prodigies fail to become exceptional adults, and why many exceptional adults were able to overcome childhood adversity, or why some people "born with a golden spoon in their mouth" go on to do great things (like Nelson Rockefeller) and why some lives begin tragically and end that way. Of course, many of the details of such stories are intimately entwined in the broader sociohistorical setting in which they unfold (Elder, 1995).

Perhaps in the most personally (if not always socially) successful and satisfying lives, autobiographical memories and concepts that reflect what the individual believes to be true and valuable about the self merge in an *aesthetic appreciation* of the self (McAdams, 1993); that is, one may come to contemplate one's own and another's life story on it own terms, as one does a work of art (Bakhtin, 1986; Baldwin, 1915). This sort of experience may be the essence of feelings such as flow (Csziksentmihalyi, 1990) that seem central to self-actualization (Maslow, 1970).

Nevertheless, as Gruber (1981) points out, even the extremely successful and creative Charles Darwin did not construct a single life story or spend his life navigating toward a single destination. Nor did he achieve his creative insight in a single burst of illumination. Rather, Darwin was engaged in a network of enterprises as a writer, pigeon breeder, and botanist (among others) that he was able to weave into a new theoretical synthesis, or point of view.

Finally, the development of exceptional individuals may not be uniquely due to an act of personal will or historical circumstance; it may also involve following a vocation, that is, literally an inner calling that draws one into tasks that are sometimes difficult and even dangerous in pursuit of something in whose value one has great faith and trust (Jung, 1934/1983). From the interaction between what one is called to do, and trying and discovering what one can and cannot do, emerges the imperative sense of what one must do (Gruber, 1981). To the extent this inner voice is strong, and one has courage, one will develop into a personality who may navigate to valued destinations (sometimes at great personal cost); to the extent that that voice is weak, one will follow others' voices and navigate beaten paths, telling well-worn tales (Jung, 1934/1983).

CONCLUSION

Taken together, the contributors to this volume suggest that as biological individuals we are aware of ourselves in the present, and experience time as extending forward and backwards even beyond our own lives. Thus, self is never constructed alone, but rather is given meanings as part of a valued community. Such meanings are constructed as part of an immediate community, and as part of a community that stretches as far forward as our hopes and as far back as recorded memory: family memory, professional memory, cultural memory—to the dawn of human memory.

Ideally, individuals should engage in activities that shape self-development in ways that enhance their natural potential by seeking out emotional and intellectual contexts that permit self-development and minimize the risk of loss of self (Ferrari & Mahalignam, 1998). And to the extent that control is possible, one should navigate toward such ends. In other words, how deliberately and wholeheartedly one joins one's present self to social activity (not always to copy the past, but sometimes to resist it)—knowing its value to one's life plans, to one's community, and to a particular cultural and professional history—to that extent one actively embraces one's culture and creates a personal synthesis between the present and the past.

As this way of framing my conclusions and my comments throughout the chapter indicate, I tend to conceive of the self as involving narratives that are constructed and change over time. However, I also believe it is important to consider individual differences, such as differences in level of cognitive development or cognitive style, that influence the type of narrative one can or prefers to tell. What is more, I would argue that such narratives themselves are put to a variety of uses; sometimes to control or navigate through our lives, but sometimes simply to affiliate ourselves with others, or for entertainment. While the ability to tell and appreciate any human narrative reflects our evolutionary heritage and our biology, I find the most meaningful dimension of self is precisely that it grounds our experience in the world in ways that both individuate us and bind us to our community.

Yet the self is also difficult to measure, and some may despair of ever really measuring something so deeply personal. More profoundly, I still find that an essential question remains unresolved after reading these chapters: Are each of us a confederacy or team of independent subselves collaborating together (or struggling) for control of behavior, or are we each led by a inner author that more or less effectively unifies the efforts of these many subselves? Given these difficulties and remaining ambiguities, one might reasonably ask, after reading this volume, if we have learned anything new or essential about self-awareness, or about the self. Perhaps the best answer to this question was given by Suzuki (1962/1971, p. 1) when he wrote, not long before his death at the age of 92, "It is the height of stupidity to ask what your self is when it is this self that makes you ask the question."

NOTES

1. All contributors to the volume will be cited directly; reference to their chapters in this volume is taken for granted.

2. A telling exception is the female gorilla raised by Francine Patterson who does show mirror recognition (Patterson & Cohn, 1994). This exception suggests that such awareness is still available to gorillas under special ontogenetic conditions (Povinelli, 1994).

3. Parker (1996b) suggests that such awareness may have been the key to developing an increasingly sophisticated ability to imitate and to be directly taught through apprenticeship, an idea that I find appealing. However, I hesitate to endorse her view that self-awareness evolved through the terminal addition of Piagetian-type stages by our primate ancestors (Parker, 1996a, 1996b; Parker & Gibson, 1979).

4. Molecular genetic studies of individual differences may be needed to conclusively determine how genes influence observed differences in individual abilities and temperament. Unfortunately, these studies are still in their infancy (see Plomin & McClearn, 1993).

REFERENCES

Achebe, C. (1996). *Things fall apart*. London: Heinemann Educational.
Agnew, N. M., Ford, K. M., & Hayes, P. J. (1994). Expertise in context: Personally constructed, socially selected, and reality relevant? *International Journal of Expert Systems, 7*, 65–88.
Alain. (1956). Des Passions. In *Propos d'Alain* (pp. 107–109) (Bibliotèque de la Pléiade). Dijon: Gallimard. (Original work published 1911)
Bakhtin, M. (1986). *Speech genres and other late essays*. Austin: University of Texas Press.
Baldwin, J. M. (1906). *Social and ethical interpretations in mental development: A study in social psychology* (4th ed.). New York: Macmillan.
Baldwin, J. M. (1915). *Genetic theory of reality*. New York: Knickerbocker Press.
Bandura, A. (1986). *Social foundations of thought and action: A social cognitive theory*. Englewood Cliffs, NJ: Prentice-Hall.
Bandura, A. (in press). *Self-efficacy: The exercise of control*. New York: Freeman.
Baumeister, R. F. (1990). Suicide as escape from the self. *Psychological Review, 97*, 90–113.
Binswanger, L. (1963). *Being-in-the-world*. New York: Basic Books.
Borges, J. L. (1967). Borges and I. In *A personal anthology* (A. Kerrigan, Trans.; pp. 200–201). New York: Grove Weidenfeld.
Bourdieu, P. (1991). *Language and symbolic power*. Cambridge, MA: Harvard University Press.
Bourdieu, P. (1997). *Méditations pascaliennes*. Paris: Seuil.
Bruner, J. (1990). *Acts of meaning*. Cambridge, MA: Harvard University Press.
Bruner, J. (1996). *The culture of education*. Cambridge, MA: Harvard University Press.
Cacioppo, J. T., Petty, R. E., Feinstein, J. A., & Jarvis, W. B. G. (1996). Dispositional differences in cognitive motivation: The life and times of individuals varying in need for cognition. *Psychological Bulletin, 119*(2), 197–253.
Cantor, N., & Kihlstrom, J. F. (1987). *Personality and social intelligence*. Englewood Cliffs, NJ: Prentice-Hall.
Cantor, N., & Harlow, R. (1994). Social intelligence and personality: flexible life-task pursuit. In R. J. Sternberg & P. Ruzgis (Eds.), *Personality and intelligence* (pp. 137–168). New York: Cambridge University Press.

Carver, C. S., Lawrence, J. W., & Scheier, M. F. (1996). A control process perspective on the origin of affect. In L. L. Martin & A. Tesser (Eds.), *Striving and feeling: Interactions among goals, affect, and self-regulation* (pp. 11–52). Mahwah, NJ: Erlbaum.

Carver, C. S., & Scheier, M. F. (1981). *Attention and self regulation: A control theory approach to human behavior*. New York: Springer-Verlag.

Case, R. (1992). *The mind's staircase: Exploring the underpinnings of children's thought and knowledge*. Hillsdale, NJ: Erlbaum.

Cassirer, E. (1944). *An essay on man: An introduction to a philosophy of human culture*. New Haven, CT: Yale University Press.

Chandler, M. J. (1987). The Othello effect: Essay on the emergence and eclipse of skeptical doubt. *Human Development, 30,* 137–159.

Chandler, M. J., & Lalonde, C. (1996). Shifting to an interpretive theory of mind: 5- to 7-year-olds' changing conceptions of mental life. In A. Sameroff & M. Haith (Eds.), *Reason and responsibility: The passage through childhood* (pp. 111–139). Chicago: University of Chicago Press.

Collins, A., Brown, J. S., & Newman, S. E. (1989). Cognitive aprenticeship: Teaching the crafts of reading, writing, and mathematics. In L. B. Reznick (Ed.), *Knowing, learning, and instruction: Essays in honor of Robert Glaser* (pp. 453–494). Hillsdale, NJ: Erlbaum.

Crawford, V. M. (1996, September). *Dialogicality in the narrative construction of identity: Conversational stories among 13-year-old girls.* Paper presented at the 2nd Annual Conference for Sociocultural Research, Geneva, Switzerland.

Csikszentmihalyi, M. (1990). *Flow: The psychology of optimal experience.* New York: Harper & Row.

Damon, W., & Hart, D. (1988). *Self-understanding in childhood and adolescence.* New York: Cambridge University Press.

Dawkins, R., & Krebs, J. (1978). Animal signals: Information or manipulation? In J. Krebs & N. B. Davies (Eds.), *Behavioural ecology* (pp. 282–309). London: Blackwell.

Dennett, D. C. (1991). *Consciousness explained.* Boston: Little, Brown.

Descartes, R. (1953). *Oeuvres et lettres* [Works and letters] (Bibliothèque de la Pléiade). Paris: Gallimard. (Original work published 1647)

Duval, S., & Wicklund, R. A. (1972). *A theory of objective self-awareness.* New York: Academic Press.

Eddy, T. J., Gallup, G. G., Jr., & Povinelli, D. J. (1996). Age differences in the ability of chimpanzees to distinguish mirror-images of self from video-images of others. *Journal of Comparative Psychology, 110,* 38–44.

Elder, G. H. (1995). Life trajectories in changing societies. In A. Bandura (Ed.), *Self-efficacy in changing societies* (pp. 46–68). New York: Cambridge University Press.

Ericsson, A. (1996). The acquisition of expert performance: An introduction to some of the issues. In K. A. Ericsson (Ed.), *The road to excellence: The acquisition of expert performance in the arts and sciences, sports and games* (pp. 1–50). Mahwah, NJ: Erlbaum.

Ferguson, T., & Stegge, H. (1995). Emotional states and traits in children: The case of guilt and shame. In J. P. Tangney & K. W. Fischer (Eds.), *Self-conscious emotions* (pp. 174–197). New York: Guilford Press.

Ferrari, M., & Mahalignam, R. (1998). Personal cognitive development and its implications for teaching and learning. *Educational Psychologist, 33,* 35–44.

Fischer, K. W., & Bidell, T. R. (1998). Dynamic development of psychological structures in action and thought. In W. Damon (Ed.), *Handbook of Child Psychology* (pp. 467–561). New York: Wiley.

Fischer, K. W., Knight, C. C., & Van Parys, M. (1993). Analyzing diversity in developmental pathways. *Contributions to Human Development, 23,* 33–56.

Fish, S. E. (1980). *Is there a text in this class? The authority of interpretive communities.* Cambridge, MA: Harvard University Press.

Fodor, J. (1983). *Modularity of the Mind.* Cambridge, MA: MIT Press.

Fogel, A. (1993). *Development through relationships: Origins of communication, self, and culture.* Chicago: University of Chicago Press.

Freud, S. (1956). *Delusion and dream: An interpretation in the light of psychoanalysis of* Gravida, *a novel, by Wilhelm Jensen* (P. Rieff, Ed.). Boston: Beacon Press.

Gallup, G. G., Jr. (1970). Chimpanzees: Self-recognition. *Science, 167,* 86–87.

Gallup, G. G., Jr. (1994). Self-recognition: Research strategies and experimental design. In S. Parker, R. Mitchell, & M. Boccia (Eds.), *Self-awareness in animals and humans* (pp. 35–50). Cambridge, England: Cambridge University Press.

Gallup, G. G., Jr., Povinelli, D. J., Suarez, S. D., Anderson, J. R., Lethmate, J., & Menzel, E. W. (1995). Further reflections on self-recognition in primates. *Animal Behaviour, 50,* 1525–1532.

Gardner, H. (1993). *Frames of mind: The theory of multiple intelligences* (10th anniversary ed.). New York: Basic Books.

Gergen, K. J. (1991). *The saturated self: Dilemmas of identity in contemporary life.* New York: Basic Books.

Gergen, K. J. (1994). *Realities and relationships: Soundings in social construction.* Cambridge, MA: Harvard University Press.

Gopnik, A., & Meltzoff, A. (1996). *Words, thoughts and theories.* Cambridge, MA: MIT Press.

Goleman, D. (1995). *Emotional intelligence.* New York: Bantam.

Gottlieb, G. (1992). *Individual development and evolution: The genesis of novel behavior.* New York: Oxford University Press.

Gruber, H. (1981). *Darwin on man: A psychological study of scientific creativity* (2nd ed.). Chicago: University of Chicago Press. (Original work published 1974)

Habermas, J. (1984). *The theory of communicative action: Vol. 1. Reason and the rationalization of society.* Boston: Beacon Press. (Original work published 1981)

Hackett, G. (1995). Self-efficacy in career choice and development. In A. Bandura (Ed.), *Self-efficacy in changing societies* (pp. 232–258). New York: Cambridge University Press.

Hala, S., Chandler, M., & Fritz, A. S. (1991). Fledgling theories of mind: Deception as a marker of three-year-olds' understanding of false belief. *Child Development, 62,* 83–97.

Hameroff, S. R., Kaszniak, A. W., & Scott, A. C. (1996). *Toward a science of consciousness: The first Tuscon discussions and debates*. Cambridge, MA: MIT Press.

Harter, S., & Monsour, A. (1992). Developmental analysis of conflict caused by opposing attributes in adolescent self-protraits. *Developmental Psychology, 28*, 251–260.

Hermans, H. J. (1996). Voicing the self. *Psychological Bulletin, 119*, 31–50.

Hermans, H. J., & van Gilst, W. (1991). Self-narrative and collective myth: An analysis of the Narcissus story. *Canadian Journal of Behavioral Sciences, 23*, 423–440.

Hinde, R. A. (1992) Developmental psychology in the context of other behavioral sciences. *Developmental Psychology, 28*(6), 1018–1029.

Humphrey, N.K. (1982, August 19). Consciousness: A just-so story. *New Scientist*, pp. 474–477.

James, W. (1983). *The principles of psychology*. Cambridge, MA: Harvard University Press.

Jung, C. G. (1964). Approaching the unconscious. In C. G. Jung (Ed.), *Man and his symbols* (pp. 3–94). New York: Dell.

Jung, C. G. (1983). Development of Personality. In A. Storr (Ed.), *The essential Jung* (pp. 191–210). New York: MJF Books (Original work published 1934)

Kagan, J. (1981). *The second year*. Cambridge, MA: Harvard University Press.

Karmiloff-Smith, A. (1992). *Beyond modularity: A developmental perspective on cognitive science*. Cambridge, MA: MIT Press.

Kuhn, D., Amsel, E., & O'Loughlin, M. (1988). *The development of scientific thinking skills*. San Diego: Academic Press.

Kunitz, S. (1995). *Passing through: The later poems, new and selected*. New York: Norton.

Langer, E. (1989). *Mindfulness*. Reading, MA: Addison-Wesley.

Leont'ev, A. N. (1978). *Activity, consciousness, and personality* (M. J. Hall, trans.). Englewood Cliffs, NJ: Prentice-Hall.

Lewis, C. S. (1967). *Studies in words* (2nd ed.). Cambridge, England: Cambridge University Press.

Lewis, M. (1992). *Shame: The exposed self*. New York: Free Press.

Lewis, M. (1995). Cognition–emotion feedback and the self-organization of developmental paths. *Human Development, 38*, 71–102.

Luria, A. R. (1972). *The man with a shattered world*. Cambridge, MA: Harvard University Press.

Marino, L., Reiss, D., & Gallup, G. G., Jr. (1994). Mirror-self-recognition in bottle-nosed dolphins: Implications for comparative investigations of highly dissimilar species. In S. Parker, R. Mitchell, & M. Boccia (Eds.), *Self-awareness in animals and humans* (pp. 380–391). Cambridge, England: Cambridge University Press.

Markus, H., & Kitayama, S. (1991). Culture and the self: Implications for cognition, emotion, and motivation. *Psychological Review, 98*, 224–253.

Markus, H., & Nurius, P. (1986). Possible selves. *American Psychologist, 41*, 954–969.

Mascolo, M., & Harkins, D. (1998). Toward a component model of emotional development. In M. F. Mascolo & S. Griffin (Eds.), *What develops in emotional development?* New York: Plenum.

Maslow, A. H. (1970). *Motivation and personality* (2nd ed.). New York: Harper.

McAdams, D. P. (1993). *Personal myths and the making of the self.* New York: William Morrow.

Mead, G. H. (1934). *Mind, self, and society.* Chicago: University of Chicago Press.

Mead, M. (1963). Socialization and enculturation. *Current Anthropology, 4*(2), 184–188.

Melzak, R. (1989). Phantom limbs, the self, and the brain (D. O. Hebb memorial lecture). *Canadian Psychology, 30,* 1–16.

Montanegro, J. (1996, September). *The presence of multiple models of optimizing processes in Piaget's theory: Redundancy or complimentarity?* Paper presented to the Centennial Conference Commemorating Jean Piaget, Geneva, Switzerland.

Mydnas, S. (1997, March, 12). Nomads of Laos: Last leftovers of the Vietnam war. *The New York Times, 146,* p. A3.

Nadel, S. F. (1957). *The theory of social structure.* London: Cohen & West.

Neisser, U. (1988). Five kinds of self–knowledge. *Philosophical Psychology, 1,* 35–59.

O'Neill, D. K. (1996). Two-year-old children's sensitivity to a parent's knowledge state when making requests. *Child Development, 67,* 659–677.

Patterson, F. G. P., & Cohn, R. H. (1994). Self-recognition and self-awareness in lowland gorillas. In S. T. Parker, R. W. Mitchell, & M. L. Boccia (Eds.), *Self-awareness in animals and humans* (pp. 273–290). New York: Cambridge University Press.

Parker, S. T. (1996a). Using cladistic analysis of comparative data to reconstruct the evolution of cognitive development in hominids. In E. P. Martins (Ed.), *Phylogenies and the comparative method in animal behavior* (pp. 361–397). New York: Oxford University Press.

Parker, S. T. (1996b). Apprenticeship in tool mediated extractive foraging: The origins of imitation, teaching, and self-awareness in great apes. In A. Russon, K. A. Bard, & S.T. Parker (Eds.), *Reaching into thought: The minds of the great apes* (pp. 348–370). New York: Cambridge University Press.

Parker, S. T., & Gibson, K. R. (1979). A developmental model for the evolution of language and intelligence in early hominids. *Behavioral and Brain Sciences, 2,* 367–408.

Parker, S. T., Mitchell, R. W., & Boccia, M. L. (1994). Expanding dimensions of the self: Through the looking glass and beyond. In S. T. Parker, R. W. Mitchell, & M. L. Boccia (Eds.), *Self-awareness in animals and humans* (pp. 3–19). New York: Cambridge University Press.

Piaget, J. (1970) *Structuralism* (C. Maschler, Trans. & Ed.). New York: Basic Books.

Piaget, J. (1974a). *La prise de conscience* [Conscious grip]. Paris: Presses Universitaires de France.

Piaget, J. (1974b). *Réussir et comprendre* [Success and understanding]. Paris: Presses Universitaires de France.

Pinard, A. (l986). "Prise de conscience" and taking charge of one's cognitive functioning. *Human Development, 29*, 341–354.

Pinard, A. (1989). *La conscience psychologique*. Québec, Canada: Presses de l'Université du Québec.

Pinard, A. (1992). Metaconscience et métacognition [Metaconsciousness and metacognition]. *Canadian Psychology, 33*, 27–41.

Plomin, R., & McClearn, G. E. (1993). *Nature, nurture, and psychology*. Washington, DC: American Psychological Association.

Povinelli, D. J. (1987). Monkeys, apes, mirrors and minds: The evolution of self-awareness in primates. *Human Evolution, 2*, 493–509.

Povinelli, D. J. (1994). How to create self-recognizing gorillas (but don't try it on macaques). In S. Parker, R. Mitchell, & M. Boccia (Eds.), *Self-awareness in animals and humans* (pp. 291–294). Cambridge, England: Cambridge University Press.

Povinelli, D. J. (1995). The unduplicated self. In P. Rochat (Ed.), *The self in early infancy* (pp. 161–192). Amsterdam: North Holland/Elsevier.

Povinelli, D. J., Landau, K.R., & Perilloux, H.K. (1996). Self-recognition in young children using delayed versus live feedback: Evidence of a developmental asynchrony. *Child Development, 67*, 1540–1554.

Povinelli, D. J., Rulf, A. R., Landau, K. R., & Bierschwale, D. T. (1993). Self-recognition in chimpanzees (*Pan troglodytes*): Distribution, ontogeny, and patterns of emergence. *Journal of Comparative Psychology, 107*, 347–372.

Riffaterre, M. (1990). *Fictional truth*. Baltimore: Johns Hopkins University Press.

Rogoff, B., Mistry, J., Göncü, A., & Mosier, C. (1993). Guided participation in cultural activity by toddlers and caregivers. *Monograph of the Society for Research in Child Development, 58*(8, Serial No. 236).

Romanes, G. J. (1882). *Animal intelligence*. London: Kegan Paul.

Salovey, P., & Mayer, J. D. (1990). Emotional intelligence. *Imagination, cognition, and personality, 9*, 185–211.

Scarr, S. (1992). Developmental theories for the 1990s: Development and individual differences. *Child Development, 63*, 1–19.

Schafer, R. (1992). *Retelling a life: Narration and dialogue in psychoanalysis*. New York: Basic Books.

Schore, A. N. (1994). *affect regulation and the origin of the self: The neurobiology of emotional development*. Hillsdale, NJ: Erlbaum.

Schrödinger, E. (1944). *What is life? Mind and matter*. Cambridge, England: Cambridge University Press.

Schunk, D. H., & Zimmerman, B. (1997). Social origins of self-regulatory competence. *Educational Psychologist, 32*, 195–208.

Searle, J. R. (1969). *Speech acts: An essay in the philosophy of language*. London: Cambridge University Press.

Searle, J. R. (1995). *The construction of social reality*. New York: Free Press.

Shore, B. (1996). *Culture in mind: Cognition, culture, and the problem of meaning*. New York: Oxford University Press.

Singer, J. A., & Salovey, P. (1996). Motivated memory: Self-defining memories, goals, and affect regulation. In L. L. Martin & A. Tesser (Eds.), *Striving and feeling: Interactions among goals, affect, and self-regulation* (pp. 229–250). Mahwah, NJ: Erlbaum.

Snow, C. E. (1990). Building memories: The ontogeny of autobiography. In D. Cicchetti & M. Beeghly (Eds.), *The self in transition: Infancy to childhood* (pp. 213–242). Chicago: University of Chicago Press.

Sperry, R. W. (1980). Mind–brain interaction: Mentalism, yes; dualism, no. *Neuroscience, 5,* 195–206.

Sperry, R. W. (1995). The riddle of consciousness and the changing scientific worldview. *Journal of Humanistic Psychology, 35*(2), 7–33.

Sullivan, H. S. (1954). *The psychiatric interview.* New York: Norton.

Suzuki, D. T. (1971). *What is Zen?* New York: Perennial Library. (Original work published 1962)

Tarde, G. (1907). *Social laws: An outline of sociology* (H. C. Warren, Trans.) New York: Macmillan. (Original work published 1897)

Taylor, C. (1989). *Sources of the self: The making of modern identity.* Cambridge, MA: Harvard University Press.

Vygotsky, L. S. (1966). Development of higher mental functions. In A. Leont'ev, A. Luria, & A. Smirnov (Eds.), *Psychological research in the USSR* (Vol. 1, pp. 11–46). Moscow: Progress Publishing. (Original work published 1930–1931)

Vygotsky, L. S. (1986). *Thought and language* (2nd ed.). Cambridge, MA: MIT Press. (Original work published 1934)

Weiskrantz, L. (1993). Unconscious vision. *The Sciences, 32*(5), 23–28.

Index

✧